Electrospun Biomaterials and Related Technologies

Jorge Almodovar
Editor

Electrospun Biomaterials and Related Technologies

Springer

Editor
Jorge Almodovar
Department of Chemical Engineering
University of Puerto Rico Mayaguez
Mayaguez, PR, USA

ISBN 978-3-319-70048-9 ISBN 978-3-319-70049-6 (eBook)
DOI 10.1007/978-3-319-70049-6

Library of Congress Control Number: 2017962303

Printed on acid-free paper

This Springer imprint is published by Springer Nature
The registered company is Springer International Publishing AG
The registered company address is: Gewerbestrasse 11, 6330 Cham, Switzerland

Preface

One challenge that biomaterials engineers face is the design of extracellular matrix mimetic structures that can replicate native tissues. The extracellular matrix of many tissues is a fibrous polymeric structure that is either randomly oriented, highly aligned, or a mixture of both. It varies significantly in composition, structure, and mechanical properties depending on the type of tissue. Techniques to create fibrous polymeric structures that replicate the extracellular matrix are limited. Electrospinning has quickly grown as a multifaceted technique to create fibrous polymeric structures with applications in healthcare, energy, and the environment. Electrospun biomaterials are attractive because their properties can be easily tuned by adjusting processing parameters, yielding fibrous structures with tunable diameter, porosity, chemistry, mechanical properties, and alignment. Since approximately the early 1990s, the number of publications producing electrospun biomaterials has grown exponentially. Now, electrospinning is transitioning from the laboratory to the manufacturing industry. Several commercial-grade electrospinning equipment are available. Moreover, multiple companies have been established for the development of medical devices engineered completely or partially using electrospinning.

This powerful technique has shown great promise, but the field is at the tip of the iceberg. There are many healthcare applications in which electrospinning can be applied that have just barely been explored. Moreover, the production of reproducible and robust electrospun materials is still a challenge. Multiple investigators, including our group (Department of Chemical Engineering, University of Puerto Rico – Mayagüez), continue researching this powerful technique with the goal of producing extracellular mimetic structures that are reproducible and robust. For example, our group specializes in the production of electrospun collagen fibers. Collagen is a difficult polymer to electrospin due to its limited solubility. However, we have been successful in producing fibrous collagen structures using a mild solvent that preserves the chemical composition of native collagen. Electrospun structures are used to investigate in vivo cell-matrix interactions for the modeling of healthy tissues or tumors. Electrospun structures can also be used as scaffold for tissue regeneration, such as nerve guide conduits or heart patches. Electrospinning also allows for the inclusion of drugs or particles for drug delivery. The number of applications of electrospun materials in healthcare is endless.

This multi-contributed book aims at providing a compendium of electrospinning strategies and related technologies to produce biomaterials for tissue engineering and regenerative medicine applications. Its purpose is to provide the reader with a broad overview of the field as well as cutting-edge research on electrospinning and how it is applied to engineer biomaterials. Its intended audience is students in the biomaterials field (bioengineers, chemical engineers, materials science, chemists, etc.), as well as biomaterials investigators. This book contains contributions from experts around the world from both industry and academia alike.

Mayaguez, PR, USA Jorge Almodovar

Contents

Reproducibility and Robustness in Electrospinning with a View to Medical Device Manufacturing

Luke David Burke, Keith Alan Blackwood, and Fabio Zomer Volpato

Abstract Informed by the numerous theoretical and practical discoveries arising from scientific literature on electrospinning over the last twenty years, coupled with the authors' considerable industry experience in the field, this chapter examines the realities of engineering reproducible and robust electrospinning manufacturing systems. The development of such systems is discussed around the framework of robust process design, and focuses on the parameters critical to translation of small-scale laboratory-based electrospinning constructs to medium or large-scale production lines. The benefits of well-defined product design specifications, appreciation of desired manufacturing scale, equipment selection, process control and finally quality management systems and regulatory concerns are discussed in such a way that is accessible to those with backgrounds in either laboratory-scale electrospinning or process engineering for medical devices. The viewpoint examined herein remains relatively under-investigated in the electrospinning field, however as the technique continues to mature from a research-based to a true manufacturing technique the challenges addressed below are anticipated to become increasingly important.

Keywords Electrospining • Robustness • Medical device • ISO 13485 • 21 CFR • Design controls • Manufacturing • Yield • Scale up • Process engineering

1 Introduction

Interest in electrospinning has seen a significant resurgence over recent years, accompanied by multiple attempts to harness the process for next-generation medical devices. Despite a range of advances in electrospinning manufacturing for application in fields such as filtration [1], industrial and commercial take-up of

L.D. Burke • K.A. Blackwood • F. Zomer Volpato (✉)
Xeltis, De Lismortel 31, Eindhoven 5612 AR, The Netherlands
e-mail: fabio.volpato@xeltis.com

J. Almodovar (ed.), *Electrospun Biomaterials and Related Technologies*,
DOI 10.1007/978-3-319-70049-6_1

medical devices comprising electrospun materials remains notoriously limited [2]. Such failures in translation of electrospinning from a research technique to a medical device manufacturing process have resulted in community scepticism, and in some cases complete abandonment of the process. Nevertheless, the electrospinning community have continued to address problems with throughput, sensitivity and flexibility of the technique, with a primary focus on research-phase operations [1, 3–8]. Despite these continuous advances it is the view of the authors of this chapter that to truly mature the electrospinning technique for medical device manufacture a re-evaluation of process bottlenecks is required from an industrial perspective. Process robustness is identified as a rarely-addressed topic in the field of electrospinning, one which is essential to commercial and industrial success. This chapter presents an engineering-focused discussion of key advances and remaining knowledge gaps identified in robustness and reproducibility from research and development to manufacturing phases of medical devices using the electrospinning process.

It is well established that electrospinning, while successful in academic research and early-stage research endeavours, has seen little commercial success [2]. Whilst research articles and patents related to electrospinning number in the thousands [9], realised products number in the low dozens, and only a subset of these are categorised as medical devices [10]. It should be noted that other products in final-phase development, and even production techniques of some existing products, may not be publically available. Nevertheless, there is a clear disconnect between the level of research and development endeavours and the marketable outcomes and products in the electrospinning field. It is the goal of this chapter to provide guidance in key decision points and parameters to allow transfer from a research-phase electrospinning process to a robust manufacturing operation. Initially, the concept of yield is discussed as it relates to electrospinning as a manufacturing technique. This brief overview is intended to familiarise those with a research background in electrospinning with the most important criteria by which a manufacturing process is measured. Subsequently, process design stages required to achieve a robust electrospinning process are outlined, with particular attention devoted to key decision points such as formal product design, desired throughput, and characterisation of both inputs and outputs of the process. Finally, relevant quality management considerations are reviewed, including the impact of European and American standards on an electrospinning manufacturing process, along with the importance of proper validation, documentation and feedback processes.

In essence, this chapter aims to give electrospinning researchers as well as process engineers an insight into the transformation from a laboratory-scale electrospinning system to a robust, high-yield, manufacturing technique. It should not be considered a how-to guide, and much further information gathering will be required to achieve this transformation. However, it is hoped that this chapter will represent the key pillars upon which a robust electrospinning process can be developed from laboratory scale.

2 Robustness, Yield, and Electrospinning

For any process, translating from a research phase to a manufacturing process, yield is critical. A range of data-driven approaches for defining and improving yield such as six-sigma, TQM, Agile, Theory of Constraints, and many others, are available and considered extremely valuable across several disciplines. Due to the wide scope of these approaches, an in-depth examination of yield metrics and analyses is considered outside the scope of this publication, and readers in need of a more detailed perspective in this field are directed to the extensive literature available [11–13]. This section will discuss yield thematically, and give an insight to the concept as it relates to electrospinning robustness.

In most cases, the total yield of a manufacturing process is the primary metric used to judge process capability, which is considered the ability of a process to produce output within specification limits. However, electrospinning yield and reproducibility are widely under-reported and must be a focus for developing the field of electrospinning manufacturing. Although the concept of yield is widely used across several fields, there is no formal definition in electrospinning. Therefore, yield definitions must be predefined and relevant, to ensure quoted yield values are truly representative of the process' robustness.

To allow better analysis of robustness, yield can be further broken down by failure modes, allowing technical personnel to identify process stages that most strongly affect total process yield. The more specific the definition of process stages the more accurate problem processes can be identified. For process engineers unfamiliar with electrospinning manufacturing processes, such separation of process stages may be more difficult than initially expected. Electrospinning is by its nature a complex and multi-stage process, which can easily lead to loss accumulation and difficulty in root cause analysis for process failures. Additionally, direct observation of the electrospinning process is difficult due to the nano- or micro-scale of the final product. These characteristics of electrospinning result in two key challenges for the electrospinning process engineer: identification of process failure points and control of yield. In the case of failure identification, separation of process stages is essential; incoming materials, solution preparation, electrospinning variables, environmental noise and final characterisation may fail themselves or result in failures in subsequent processes. Incorrect attribution of failures, such as to the electrospinning process itself, rather than preceding process stages such as solution preparation will lead to poor prioritisation of process improvements, and slow overall process development. Second, limitation of losses from process errors in electrospinning can be extremely difficult, as the process cannot be easily characterised during operation, preventing early identification and interventions. In most cases, an entire device or component must be completed before characterisation steps can be performed which identify process failures. Such difficulties make the areas of loss control and failure identification key for process engineers attempting to improve yield in an electrospinning system.

3 Process Design Stages

The design and implementation of a robust manufacturing process based around the electrospinning technique can be divided broadly into three stages; Product Design, Process Scale and Process Control. The following sections aim to outline the key requirements of each of these phases as they relate to the electrospinning technique.

3.1 Product Design

Due to the highly-regulated nature of medical device design and manufacturing processes a significant degree of legislation and guidance is available concerning product design approaches and processes [14, 15]. One of the most familiar product design processes is the "Waterfall model", as featured in numerous regional and international standards. An overview of the design concept can be seen below in Fig. 1.

The waterfall model of product design illustrated above is intended to be applicable to any product development process. The concept focuses on a logical sequence of design stages; first, a clear definition of end user needs and product requirements leads to initial device design. Subsequent stages of evaluation, transfer to production and manufacture follow sequentially, along with continuous feedback, evaluation and iterative development throughout. While the outline of the waterfall model shown in Fig. 1 is general, the influence of proper design controls on the design process is illustrated, along with the appropriate timing and usage of verification and validation processes. The importance of the cyclic nature of the

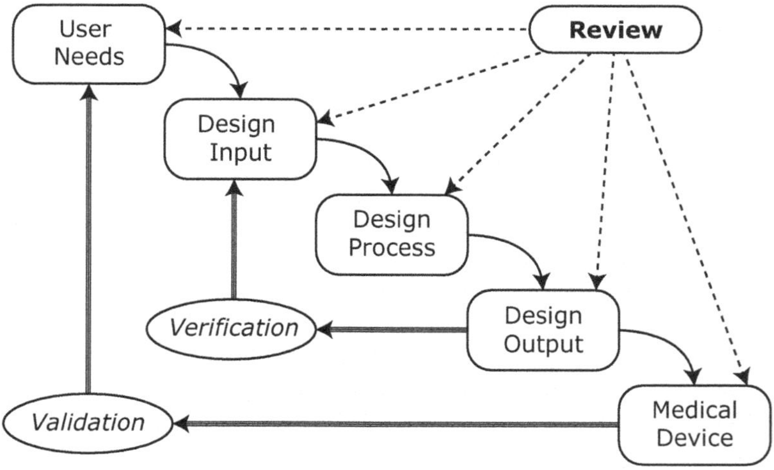

Fig. 1 Basic overview of the waterfall diagram

design process is also highlighted. Initially high-level requirements are developed from Design Inputs to yield product specifications. These specifications thereby become the Design Outputs, which are in turn verified to meet the requirements outlined by the Design Input. The outcomes of these verifications, both positive and negative, become Design Inputs for the subsequent development of the process, completing the cycle.

Of the phases described by the waterfall model, Design Input and Design Process are particularly key for the successful translation of an electrospinning process from R&D to a manufacturing scale. These two phases should define both the product and process specifications in a non-ambiguous and achievable fashion. Initially, both Design Input and Design Process are a rapid, iterative process, and will develop both product and process specifications by the process of frequent technical reviews. The importance of a robust, well-defined and accurate product specification of an electrospun product cannot be overstated, and is best facilitated by strong early-phase research data to accurately define the widest acceptable output parameters which still robustly meet the physical and performance requirements.

Alongside development of product specifications it is essential to develop and define process specifications. Although these process specifications will in almost all cases be dependent on product design and specifications, considerations of total throughput, potential for product design changes and/or manufacture of multiple unique products into the process should be considered as soon as possible in the development stage to avoid conflicts and process bottlenecks later in development.

3.2 Throughput and Scaling Considerations

Having considered the effect of product design and specifications on process requirements, the next key decision point in process design is that of manufacturing scale and throughput. Due to moderate research efforts in these areas, a range of options around electrospinning technique and equipment are available, the choice will be highly dependent on product and process specifications as well as desired throughput. Making these choices as early as possible in process design is strongly recommended, as downstream disruption may completely preclude modifying a fundamental electrospinning technique. Therefore, the process engineer must attempt to foresee not only the current scale and important product specifications, but also the adaptability of the chosen system. For example, unlike a research and development project, the manufacturing process of a well-defined product can afford to sacrifice flexibility of equipment to achieve a higher level of reproducibility, throughput, or both.

There are a number of factors that allow process engineers to predict the most suitable electrospinning techniques for manufacture of a medical device. Considerations of throughput along with product and process design specifications are common key attributes not usually considered at the research and development phase.

With product design specifications well defined, it is important early in the development to consider the targeted total annual throughput of the process. Medical device manufacturing, as a highly regulated field of high value-density products, often results in lower throughput when compared to other industries employing electrospinning manufacture. This lower total throughput frequently results in a high cost-per-device, which can be strongly affected by higher throughput systems.

Flexibility is a key strength of the electrospinning technique, and electrospun materials can comprise either entire medical devices and/or their subcomponents. Products themselves range from the simplest two-dimensional sheets, through simple three-dimensional shapes such as tubes, to complex bespoke shapes such as valves and/or other physically active devices [16]. In the case of entire complex medical devices comprised entirely of electrospun components, tight conformity of produced materials to design specifications will be imperative. Conversely a manufacturing process for simple sheets will likely place greater emphasis on throughput and post-processing of materials. In either case, process specifications, which are well matched with requirements, are imperative to ensure process robustness.

In addition to the considerations above, manufacturing continuity strategies shall be set to best suit the manufacturing scale and post-processing requirements of the product. As with other processes, batch versus continuous manufacturing have a range of costs and benefits that should be assessed. The production of large batches, with considerable downtime between runs, may allow greater focus on complex post processing techniques, whereas smaller semi-continuous processes where several products exist in different stages of completion may allow personnel focus on individual process stages, reducing operator error. Further, the use of large batch processing has the advantage of reducing variability, for example the reduction in the number of polymer solutions produced for electrospinning and the associated irregularities. However, such advantages can be offset by the increased risk should a batch fall out of specification, combined with the increased cost and reduced flexibility of large scale electrospinning systems.

Finally, truly continuous processes have the potential advantage of rapid feedback if product is characterised as it is produced, thereby removing significant losses caused by identifying defective batches post-production if feedback methods allow. On the contrary, batch production can offer a higher degree of control by reducing the effect of changing daily conditions that must be monitored, regulated or controlled in semi-continuous production processes. Additionally, characterisation of continuous processes can be both costly and expensive to implement, as discussed in detail below in Sect. 3.3.

As with product design, process design should be an iterative process, subjected to review and development at regular intervals. Effective bottleneck analysis techniques, such as Lean Six Sigma, can be used to determine needs in terms of equipment, manpower and other resources. Such analysis techniques are well documented and apply well to electrospinning processes; as such they are not discussed in detail here.

3.3 *Equipment Selection and Process Automation*

Having completed the design and specification of the manufacturing process, the identification of suitable electrospinning equipment is possible. Currently, several commercial electrospinning devices are available with a range of proprietary features such as environmental control, multiple nozzles, and high throughput, amongst others. As discussed in Sect. 3.2, choice of electrospinning equipment is strongly linked to total desired throughput and, in most cases of medical device manufacture, product and process specifications. In addition to these considerations, the ability of equipment to adapt to changing process specifications should be considered, particularly in the case of total throughput. Such decisions must account for product complexity; for example whereas scale up of equipment producing two-dimensional sheets may simply require a larger collection area for the electrospun fibres, complex three-dimensional or multi-component devices are unlikely to be scalable in this fashion.

The variety of electrospinning systems available today facilitates control of a wide range of process parameters for a range of total throughputs and price points. The range of options available further emphasises the requirement for well-defined specifications for both product and process, as described above in Sects. 3.1 and 3.2, respectively. Further, in the event that a particular specification cannot be met by off-the-shelf systems, bespoke systems can be considered. Depending on the particular modifications required, such systems frequently can be requested from existing suppliers, or developed in-house. Advantages of bespoke equipment should be weighed against increased financial and time costs, as well as the loss of experience and information from suppliers and users that exists with standard equipment.

The limited number of commercially available electrospun products in the highly regulated field of medical device development has contributed to the slow development of equipment automation in the electrospinning field. Due to the high level of investment required from primarily academic research funding, which focuses on high degree of flexibility over process automation and large-scale reproducibility, such automated systems are not widespread. Despite the restricted amount of commercial equipment available that can provide automated electrospinning manufacturing, automation is feasible for many stages of the manufacturing process.

The most common application for automated electrospinning equipment is manufacturing of 2D sheets, where a semi-continuous/high throughput process is used to maximize throughput. In this sector, largely linked to the production of filtration devices, automation focuses on continuous delivery of polymer solutions, and collection of the electrospun fibres on large conveyor belt collectors. However the number of products of complex geometries and functionalities is still relatively small, therefore, process and product characterization automation is limited to modification of equipment available for different purposes and industries, or custom upgrades of existing commercial equipment which are client specific, and often produced to bespoke requirements which do not enter the general market.

In-line product characterization is an area which is highly valuable in robust processes since it allows real-time feedback of the process during manufacturing, which can be modified to ensure appropriate yield. A variety of in-line methods widely used in other industries are becoming available to the electrospinning market e.g. viscosity measurements as verification of correct solution preparation, optical measurements of fibre diameter and/or scaffold thickness as primary outputs of the electrospinning process, allowing for real-time adjustment of parameters to maintain manufacturing within production limits. Once key requirements of in-line processes are identified, specific characterisation systems should be identified and set-up in coordination with experts in the appropriate field.

The decision of automating the manufacturing process of electrospinning medical devices is based on product complexity, production volumes and business strategy. Implementation timeline for processes, systems and/or subsystems is frequently a business decision as opposed to a technical one. Early implementation of automation systems, whilst initially expensive, can serve to illustrate the potential of scalability to greater production volumes. Such automated systems are likely to streamline production ramp-up as the business matures, ensuring maintenance of quality and product conformance through manufacturing volume increases. Conversely, processes that never require scale up or require significant alterations are likely to be highly expensive to implement while offering little or no observed benefit from the investment.

Process and product characterization automation has its advantages and disadvantages, being increased product reproducibility, cost of goods sold (COGS) reduction, improved time efficiency and increase of process throughput the key drivers to automate the process; and requirement for high investment and customization, lack of expertise on the field the drawbacks of automation. Therefore, automation currently needs to be judged on case-by-case basis, reiterating the advantages and disadvantages described in this chapter.

3.4 Process Controls

The lack of electrospinning as a manufacturing technique in commercial and industrial endeavours results in a clear lack of “standard” electrospinning systems. The variety in electrospinning set-ups, particularly when considering the range of available variations of polymer chemistry, solvent selection, environment, electrospinning system, and the purpose of the final product, makes generalised comparisons between processes difficult, if not impossible. Ideally, electrospinning processes should be analysed individually using a clear, effective and reliable framework. The subsequent sections aim to provide the foundations of such a framework by illustrating universally applicable stages of process control for electrospinning processes, from specification of incoming materials, through manufacturing control and finally output characterisation.

3.4.1 Material Specifications and Controls

As previously discussed in this chapter, the electrospinning process is difficult to characterize during operation. It is therefore extremely important that inputs to the system are tightly controlled, to both maximize yield and facilitate traceability of losses and/or failures. This section outlines the inputs to solution-electrospinning systems generally, including properties that should be tightly controlled and/or accurately characterized to best improve robustness of the process.

Incoming Polymer

Two key decision points for any electrospinning manufacturer are polymer chemistry, and source. Polymer solutions represent the vast majority of fluids employed in electrospinning experiments and as such an extensive catalogue of polymer/solvent combinations are available in the current scientific literature [1, 5]. The choice of polymer is often defined by, or at the very least strongly linked to, the final product application. Well-established polymers such as poly(caprolactone) (PCL), polylacatide-glycolide (PLGA), polylactic acid (PLA) and others offer an extensive background in the medical device field [17]. Alternatively, innovative polymer chemistry may contribute essential or unique properties to the final product, improving device performance and/or value. Having selected the desired polymer chemistry it is necessary, for the majority of polymers, to specify a molecular weight range suitable for manufacturing within the product specifications. For most electrospinning polymers the possible molecular weight range is extensive, and prone to batch-to-batch variability due to the nature of synthesis techniques. Naturally, such variability in the molecular weight of individual polymer batches must be accounted for by either adaptations to solution preparation or electrospinning processing variables in order to achieve consistently reproducible output. Such adaptations generally take the form of solution concentration alterations, accelerating voltage compensation, tuning of separation distance or a combination of these and other processing parameters. Details of the effects of these parameters have been both predicted theoretically, and experimentally verified by many groups over the last twenty years and as such will not be discussed here [4, 7, 17].

This range of electrospinning parameters supported by strong and extensive research allows processing of polymers with a wide range of properties via electrospinning For many small companies or research labs entering into the electrospinning field with a potential of future to scale-up, the specification of incoming polymer represents a highly important decision point. Reduction of polymer specification at the expense of greatly increased costs may not appear justified at the research phase, as acceleration voltage, or separation distance, or other process parameters can be easily adapted to achieve required product attributes. However, as the process and/or product matures such adaptations become increasingly difficult, and savings associated with a wide polymer specification are quickly outstripped by increasing cost-per-device resulting from low production yields, or the requirement for multiple highly-skilled electrospinning operators able to adjust parameters in real time. This risk can be offset by in-house polymer synthesis, which represents a significant initial investment in both facilities and personnel, but may pay dividends to stability during later phases. Conversely, overly-rigid polymer specifications will increase cost exponentially, potentially crippling early-stage research efforts. Such decisions must, therefore, be reviewed continuously as the product matures and the design control phases progress.

Solvents

Having considered the specifications of the polymer to be electrospun, the remaining primary component for consideration is the choice of solvent. In the majority of cases the decision of solvent will be limited by the selected polymer and desired properties of the electrospun materials [4, 17]. Despite these restrictions there remain several important decision points for the electrospinning process engineer, particularly in the case of medical devices. Regulatory considerations, or the function of a final medical device may immediately limit both choice of solvent and purity. For example, low volatility solvents leading to high residual levels in the final product are unlikely to be suitable for implantable devices, just as transfer of trace levels of contaminants may be acceptable for filtration or research applications.

It should be noted that the choice of solvent should always be reassessed when changing production scale from R&D to manufacturing phases. Solvents selected for R&D uses are frequently aggressive, highly volatile organic solvents selected due to electrospinning performance and ease of solution preparation. These solvents have a wealth of research from the academic field [17] but can be precluded from use if excessive levels remain in the final product.

Having selected an appropriate solvent based on process requirements and in line with relevant regulatory conditions, solvent purity specifications should be considered. Generally, the choice of solvent purity is reduced to a cost-benefit ratio; as, although more expensive, the purer the solvent the more repeatable the solution preparation and electrospinning process, as variation in contaminant levels between solvent lots affecting conductivity of solutions can have a profound effect on the electrospinning process, even at very low differences in concentrations. The challenge then for the process engineer is to elucidate the effect of solvent impurity on the yield of the specific electrospinning setup, source and select the lowest priced acceptable purity grade, if available.

3.4.2 Manufacturing Controls

The process of electrospinning has, over the last 25 years, been extensively researched and developed [7]. Processing parameters such as acceleration voltage, polarity, separation distance, solution supply flow rate, solution conductivity, viscosity, surface tension, peak current and many others have been well documented [4, 5, 7, 17]. As such these parameters are extremely well understood and are easily controlled by commercially available equipment. More recently, the effect of environmental parameters on the electrospinning process has been an increasingly common focus of research endeavours [4]. The control of these environmental variables (temperature, humidity, electrostatics, air flow, sterility and similar) is the contemporary challenge faced by a robust electrospinning manufacturing process.

While, a great deal of research around parameters and process aspects of electrospinning has been conducted, reproducible electrospinning production remains a complex production process governed by multiple parameters that are complex to fully monitor and control. For the researcher this presents a wide potential for technical discovery, however for the process engineer the challenge lies in defining and controlling an inherently chaotic process. As such, it is important to set tolerances and specification windows which facilitate compliance with both regulatory requirements, discussed in more detail in Sect. 3.5.1, and product design specifications by minimising the effects of poorly-understood and/or poorly-controlled variables. As discussed above in Sect. 3.4.1 establishing well-defined parameter ranges, developed from knowledge gained during research and development stages, allows a reproducible and robust electrospinning process to be designed. To assist the definition and development of such ranges, the following sections detail process parameters strongly affecting the reproducibility and robustness of a manufacturing-scale electrospinning process.

Solution

A major source of process variability in solution-electrospinning systems is the solution to be electrospun, which is the main external input to the process. Variation can occur from differences in actual versus targeted concentration, impurities in materials, and inhomogeneous or incomplete mixing profiles. Mostly not relevant during research phase, where there is rarely the requirement for high reproducibility of scaffold between experiments. For this reason, there are very few examples in scientific literature where mixing conditions and solvent suitability for the solvent/solute combination are discussed in detail. This can lead both experienced process engineers and those familiar with electrospinning in a research environment to overlook the importance of solution robustness, or to prioritise other aspects of the electrospinning process for optimisation or development. This is of particular importance in situations where long-term, repeatable production is a key goal of the process.

A logical first consideration in any solution preparation process is the volume to be prepared, as this will dictate a number of requirements for the process design. Generally, the minimum amount of required solution per device will be clear from research-phase data, however it should be noted that unavoidable operation losses, such as delivery tubing dead-space, may change when translating between R&D and manufacturing systems. Conversely, maximum levels of solution can be significantly harder to define. Larger solution volumes generally lead to lower proportional error, and using a single solution to electrospin multiple products greatly reduces the effect of solution variation on the process. However, the use of large solution volumes is rarely a simple scale-up transition. Solution shelf-life, total mixing energy of the system, safe storage of polymer solutions containing volatile organic solvents, and measurement device capacity must all be considered to ensure that the benefits created by solution volume scale up are not lost.

Regardless of scale, the key goal in solution preparation techniques is repeatability and reproducibility. In the simplest form, such a robust process is one that ensures the complete dissolution of an accurately controlled amount of solute in a specified volume of solvent without contaminants.

Environmental Control

In addition to controlling the classical electrospinning process parameters used for maintaining process stability and robustness, the control of environmental conditions is highly important in a manufacturing setting. Numerous environmental conditions have an impact upon the electrospinning process, thus variation in environment, such as seasonal changes, humidity fluctuations, have a distinct impact upon the overall robustness of the process as a whole.

Of the environmental conditions, both temperature and humidity are well documented to have impact upon electrospinning [18–20], and both have robust control systems available, including within multiple commercially available electrospinning devices. However while these two conditions are both easy to monitor and control, there are a number of others that, while less understood, are known to have impact upon the electrospinning process. Therefore the monitoring, and preferably the active controlling of these conditions should be a significant consideration for improving robustness. The additional environmental conditions known to impact electrospinning include airflow, atmospheric pressure and static charge of the air, in addition charge build-up on surfaces within proximity to the spinning environment.

These environmental conditions are particularly important in the medical device industry, since most of the manufacturing occurs inside cleanrooms. Cleanrooms are designed to monitor and control particle count, bioburden temperature and humidity within a confined area. Frequently, the use of high efficiency particulate arrestance (HEPA) filters are integral to control of incoming air in a cleanroom system, along with heating and air conditioning systems to maintain temperature and humidity within specified parameters. Control of these parameters is considered essential to medical device manufacture, and a dedicated industry for both bespoke and retrofitted cleanroom systems exists for this reason. However, the parameters controlled in traditional cleanroom systems do not include electrostatic charge, which as discussed may significantly affect electrospinning phenomena. This lack of control is exacerbated by the increase in number of air exchanges, air filtration and humidification controls required to maintain other parameters. Therefore, at the very least monitoring, and preferably control, of electrostatic charge within the environment is strongly recommended to improve electrospinning reproducibility, and thus robustness and yield.

Ideally, the development of a bespoke environment suited to full control over the environmental parameters discussed above is recommended. However, cost and footprint of such facilities may prevent this process altogether, or preclude total control of every parameter. In these situations it is recommended to understand the environmental controls available, along with their effect on each of the environmental parameters outlined above, as they relate to the electrospinning technique. Wherever possible, the monitoring and tracking of these environmental parameters should be undertaken at the very least, if only to allow proper failure identification and feedback for process development, as discussed in Sect. 3.5.3.

3.4.3 Output Characterisation

The definition of the output characterisation of the electrospun devices or device components should be based on the relevant product functionalities and specifications identified during the Design Input and Output phases, respectively. A comprehensive understanding of required product characterization at the design phase is crucial to avoid developing a non-capable process, which may lead to a low yield. As outcome of Design Output, product specifications will indirectly determine the required level of control over the process, e.g. if strict fibre diameter and standard deviation are relevant for the functionality of the product, special attention has to be taken on input variables like solution homogeneity and relative humidity of the spinning environment, as well as accuracy of measurement tools. Therefore, it is highly recommended that process experts take part on the development phase to ensure a proper process capability.

Prior to making decision of the level of process controls it is advisable to identify the most relevant attributes of the product to maintain quality and functionality. In electrospinning manufacturing, the most relevant attributes that directly affect product performance via mechanical, chemical and biological properties are fibre diameter, interconnectivity and alignment, thickness and thickness distribution, porosity, average pore size, fibre surface roughness and polymer molecular weight. Due to the vast literature on the effect and control of these attributes, an in-depth examination is considered outside the scope of this publication, and readers in need of a more detailed perspective are directed to the extensive literature available [1, 4–6, 8, 18–20].

In combination with product attributes and specifications, the identification of reliable and accurate test methods will affect the product yield and/or quality. Regulations and guidance are available in order to aid the selection of test method based on product specifications and accuracy of equipment [14, 15]. When defining the test methods to characterize the product outputs, it is recommended to consider in-line tools, e.g. fibre diameter, thickness and Taylor cone characterization. These measurements can be performed in-line ensuring process feedback and increased product yield. Such tools are although mostly employed at later stages of the product development, e.g. design verification and validation, and commercialization phases.

During the research and development phases it is acceptable to characterize the product after the process is concluded, this is possible due to the fast turnaround of post processing activities and the low volume of production. At this stage product yield is not a relevant concern; however, at higher output manufacturing stages the process chain becomes longer, volume output, product yield and traceability relevance increases. Therefore, collecting information on product quality at the end of the process becomes expensive and usually non acceptable from business reasons.

3.5 Quality Management

The field of medical device manufacture is, by necessity, highly regulated. Compliance with these regulations and standards from the earliest possible phase is essential to the success of a medical device manufacturing process. To this end the following section discusses both the regulatory requirements and information provided by the most commonly encountered regulatory bodies, along with the usefulness of well-defined process validations and feedback systems. The application of these three topics will not provide the reader with a recipe to a perfectly validated, regulatory acceptable process, but will provide an excellent basis upon which to begin the development of such a process.

3.5.1 Regulatory

Regulatory considerations should be a key focus for any medical device manufacturing process from inception. A lack of focus or prioritisation around regulatory concerns, regardless of device purpose or manufacturing technique, will doom any process to failure before beginning. This is, arguably, more applicable to electrospinning manufacturing than many other techniques. As a research-driven technical process with very few examples of existing products the pitfalls of regulation are numerous and engineers working with electrospinning must be acutely aware of this fact and, ideally, what is and what is not acceptable when considering process changes and developments.

To assist the reader in beginning to understand the broad strokes of medical device regulation, this section will focus on the impact regulatory affairs on process robustness. It should be acknowledged up-front that specific product attributes and their effect on classification by regulatory bodies will not be discussed. Instead a view of both general and certain specific regulations based on similar processes which apply to electrospun devices, will be presented.

In the European and North American markets, the two main regulatory authorities, Food and Drug Administration [FDA] (USA) and European Commission (EU), do not have specific regulations and/or harmonized standards available for electrospun products. Therefore, products manufactured via electrospinning are expected to adhere to quality systems requirements in order to ensure devices perform as intended. Additionally, the FDA has released in 2016 a draft guidance for additive manufactured devices: "Technical Considerations for Additive Manufactured Devices" [21]. Although this guidance is based around more traditional layer-by-layer 3D printing rather than electrospinning, there are significant similarities between the two techniques, e.g. inherent variability product quality in controlled systems, which allow the application to electrospun medical device manufacturers.

In reference to the draft guidance, the FDA states that since there are multiple additive manufacturing technologies, printers and materials, it is important for manufacturers to "clearly identify each step in the printing process... from the initial device design to the post-processing of the final device" [21]. Furthermore, companies should make sure they understand the upstream effects different steps in manufacturing process might have on a device. Such considerations and statements related to 3D printing technologies can be extrapolated to electrospun devices manufacturers once as mention above are technologies with similar degree of process variation.

Medical devices manufactured from electrospun materials have to comply to the code of regulations 21 CFR part 820 (USA) [14] and the Medical Device Directive MDD 93/42/EEC (EU), which require that manufacturers have to demonstrate that the process is capable of reproducible and controlled commercial manufacturing. Despite not being a requirement by the European Union, International Organization for Standardization (ISO) 13485 is recognized as a critical aid in supporting compliance with the MDD and it is therefore considered to be the de facto standard for the medical device industry in Europe. Furthermore, Article 5 of the MDD indicates that compliance to a harmonized standard is sufficient to assume conformity to relevant essential requirements of the MDD.

Independent of the regulatory body, it is a common requirement that prior to products being commercially distributed, the manufacturers must have gained a high degree of assurance in the performance of the manufacturing process such that it will consistently manufacture products meeting the product specifications. The assurance should be obtained from objective information and data from laboratory-, pilot-, and/or commercial-scale studies. Information and data should demonstrate that the commercial manufacturing process is capable of consistently producing acceptable quality products within commercial manufacturing conditions. Furthermore, focusing exclusively on qualification efforts without also understanding the manufacturing process and associated variations may not lead to adequate assurance of quality.

3.5.2 Process Validations

The lack of specific regulations and standards for electrospun devices, further emphasize the need to create deep knowledge on how each input parameter and process step affects the quality of the finished product. Improving the control of the input parameters minimize process variability improving process capability whilst facilitating validation activities, e.g. improving robustness of the solution preparation process can minimize variability on fiber diameter. An additional method to facilitate the validation activities and improve yield is to take advantage of the product specifications, which depending on the requirements and functionality of the device, can be widened allowing a more capable process, common examples are device thickness, fiber diameter distribution, etc.

Where the product output cannot be verified, process validations is required by regulations and standards. As clearly described by the FDA and ISO standards, Process validations shall occur for processes and services where the output attributes cannot or may not be verified by subsequent inspection [15]. This includes any processes where deficiencies become apparent only after the product is in use [15]. Process validation must be performed, prior to product commercialization, in order to ensure and maintain quality of the manufactured device [14]. Briefly, validation is composed by three phases Equipment Installation Qualification (EIQ), Operational Qualification (OQ) and Performance Qualification (PQ), where EIQ aims to ensure the equipment is installed and performs accordingly to the specifications required to manufacture the product under a controlled environment; OQ focus on challenging the process specifications by testing wider ranges than the specified in the process in order to demonstrate that even at wider process windows the product still conforms to the specifications; whilst PQ aims to confirm that whenever manufactured with the nominal process specifications the product will conform to the specification within a statistical significance level.

Process validations can also be performed on process where the output can be measured but decided not to for business reasons. Mostly, it is used where the attribute inspection has high costs, involves high degree of variability and/or requires a significant amount of time.

The correct timing to start validation activities may vary depending on business and/or development strategies, however it is mandatory that processes requiring validation are completed by the submission of the request for product approval to the regulatory bodies. The authors of this chapter, although, recommend to initiate as soon as the process is at final stage of development in order to avoid multiple revalidation activities since process re-validation is required whenever a process change that may affect product output is implemented. It is important to remember that processes like electrospinning, which have not seen widespread implementation in large scale manufacturing and most likely will be customized to specific products may impose a higher degree of complexity during validation.

3.5.3 Process Feedback

Process feedback and the analysis of the resulting data is a key step towards establishing a continuously robust process. Quality management systems allow for a high degree of traceability and ease of access to key data, including product traceability, which is crucial for effective root cause analysis and corrective and preventative action implementation. Furthermore, it should be a direct input for the process failure analysis and product risk assessment.

The implementation of process feedback in electrospinning manufacturing is even more crucial than other techniques since there are numerous variables that may result in the same failure mode. Furthermore, noise variables are usually not well understood/controlled and can influence the final product. By adding proper feedback stages along the process, e.g. solution characteristics, record of input and real value of parameters in electrospinning machine, environment readings, etc., it allows more robust isolation of possible failures and therefore act on process modifications whenever necessary.

4 Summary

Electrospinning has long been recognised as a powerful technique for the production of highly porous non-woven materials. However, the concept of process reproducibility and robustness in electrospinning is widely under-reported, despite the large body of work devoted to the technique. This has resulted in a substantial disconnect between research success compared to commercial application, and must be a focus of any endeavour to utilize electrospinning in a manufacturing capacity. To effectively address the robustness and yield requirements of an electrospinning manufacturing process for medical devices, a combination of scientific, engineering and regulatory expertise is essential. Due to the size of the field, coupled with the variety and complexity of products, an in-depth analysis of all parameters is not possible in the space available here. Therefore, the preceding sections are intended to highlight key decision points and parameters not widely reported by the scientific community, which have a significant effect on process robustness.

Beginning with the effective application of Design Controls to electrospinning, the use of well-defined structures available to medical device engineers is highlighted. Decision points during the development phase will have wide-ranging implications on product suitability, process requirements, characterisation and, ultimately, process yield. Verification of initial process and product design specifications will give a strong framework around which to base further process design decisions, as outlined by the waterfall model of process design.

The field of electrospinning as a manufacturing technique has been developing over the past several decades. During this time the wealth of information from fundamental research has been harnessed to propose and develop a range of novel and innovative medical devices. Such devices have the potential to be truly paradigm shifting in terms of their interaction with the body, with the goal of improving a variety of clinical outcomes across a range of medical disciplines. To achieve this, it is imperative that these goals are supported by knowledge of both the electrospinning technique and the strong frameworks that exist for medical device product and process development. Only through this multidisciplinary approach will electrospinning truly mature into a reproducible and robust technique for medical device manufacture.

References

1. Ahmed FE, Lalia BS, Hashaikeh R (2015) A review on electrospinning for membrane fabrication: challenges and applications. Desalination 356:15–30. https://doi.org/10.1016/j.desal.2014.09.033
2. Hayes TR, Hosie IC (2015) Turning nanofibres into products: electrospinning from a manufacturer's perspective. In: Electrospinning for high performance sensors. Springer, Berlin, pp 305–329
3. Dai Y, Liu W, Formo E et al (2011) Ceramic nanofibers fabricated by electrospinning and their applications in catalysis, environmental science, and energy technology. Polym Adv Technol 22:326–338
4. Haider A, Haider S, Kang IK (2015) A comprehensive review summarizing the effect of electrospinning parameters and potential applications of nanofibers in biomedical and biotechnology. Arab J Chem. doi: https://doi.org/10.1016/j.arabjc.2015.11.015
5. Khorshidi S, Solouk A, Mirzadeh H et al (2015) A review of key challenges of electrospun scaffolds for tissue-engineering applications. J Tissue Eng Regen Med 10:715–738. https://doi.org/10.1002/term.1978
6. Miao J, Miyauchi M, Simmons TJ et al (2010) Electrospinning of nanomaterials and applications in electronic components and devices. J Nanosci Nanotechnol 10:5507–5519. https://doi.org/10.1166/jnn.2010.3073
7. Pillay V, Dott C, Choonara YE et al (2013) A review of the effect of processing variables on the fabrication of electrospun nanofibers for drug delivery applications. J Nanomater 789289:22
8. Wu H, Pan W, Lin D, Li H (2012) Electrospinning of ceramic nanofibers: fabrication, assembly and applications. J Adv Ceram 1:2–23
9. Persano L, Camposeo A, Tekmen C, Pisignano D (2013) Industrial upscaling of electrospinning and applications of polymer nanofibers: a review. Macromol Mater Eng 298:504–520
10. Electrospintech (2016) Electrospun applied products. http://electrospintech.com/products.html. Accessed 16 Dec 2016
11. Abrahamsson P, Warsta J, Siponen MT, Ronkainen J (2003) New directions on agile methods: {A} comparative analysis. 25th Int Conf Softw Eng Proc, pp 244–254. doi: https://doi.org/10.1109/ICSE.2003.1201204
12. Bozdogan K (2010) Towards an integration of the lean enterprise system, total quality management, six sigma and related enterprise ESD-WP-2010-05 ESD working paper series towards an integration of the lean enterprise system, ESD-WP-2010-05. ESD Work Pap Ser 5:1–23
13. Cagnazzo L, Taticchi P (2009) Six sigma: a literature review analysis. In: Bulucea CA, Mladenov V, Pop E et al (eds) Recent advances in e-activities, information security and privacy. Puerto De La Cruz, pp 29–35
14. FDA (2016) Code of federal regulations. In: 21 C.F.R. 21. US Government, p § 820
15. ISO 13485:2003(E) (2003) Medical devices – quality management systems – requirements for regulatory purposes. ISO 134852003(E)

16. Chen R, Morsi Y, Patel S et al (2009) A novel approach via combination of electrospinning and FDM for tri-leaflet heart valve scaffold fabrication. Front Mater Sci China 3:359–366. https://doi.org/10.1007/s11706-009-0067-3
17. Leach MK, Feng Z-Q, Tuck SJ, Corey JM (2011) Electrospinning fundamentals: optimizing solution and apparatus parameters. J Vis Exp:1–5. https://doi.org/10.3791/2494
18. De Vrieze S, Van Camp T, Nelvig A et al (2009) The effect of temperature and humidity on electrospinning. J Mater Sci 44:1357–1362. https://doi.org/10.1007/s10853-008-3010-6
19. Nezarati RM, Eifert MB, Cosgriff-Hernandez E (2013) Effects of humidity and solution viscosity on electrospun fiber morphology. Tissue Eng Part C Methods 19:810–819. https://doi.org/10.1089/ten.TEC.2012.0671
20. Pelipenko J, Kristl J, Janković B et al (2013) The impact of relative humidity during electrospinning on the morphology and mechanical properties of nanofibers. Int J Pharm 456:125–134
21. FDA (2016) Technical considerations for additive manufactured devices draft guidance for industry and food and drug administration staff. US Government

Electrospun Collagen Scaffolds

David A. Castilla-Casadiego, Carol A. Rivera-Martínez, Beatriz A. Quiñones-Colón, and Jorge Almodóvar

Abstract Nanofibrous collagen scaffolds developed via electrospinning have revolutionized the field of designing useful biomaterials for regenerative or tissue engineering. Electrospun collagen scaffolds allow for the replication of the extracellular matrix of tissues with regards of their chemical, physical, and mechanical characteristics. Because collagen is the most abundant protein found in tissues, it can be the base for an ideal scaffold to mimic the majority of soft or hard tissues such as bone, which contains an organic component composed of collagen and a mineral component. The physical and mechanical properties of the collagen nanofibers are of vital importance to promote the necessary and specific signals of the cellular or tissue environment supporting different cellular processes. This chapter describes the fabrication and modulation of the physical and mechanical properties of electrospun collagen nanofibers and their biomedical applications.

Keywords Collagen • Electrospinning • Nanofibrous scaffolds • Nanofiber alignment • Diameter • Mechanical properties • Crosslinked nanofibers

D.A. Castilla-Casadiego
Department of Chemical Engineering, University of Puerto Rico Mayaguez, Mayaguez, PR, USA

C.A. Rivera-Martínez
Bioengineering Graduate Program, University of Puerto Rico Mayaguez, Mayaguez, PR, USA

B.A. Quiñones-Colón
Department of Biology, University of Puerto Rico Mayaguez, Mayaguez, PR, USA

J. Almodóvar (✉)
Department of Chemical Engineering, University of Puerto Rico Mayaguez, Mayaguez, PR, USA

Bioengineering Graduate Program, University of Puerto Rico Mayaguez, Mayaguez, PR, USA
e-mail: jorge.almodovar1@upr.edu

J. Almodovar (ed.), *Electrospun Biomaterials and Related Technologies*,
DOI 10.1007/978-3-319-70049-6_2

Abbreviations

AcOH	Dilute acetic acid
AFM	Atomic force microscopy
BDDGE	1,4-Butanediol diglycidyl ether
CD	Circular dichroism spectroscopy
Col/PCL	Collagen/polycaprolactone
DMSO	Dimethyl sulfoxide
ECs	Endothelial cells
ECM	Extracellular matrix
EDC	1-Ethyl-3-(3-dimethyl-aminopropyl)-1-carbodiimide hydrochloride
EtOH	Ethanol
FTIR	Fourier transform infrared spectroscopy
GAG	Glycosaminoglycan
GFAP staining	Glial fibrillary acidic protein staining
HA	Nanohydroxyapatite
HAc	Acetic acid
HFIP	Hexafluoroisopropanol
HFP	1,1,1,3,3,3-Hexafluoro-2-propanol
MSCs	Mesenchymal stem cells
NHOK	Normal human oral keratinocytes
NHS	*N*-Hydroxysuccinimide
NOI	Normalized orientation index
NRVCM	Ventricular cardiomyocytes of primary neonatal rats
PHBV	Poly(3-hydroxybutyrate-co-3-hydroxyvalerate)
PIECs	Porcine iliac artery endothelial cells
PLGA	Poly(lactide-co-glycolide
PLLA	Poly(L-lactic acid)
SCs	Schwann cells
SCI	Spinal cord injury
SEM	Scanning electron microscope
SMCs	Smooth muscle cells
TFE	Tri-fluoroethanol
TPU	Thermoplastic polyurethane
vWF	Von Willebrand factor

1 Introduction

In the field of tissue engineering, many researchers have focused their studies on fabricating biodegradable polymeric scaffolds, attractive to the cellular environment, and that would have the capacity of promoting tissue regeneration [1]. In the beginning, their efforts were focused on mimicking the shape, structure and composition of the cellular environment, developing scaffolds in the form of hydrogels [2, 3], sponges [4], nanolayers [5] and nanofibers [6], composed of polymers such as chitosan [7], hyaluronan [8], pullulan [9], dextran [10], among others. Elaborated scaffolds based on these polymers and presented in the different known models, have evidenced positive results at a cellular level. Demonstrating that these can overcome cytotoxicity analysis, adhesion, proliferation and cellular differentiation, indicating the importance for them being considered useful for biomedical applications [11].

An important challenge in this field of research is the mimicking of the physical and chemical characteristics of the extracellular matrix (ECM) of tissues. In general, the ECM is composed of a variety of polymeric biomacromolecules, among these are glycoproteins, proteoglycans, and polysaccharides [12]. Being collagen the most abundant component of the ECM [13], which contributes to the strength and structure of the ECM, directly interacting with cells and other ECM molecules [14].

The electrospinning technique has the capacity of creating scaffolds with the ability to replicate the physical characteristics of the ECM, such as the fibrous morphology, the diameter of the fibers, and their orientation [15]. Compositional characteristics have also been abundantly studied with the use of polymers and proteins found in the ECM to fabricate nanofibers, allowing for the modulation of other properties such as mechanical properties. These have been supported with the use of chemical agents for crosslinking [16], in addition to fabricating collagen scaffolds [17].

Collagen fibrous scaffolds fabricated by the electrospinning process have proven to possess excellent biocompatibility and the mechanical properties necessary to be used in the field of tissue engineering [18]. Electrospun nanofibers have shown to have a high surface area to volume ratio, mimicking the ECM of native tissue and therefore facilitating cell proliferation and attachment [19]. These can be applied for the regeneration of nerve [20], bone [21], skin [6], tendon/ligament tissue [22], and vascular grafts [23], among others. They provide promotion of function and tissue repair, both in vivo and in vitro. In addition to serving as an artificial ECM for growing cells, these scaffolds can be used to deliver bioactive agents to promote regeneration of tissue [24, 25].

This chapter describes the history of the fabrication of collagen nanofibers, highlighting from the conventional electrospinning method to their fabrication through the green electrospinning process to preserve the secondary structure of collagen. Because the physical characteristics are of vital importance to recreate a cellular environment attractive for the native cells in the tissue, in this section different works that have been published that modulate fiber diameter, orientation, and mechanical properties are highlighted. Finally, several biomedical applications of collagen nanofibers useful for the regeneration of tissues in the human body are presented as well as the process towards scale-up and industrialization of electrospun collagen nanofibers.

2 Collagen Nanofibers: Use of Toxic Solvents to Benign Solvents

The fabrication of nanofibrous scaffolds have gained a greater reputation and importance in the field of bioengineering, this because these types of scaffolds have demonstrated that they have the capacity to replicate the structural characteristics and composition found in the ECM of tissues [26]. Other scaffold presentations, such as gels and polymeric nanolayers, also offer characteristics and properties to promote tissue repair, but these types of scaffolds have shown limitations and disadvantages in comparison with nanofibrous scaffolds [11]. The biggest advantage of nanofibrous scaffolds lies on the fact that it is possible to replicate the scale, morphology, geometry and orientation of the nanofibers, composition, while also modulating the mechanical properties of tissues, which is known that they vary depending on the type of tissue in the human [27–29].

By early 2000's, Bowlin and his research group fabricated electrospun collagen nanofibers [30] to be used as a scaffold mimicking; (1) the physical characteristics of the extracellular matrix (nanofibrous structure) and (2) the composition. When overcoming the challenge of finding a solvent that would dissolve collagen and would be useful for the electrospinning process, they focused on finding an optimum ratio of collagen type I dissolved in 1,1,1,3,3,3-hexafluoro-2-propanol (HFP) solvent. Experimenting with different concentrations of collagen, they found that 0.083 g/mL resulted in being the appropriate concentration for nanofiber formation, operating the electrospinning equipment at 25 kV, 125 mm distance from the collector, and 5 mL/h of injection flow. Electron microscopy images demonstrated the successful obtainment of collagen type I nanofibers, which possessed a diameter of 390 ± 290 nm [30]. Experimenting with collagen type III, they were also able to obtain nanofibers with a solution composed of 0.04 g collagen III/mL dissolved in HFIP. They could obtain an average nanofiber diameter of 250 ± 150 nm, maintaining the same conditions of operation of the electrospinning equipment used for the obtainment of collagen type I nanofibers [30].

When examining the different parameters of the equipment, they found that it might be possible to obtain random and aligned nanofibers by adjusting the rotation velocity of the cylindrical collector of the equipment [30]. Using a high velocity of rotation of 4500 RPM, it was possible to have a control over the alignment of the nanofibers, opening the doors to the possibility of mimicking not only tissues that have random collagen nanofibers, but also other tissues such as cardiac and nervous, which contain aligned nanofibers [31–33].

Bowlin's group developed an in vitro study to prove the viability of the prepared scaffold, using aortic smooth muscle cells. The cells were seeded on the nanofibrous membrane after being crosslinked in glutaraldehyde vapor for 24 h and sterilized. This crosslinking process was necessary to maintain the morphologic stability of the nanofibers obtained because when contacting water they would dissolve. The in vitro results demonstrated that the nanofibrous scaffold obtained, using HFIP as a solvent, promoted the adhesion and cellular growth. This immediately resulted in being able to say that this scaffold would be ideal and potentially useful for tissue engineering, since the collagen nanofibers could replicate in large proportions the structural and compositional properties in addition to the biological functions of the ECM [30]. Shields et al. [34] fabricated collagen type II nanofibers from chicken sternae which was also dissolved in HFIP. Their results showed that a scaffold cultured with chondrocytes could promote the cells' ability to infiltrate the scaffold surface and interior. By this time, it was possible to fabricate nanofibers of all three main types of collagen mainly found in the tissues of the human body.

Once it was demonstrated that it was possible to fabricate collagen nanofibers, researchers such as Byung-Moo Min and his group in 2006 [6], decided to continue investigating the potential of collagen nanofibers. They obtained nanofibers from a solution of 8% w/v collagen in HFIP. Their investigation was focused on studying the cytocompatibility and cell behavior including cell attachment and spreading of normal human keratinocytes seeded on the collagen nanofibrous matrix, and other ECM components such as fibronectin and laminin. They also evaluated in vivo the effect of the collagen scaffold on open wound healing in rats. In their in vitro studies, they seeded cells over collagen only, collagen nanofibers, and nanofibers covered with: type I collagen, bovine serum albumin, fibronectin, and laminin. Their results showed the ability of the collagen nanofibers covered with collagen type I, which promoted cell adhesion and spreading of normal human keratinocytes in comparison to the "collagen nanofibers" obtained and the other coatings of different proteins evaluated. Their results determined that the components generated the highest promotion for cellular processes following the ratio: type I collagen > laminin > fibronectin > uncoated collagen nanofibers. Observing a level of cellular adhesion of almost double in collagen nanofiber covered with type I collagen in comparison with "collagen nanofibers" fabricated by them without any coating. These results opened the doors to a variety of uncertainties. Why collagen nanofibers without any additional coating do not have the same effects than those with the additional collagen coating? Is the use of the HFIP solvent having a negative effect over the chemical structure of collagen? Is the crosslinking process ideal?

A year later, in 2007, Lin Yue Lanry Yung and his team [35], wanting to keep replicating the ECM of tissues decided to fabricate collagen nanofibers containing glycosaminoglycan (GAG), which are also part of the native tissues in humans. In their research, they used chondroitin-6sulfate as the GAG. The fabrication of collagen/chondroitin-6sulfate nanofibers was successful, with a concentration of 4 wt. % of chondroitin-6sulfate and 10 wt. % of collagen, dissolved in a mixture of solvents composed of 2,2,2-trifluroethanol and water in a ratio of 4:1. Their in vitro study reported that the proliferation of rabbit fibroblasts exhibited high density of cells in the crosslinked scaffold (glutaraldehyde) in comparison to non-crosslinked. The collagen scaffold with GAG crosslinked reported a higher effect over the proliferation than collagen nanofibers without GAG, demonstrating that the more you can replicate the cellular environment, cellular activities are favored.

By the year 2008, Jan Feijen and his group reported a problem over the collagen nanofibers fabricated with fluorinated solvents. They demonstrated, from their study with circular dichroism (CD) spectroscopy, a decrease of 45% of collagen's triple helical structure after electrospinning using HFIP as the collagen dissolvent [17]. Works where fluorinated solvents such as trifluoroacetic acid, [68] trifluoroethanol, [35] and 1,1,1,3,3,3-hexafluoro-2-propanol (HFIP) were used [30, 34] to produce "collagen nanofibers" did not preserve the chemical structure of collagen. These fluorinated solvents disrupt collagen's secondary structure, converting the electrospun collagen into gelatin [36].

Because of these findings, by the year 2009, Wnek and his group [37], proposed the fabrication of collagen nanofibers using benign solvents to replace toxic solvents that demonstrated the destruction of the chemical composition of collagen. They were able to dissolve collagen type I and III in a mixture of phosphate-buffered saline and ethanol, effectively generating a useful solution for the electrospinning process, which helped them obtain the collagen nanofibers while conserving the chemical structure of native collagen. Following this, several groups present alternatives to prepare collagen nanofibers using benign solvents. In 2010, Sing Yian Chew and his group [38], reported the production of collagen nanofibers using acetic acid, with a solution that consisted of 25 wt. % collagen in 40% (v/v) acetic acid. This solvent also prevented the degradation of collagen, demonstrating the advantage over toxic solvents.

V.R. Giridev [39, 40] with the purpose of preparing other formulas, proposed the alternative of preparing a collagen solution mixed with other polymers, to facilitate the electrospinning process. V.R. Giridev in his study used a mixture of COL/PCL and Li-Ming Zhang, on the other hand, used a mixture of collagen and zein, using acetic acid as the dissolvent. In both studies, they could obtain nanofibers but the possibility of facilitation the electrospinning process was not significant, because of difficulties in the preparation, in the morphology of the nanofibers, and in the use of approximately 50% of another polymer.

By 2014 new recipes supported by the solution preparation of Lin et al. [40] with the use of acetic acid, were reported. Andrea Maria Letizia Focarete and her research group [41], generated electrospun mats dissolving collagen in acetic acid/water (20/80 v/v) at a concentration of 30% (w/v), being able to achieve and conserve the chemistry of collagen type I. Kazanci [42], who was able to prepare collagen nanofibers with 40% w/v collagen in 40% v/v acetic acid.

In 2015, Anuradha Elamparithi, demonstrated the fabrication of nanofibers using glacial acetic acid and DMSO to dissolve collagen at a ratio of 93/7 (glacial acetic acid/DMSO) with a concentration of 10% collagen [43]. In their work, they did a biocompatibility study of the nanofibrous scaffold obtained, using rat skeletal myoblasts (cellular line L6). Their results showed that the scaffold could support the cells, allowing them to carry out their normal functions. This is due to the evidence ventricular cardiomyocytes of primary neonatal rats (NRVCM) seeded were able to maintain their contractile function of a 17-day period. Indicating that it could be a promising scaffold for the field of tissue engineering, when learning about its potential, demonstrated in the evaluation with different types of cells.

By 2016 [15], in our most recent work we were able to report the obtainment of collagen nanofibers using collagen in acetic acid (90% v/v in water) to a final concentration of 20% w/v (Fig. 1). This was done with the purpose of demonstrating the

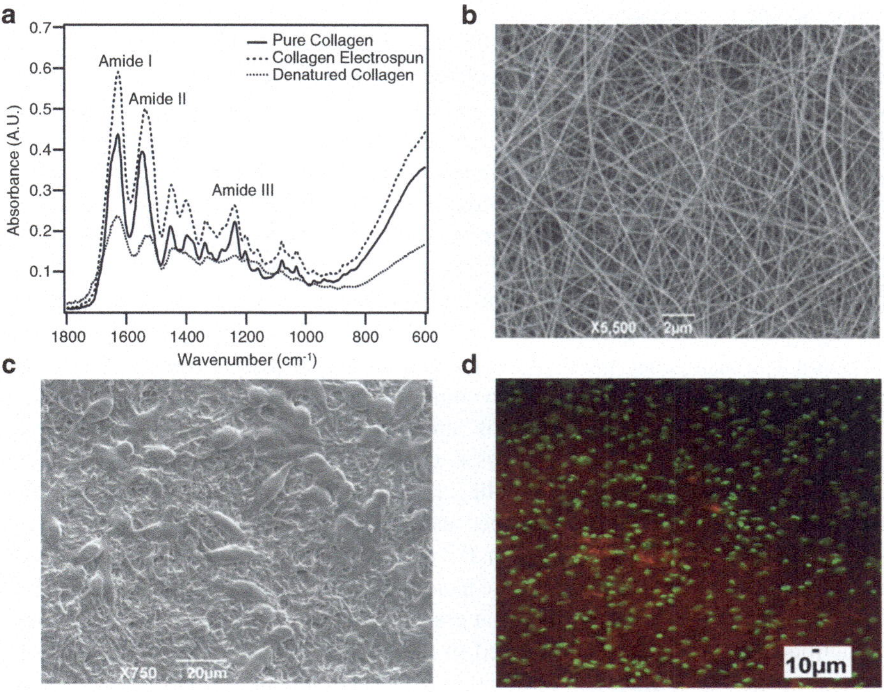

Fig. 1 (**a**) FTIR spectra of pure type I collagen sponge, electrospun collagen, and denatured (200 °C treatment) collagen and in vitro cellular evaluation of nanofibers. (**b**) SEM image of collagen nanofibers. (**c**) SEM image of fibroblasts cultured on collagen nanofibers after 24 h. (**d**) Fluorescent microscopy images of 3T3 fibroblasts (green: nuclei) cultured on collagen nanofibers (red). Reprinted with permission from [15]. Copyright 2016 Wiley-VCH Verlag GmbH & Co. KGaA

preservation of the chemical composition and secondary structure of collagen in the obtained nanofibers, with the use of benign solvents. FTIR spectroscopy was used to study changes in the secondary structure of type I collagen and the results are presented in Fig. 1.

Figure 1a shows the spectra of pure collagen, electrospun collagen, crosslinked and denaturalized collagen nanofibers. Indicating that the characteristic bands of type I collagen were clearly observed. Obtaining that for all the spectra evaluated the amide I (1700–1600 cm^{-1}), amide II (1600–1500 cm^{-1}), and amide III (1300–1180 cm^{-1}) are present. Demonstrating that the chemistry of collagen in the nanofibers with and without crosslinking is stable, when observing the spectra showing that these are identical to the native collagen without processing.

A study about the secondary structure of the obtained nanofibers consisted in evaluating the changed of the comprised bands in the amide I region from 1700 to 1600 cm^{-1}. The second derivative of the amide I data reported the location of four intense peaks in 1683 cm^{-1}, 1652 cm^{-1}, 1624 cm^{-1}, and 1617 cm^{-1}, which have been reported in the literature to belong to the bands (β-sheets), (α-helix), (triple helix), and (side chains), respectively [44–46]. Indicating that the secondary structure of collagen in the collagen nanofibers with and without crosslinking is maintained, in comparison to the denaturalized nanofibers, in which changes in the absorption were shown from 1620 to 1640 cm^{-1} indicating a loss in side-chain structures (at 1617 cm^{-1}) and in parallel β-sheets (1624 and 1628 cm^{-1}). Once the stability of the chemical and secondary structure of the nanofibers was confirmed, we performed an in vitro study using 3T3 fibroblasts cultured over crosslinked nanofibers, to evaluate the capability of the scaffold to promote cellular adhesion. Figure 1c, d shows a large number of cells that adhere and spread on the produced electrospun collagen fibers. Because of the rise of using benign solvents to prepare nanofibers through electrospinning, a new concept has risen called "green" electrospinning [47, 48]. This concept has opened the doors to the preparation of nanofibrous scaffolds that replicated the different tissues using benign solvents.

Focused on applying the concept of "green" electrospinning, we decided to prepare nanofibers that would replicate osseous tissue. By mid-2016 [15, 49] we were able to report the obtainment of collagen nanofibers containing hydroxyapatite, which have been previously developed using toxic solvents. The electrospinning solution consisted of 0.6 g of collagen, 0.3 g of hydroxyapatite dissolved in 3 mL of acetic acid/water at 90%, obtaining the nanofibers under an injection flow of 3 mL/h and a voltage of 47 kV. Nanofibers that presented a morphology and rough surface, by action of hydroxyapatite immersed in the collagen, in comparison to the pure collagen nanofibers. Since the fabrication of the nanofibers was based in the principles of "green" electrospinning, this assures the preservation of the chemical structure, in addition to the capacity to promote cellular adhesion of the seeded preosteoblast [15, 49].

In summary, through the application of "green" electrospinning, it is possible to generate a collagen nanofibrous that mimic to a large degree the characteristics of the extracellular matrix of tissues, preserving the chemical and secondary structure of collagen. This way allowing to offer a promising scaffold to the field of engineering, with the ability of supporting the different cells that compose tissue in the human body.

3 Properties of Collagen Nanofibers

The components of a polymeric solution are key elements towards the outcome of nanofibers, mainly because of the influence of electrospinning parameters on the polymeric solution [28]. Many researchers have focused on variable compositions, solvents, and components that affect solution conductivity to understand how each factor helps tune the diameter of the nanofibers [50]. For example, polymer concentration in the solution affects the solvent evaporation rate and viscosity which is linked to the fiber stretching, diameter and morphology [50]. The alignment of the electrospun nanofiber has high correlation with cell adhesion, growth direction, proliferation and mechanical properties of the tissue [51, 52]. This has aroused interest on figuring out how to control the development of aligned electrospun fibers through the electrospinning parameters [53]. Once the collagen scaffold is developed through the electrospinning process, in order to preserve its integrity, the crosslinking process is implemented. Crosslinking serves to control the mechanical properties, preserve its nanofiber morphology and improves degradability, all to mimic body conditions. Up next a discussion focused on diameter tuning through concentration variation of the polymeric solution; fiber orientation, and diameter tuning through electrospinning parameters and preserve collagen scaffold integrity via crosslinking.

3.1 Nanofiber Diameter Tuning Using Polymeric Solution Blends

As many efforts to develop an environment capable of mimicking the ECM to foment cellular activity, a variety of polymeric blends are an option to modify scaffolds morphology. Collagen plays an important role on the ECM and its biological properties such as cell recognition, making it an ideal material to use in the development of nanofibrous scaffolds but with limited mechanical properties [54]. Several works present how the addition of collagen influences solution composition and therefore, nanofiber diameter, as is the case of a blend of two biopolymers, such as thermoplastic polyurethane (TPU) and collagen. This blend overcomes the limitations of the collagen mechanical properties with the addition of the excellent mechanical properties of TPU as shown by Chen et al. [51]. Before the scaffold preparation TPU and collagen (80–100 kDa) were dissolved in 1,1,1,3,3,3-hexafluo ro-2-propanol (HFP) separately, and then blended at the desired weight ratios and stirred at room temperature. The variables studied were weight ratio and concentration, and the results indicate that the fiber diameter is influenced by both. An interesting trend is that weight ratios with higher quantity of collagen develops smaller average diameter fiber, with diameter distribution of 0.10–0.20 μm compared to pure TPU with a diameter distribution between 1.2 and 1.6 μm. The reason for diameter distribution is that collagen increases conductivity, which increases the stretching of the solution, resulting on smaller diameter fibers [51, 55]. To investigate the concentration factor (from 1.5 to 9 wt. % at equal intervals) a fixed weight ratio (TPU/collagen 80:20) was established. The outcome was that diameter increases as the concentration factor increases: 1.5 wt. %: 0.08–0.12 μm, 3 wt. %: 0.15–0.20 μm, 4.5 wt. %: 0.2–0.3 μm, 6 wt. %: 0.3–0.4, 7.5 wt. %: 0.3–0.4 μm, and 9 wt. %: 0.3–0.4 μm. In the two limits, high and low concentration, beads were developed because of chain entanglements. The appropriate concentration, after experimentation was 4.5 wt. %, with a diameter range from 0.2 to 0.3 μm, for a homogeneous nanofiber. This blend solution could produce an effective tissue engineered scaffold, with improved porosity and hydrophilicity, and with a diameter dependent on the concentration.

Another type of polymeric solution is a blend between type I collagen and polycaprolactone (PCL) [39]. The ratio between the collagen type I and PCL in the solution was used to analyze its effect on the nanofiber diameter, varying ratio percent from 100 to 0. The average diameters for the collagen percent in the solution blend are: 75%: 115 ± 38 nm, 50%: 130 ± 30 nm, 25%: 148 ± 30 nm, 0%: 160 ± 35 nm, indicating an increase in diameter as the content of collagen decreases, in other words the content of PCL increases. SEM analysis shows that the addition of PCL adds a globule structure to the nanofibers while also increases the diameter. The explanation for this diameter increment is attributed to the increase in viscosity of the solution that forms the electrospun nanofibers. Zhang and his group, in a similar work, corroborates that the scaffold diameter decreased from 987 ± 274 to 689 ± 299 nm as the collagen content increased compare with PCL [56]. These approaches served to understand how a balance of natural and thermoplastic material are mixed to achieve a biodegradable substrate. An example was the strength contributed by PCL and natural structural composition by collagen. Different blends of polymers give a variety of results and all of this is indicative of the sensitiveness of the electrospinning process.

An approach for vascular regeneration is using a poly-L-lactide (PLLA) and collagen type I blend (1:1) blend. This scaffold proved to differentiate mesenchymal stem cells into endothelial cells by the expression of proteins such as platelet endothelial cell adhesion molecule-1 (PECAM-1 or CD31) and Von Willebrand factor (vWF) [57]. The blend of PLLA/collagen type I (1:1), produced fibers of 210 ± 160 nm, in a different proportion, PLLA/collagen type I (3:1) produced fiber of 270 ± 120 nm when compared with PLLA fibers 430 ± 90. The reason for a decrease of diameter with presence of collagen is due to an increment on solution conductivity, which plays an important role in the electrospinning process. Pore analysis resulted in 0.273, 0.340 and 0.721 μm for PLLA/collagen type I (1:1), PLLA/collagen type I (3:1) and PLLA, respectively. Pore diameter is proportional to fiber diameter, and SEM images are shown in Fig. 2. The presence of collagen on the PLLA and collagen scaffold decreased the diameter and was found to have 256% higher proliferation than PLLA scaffold and prove the potential of mesenchymal stem cells to differentiate in vitro in the presence of differentiation media.

The approach of tuning the nanofiber diameter through the solution composition is proved to have a direct impact. A balance between biological and mechanical properties was reached by the addition of collagen and synthetic polymers respectively, to develop an optimal scaffold. The most common synthetic polymers are: TPU, PCL, PLLA, and tend to blend homogenously with collagen. The overall tendency is that the presence of collagen decreases the diameter of the electrospun nanofiber and is attributed to the increases in conductivity. The presence of collagen on the blend increased proliferation and the scaffold proved to be a viable solution for in vitro experiments and could further serve as ideal scaffolds for tissue regeneration.

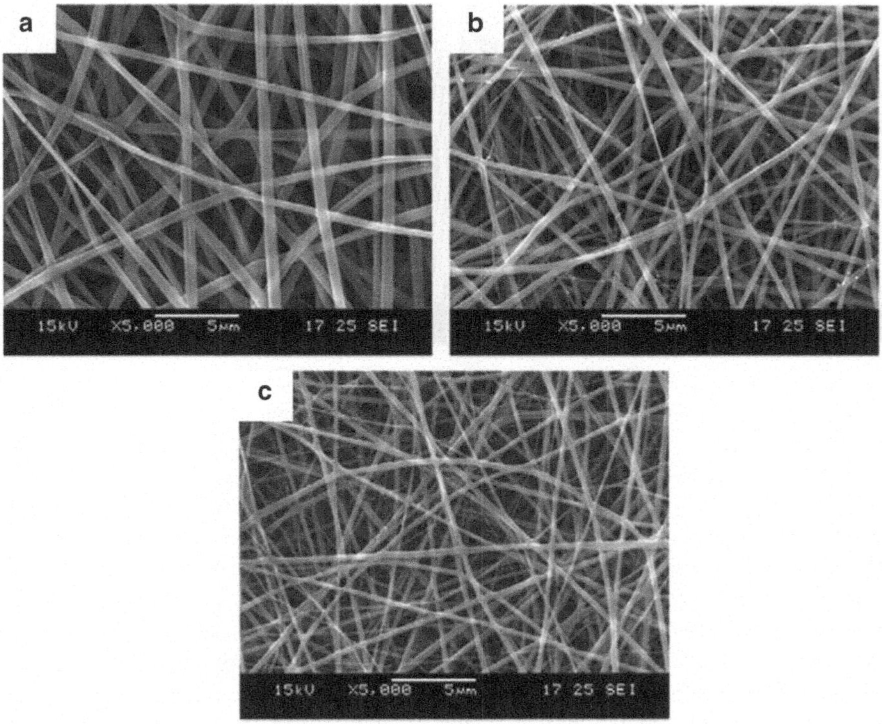

Fig. 2 SEM images of electrospun (**a**) PLLA, (**b**) PLLA/COL (3:1) and (**c**) PLLA/COL (1:1). Reprinted with permission from [57]. Copyright 2013 Elsevier

3.2 Electrospinning Parameters Impact on Nanofiber Alignment and Diameter

A variety of approaches have been investigated to obtain aligned electrospun fibers with collagen nanofibers and hexafluoro-2-propanol, such as: spinning onto a rotating drum, spinning onto a sharp edge of thin rotating wheel, introducing an auxiliary electrode or electric field, and others are proposed to develop aligned nanofibers [33]. Nanofiber orientation is mostly controlled by the type of collector, for example, a rotating collector for aligned and flat collector for random orientation [31–33, 52, 58]. In addition to the type of collector, tuning of the electrospinning parameters can also directly influence nanofiber orientation [15, 59, 60].

The solution composition, as mentioned above, contributes to the diameter of the scaffold. Nevertheless, the complexity of the solution also influences fiber orientation making the outcome of the nanofibers less predictable. A blend of collagen-chitosan-thermoplastic polyurethane (TPU), can be categorized as a complex mixture which means that the morphology of the scaffold is not only controlled by the electrospinning parameters, but also strongly by the conductivity imparted by components of the solution [58].

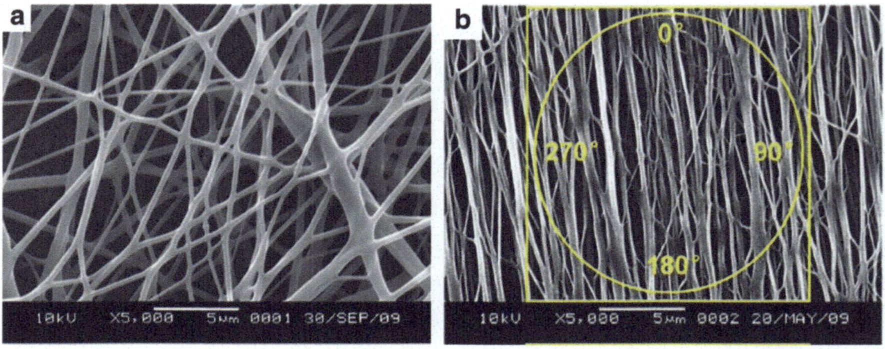

Fig. 3 SEM micrographs of (**a**) random oriented collagen-chitosan-TPU and (**b**) aligned oriented pure TPU (control) 5000× magnification. Reprinted with permission from [58]. Copyright 2011 Elsevier

The conditions to generate aligned nanofibers by Huang et al. [58] was a rotating drum (6 cm diameter) at a rotating speed of 4000 rev/min and 12 cm away from tip. The nanofibers resulted with average diameter of 256 ± 145 nm. SEM images of scaffold obtained with the complex mixture confirmed that a fully aligned scaffold is not obtained, when compared with TPU control, Fig. 3. The cause, for not aligned nanofibers, is attributed to all the interaction between the components of the complex mixture of collagen-chitosan-thermoplastic polyurethane. The random orientation was obtained using a flat collector plate wrapped with aluminum foil at 12–15 cm, with nanofibers average diameter of 360 ± 220 nm. This work highlights the fact that electrospinning is a complex system and that solution components play an indispensable role in the determination of fiber morphology by means of contribution to molecular weight, conductivity, viscosity and others. The generated scaffolds were cultured for 3 days with porcine iliac artery endothelial cells (PIECs) and Schwann cells (SCs). The PIECs cells showed a slight orientation on aligned fibers, compared with SCs that seem to be more responsive to the seeding matrix, with a spindled-shaped morphology on aligned. This confirms that cells are influenced by the orientation of the scaffold and sensitive to the morphology.

A flexible wheel with high rotational speed 10–20 m/s is the proposed assembly by Zhong et al. [33] to achieve aligned nanofibers. The focus was to study the effect of alignment on cell orientation and cell-scaffold interaction. The control group were random nanofibers prepared by a nonmoving metal collector. The average diameter for aligned was 180 and 250 nm for random and the morphology is confirmed with SEM micrographs. The aligned scaffolds have lower porosity due to less overlaying of fibers compared with the random. Is important to mention that this research group did an optimization experiment and concluded that a rotation speed of 5 m/s does not generated aligned fibers and high rotational speed of 25 m/s is too strong of a force and breaks the fibers. Therefore, it was established that 15 m/s as an appropriate rotational speed. This established rotational speed generates thinner aligned fibers because it exerts a pulling force on the solution jet when compared with a static collector.

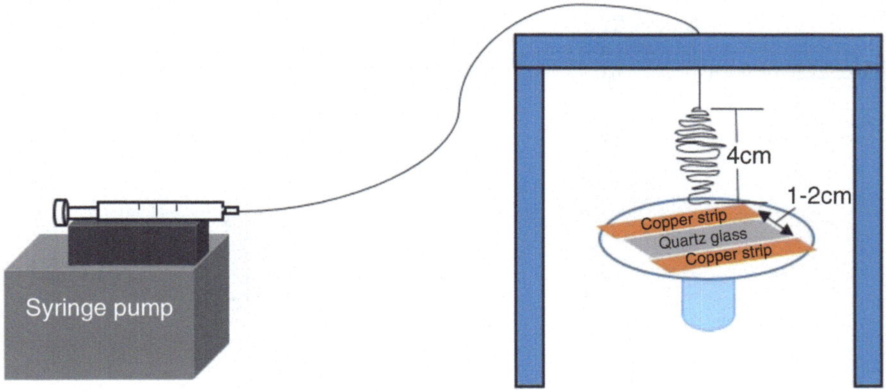

Fig. 4 Electrospinning assembly. Adapted from [61]

Collagen's versatility is taken in advantage by developing a composite collagen structure [61]. The combined technologies are gel and electrospinning to achieve a three-dimensional hybridized collagen implant to mimic tendon architecture. The reason for the composite is to obtain a more voluminous collagen scaffolds, a requirement for tendon implants. This is achieved by the addition of a collagen gel to collagen fibers both being promising matrixes for tissue engineering [62]. The polymeric solution consists of acetic acid and collagen type I, 7.5 w/w % collagen. The aligned collagen fibers were constructed via an assembly consisting of two dual plates device, 2.5 × 0.5 cm copper strips, 4 cm away from needle tip, attached with Gluseal to 1 cm gap of quartz glass [61] Fig. 4. The fibers alignment is attributed to the electrical resistivity of the gap substrate, quartz glass with 10^{20} Ωm, in which the fibers were collected. Also, the distance between the gap contributes to fiber alignment with the tendency that narrow plates developed more aligned fibers. The potential difference is incremented with the additions of an insulating material between the copper strips and is increased by decreasing the distance of the gap which leads to the align fiber organization along the quartz glass surface. The aligned fiber resulted with a diameter of 272 ± 183 nm, and later was mixed in a chamber with collagen solution to develop the collagen composite, overall to induce proper healing and recovery, attributed to the presence of the types of collagen, gel and fibers.

Nanofibers develop by Prabhakaran et al. [63], consisted of poly(3-hydroxybutyrate-co-3-hydroxyvalerate (PHBV) and collagen. Two different ratios were evaluated PHBV:collagen solution 8 w/v. % at 75:25 and 50:50. The solution preparation is not directly discussed only that is blended and deposited on syringe. Random fibers were produced by a flat aluminum collector (Fig. 5a) and the aligned fibers by the collection of the fiber on a rotating drum set to 2400 rpm, (Fig. 5b). The electrospinning parameters consisted of 12 cm as the distance between flat tipped spinneret, flow rate 1 mL/h, and 15 kV. The syringe containing the polymeric solution as assembled in a vertical position, ejecting down towards the collector. The aligned fibers using a rotating cylindrical drum, had diameter between 205 ± 50 nm for

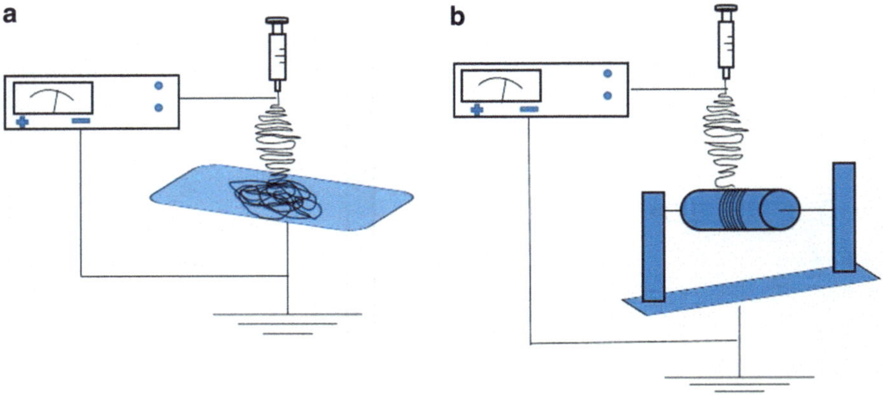

Fig. 5 Electrospinning assembly; (**a**) random fibers on flat aluminum collector; (**b**) aligned fibers on rotating drum collector. Adapted from [63]

75:25 ratio and 229 ± 65 nm for 50:50 ratio and the random fibers had an overall diameter of 472 ± 85 nm. The cell-scaffold interaction was analyzed using PC-12 cells at day 8, higher proliferation of 34.30% for random and 40.00% for aligned fibers relative to PHBV nanofibers.

Ouyang and his group were able to develop a seamless 3D nanofibrous nerve conduit due to collagen's high biocompatibility and the strength of the synthetic polymer poly(lactic-co-glycolic acid) [64]. The focus was aligned nanofibers produced via a new approach of the electrospinning technique. The polymeric solution is a blend 25/75 collagen/PLGA; 8% collagen and 12% PLGA were first dissolved separately overnight on 1,1,1,3,3,3-hexafluoroisopropanol (HFIP). The mandrel that served as the collector for the aligned nanofibers was composed of three parts: two metal sides and an insulator that separated the two metal sides. The general mechanism that made possible the fibers alignment is by the potential difference developed by the insulating material and the two metal sides causing the deposit fibers to align along the longitudinal axis of the mandrel as it rotates (Fig. 6) [64]. The advantage of this assembly is that diameter of the nanofiber can be modified through mandrel sizes and nanofiber diameter can range from 1 to 10 mm. The obtained average diameter of aligned and random collagen:PLGA was 766 nm and 834 nm, respectively. Once again this highlights the versatility of electrospinning which eliminates the process of conduit rolled that provoked discontinuous joint and inconsistent size. A brief in vivo experiment was developed to regenerate 13 mm sciatic nerve defects and these were compared between randomly, aligned conduits and autograft. Results confirmed that alignment promotes better motor function, axonal conduction and Schwann cell morphology when compared with random. The autograft resulted better on axon regeneration, myelination, action potential propagation, neuromuscular transmission and functional recovery, when compared with aligned seamless 3D nanofibrous nerve conduit.

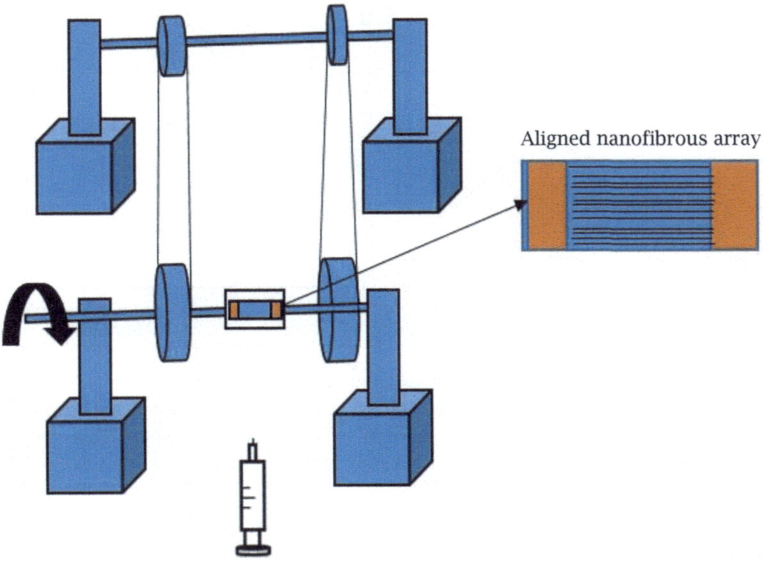

Fig. 6 Electrospinning assembly. Adapted from [64]

In our most recent work, the objective was to evaluate the effect of the flow rate and voltage in the diameter and orientation of the electrospun fibers [15, 49]. The particularity of this work when compared to others is the composition of the polymeric solution. The components of the solution are: pure collagen type I and a non-toxic solvent, acetic acid. Previous presented literature had the necessity to use a blended polymeric solution of synthetic and natural polymers to produce electrospun nanofibers and achieve promising mechanical properties.

The experiment consisted on varying the flow rate from 0.5 to 3.0 mL/h, at a constant voltage of 45 kV and another round varying the voltage from 25 to 45 kV, with a constant flow rate of 1.5 mL/h. The observed trends were decrease in diameter, from approximately 400 to 200 nm, as the voltage increases and increase in diameter, from approximately 160 to 350 nm, proportional to the flow rate (Fig. 7). A trend for fiber orientation was observed: for a constant voltage of 47 kV, aligned fibers were obtained with high flow rate 5 mL/h and random fibers with low flow rate of 1 mL/h, both validated with SEM. The orientation is confirmed with orientation histograms and normalized orientation index (NOI). A 58.9% NOI correspond to random orientation and 90% NOI to aligned orientation. The average fiber diameter for aligned orientation was 202.47 nm and for random 117.52 nm. The aligned nanofiber had a larger diameter because it was produced with a high flow rate which means more solution available at the tip of the syringe for the same voltage to stretch. Is important to mention that the collector used was a rotational drum, what suggests that random and aligned fibers can be obtained with a rotation drum by controlling the flow rate, even though other researchers used rotational drum only for aligned electrospun fibers.

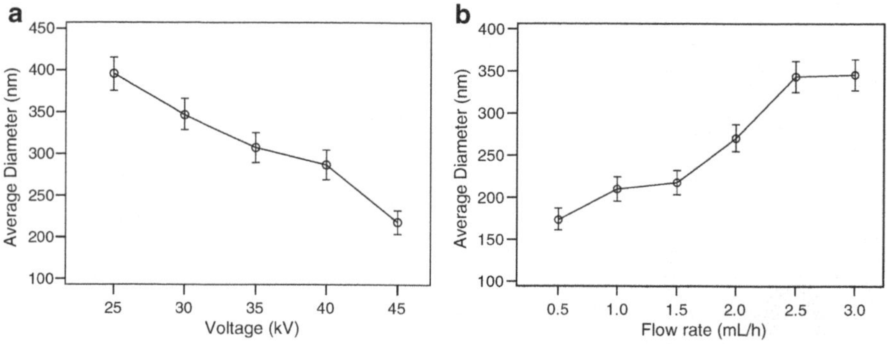

Fig. 7 SEM (**a**) Average fiber diameter versus applied voltage for fix flow rate of 1.5 mL/h. (**b**) Average fiber diameter vs. flow rate for fixed voltage at 45 kV. Values plotted and error bars in (**a**, **b**) correspond to the average and standard error of three independent experiments, respectively. Reprinted with permission from [15]. Copyright 2016 Wiley-VCH Verlag GmbH & Co. KGaA

The overall electrospinning parameters based upon recent literature are the following: flow rate of 1 mL/h, voltage range from low as 15 kV and a maximum of 47 kV, tip-collector distance range from 10 to 15 cm, needle-inner diameter 21-G and 22-G. The type of polymers used goes from natural to synthetic and all contribute in different ways to the end product and each of the components interacts differently with the electrospinning parameters [65]. Variation in polymer concentration along with the solvent, tunes the diameter size of the electrospun fibers [39, 56, 57]. It is important to emphasize that fiber orientation and diameter define the scaffold morphology which provides the biochemical and biomechanical cues for the cell environment, and further stimulate tissue regeneration [65]. Fiber orientation is mainly achieved through a rotational drum as a fiber collector, although our most recent work serves a platform to evaluate that alignment can be modified through the flow rate [15, 31, 33, 49, 52]. Electrospinning parameters clearly impacts fiber diameter and alignment along with polymeric solution concentration. This technique along with natural polymer collagen combined to truly innovate and contribute to the future of biomaterial to mimic ECM.

3.3 Mechanical Properties in Relation to Crosslinking

The structural integrity of collagen is key to simulate the native collagen and its morphology is preserved through the crosslinking process, which allows collagen scaffolds to withstand body conditions [66]. Crosslinking agents also known as crosslinkers are molecules that form covalent bonds with functional groups (primary amines, sulfhydryls, etc.) [67]. Some of the ones often found in literature are: glutaraldehyde, formaldehyde, and *N*-(3-dimethyl-aminopropyl)-*N*′-ethyl carbodiimide hydrochloride (EDC) [33, 68]. The products developed by electrospinning possess high degradability, especially the ones in which the main component is collagen, due to the low melting point and quick denaturation [33]. This high degradability limits the product to endure an aqueous environment, like in vitro conditions, and body temperature [69]. The crosslinking technique not only improves the degradability of nanofibers; it also impacts the mechanical properties. The tensile strength of a specific tissue can be simulated on a biomaterial by variation on the type of crosslinkers, method of application and time of reaction [70]. The assembly for the crosslinking technique is not mentioned in most of the literature. Presented below are several crosslinkers, mechanical properties, and possible assemblies that have been found in literature.

Zhong and colleagues developed electrospun fibers with collagen nanofibers and hexafluoro-2-propanol (HFIP). The scaffold was crosslinked with 30% glutaraldehyde vapor for 48 h at room temperature to obtain a physically stable product and study its effects in vitro with fibroblast culture [33]. To remove the unreacted glutaraldehyde, an additional solution of 0.002M glycine is added, followed by a wash with deionized water for 20 min. Glycine is an amino acid which is used to quenched the residuals of crosslinkers such as, glutaraldehyde and formaldehyde [71]. SEM images confirmed the integrity, meaning unchanged orientation and overall porous morphology of the fibers. The surface roughness of the nanofiber was analyzed by atomic force microscopy (AFM), in Ra which is the arithmetic mean value of the surface roughness. The results for crosslinked random fibers was 60 Ra, and for uncrosslinked random fibers, 90 Ra. The water content, which is measured in water contact angle, was increased from 13° to 30°, which confirms the effectiveness of the crosslinking process.

The surface roughness analysis shows that the crosslinked nanofiber had smoother surface compared with uncrosslinked nanofiber, due to the washing steps needed on the crosslinking method, for both random and aligned [33]. Increase in water contact angle measurements helped confirm the reduction of the hydrophilicity, and prove the reaction of the glutaraldehyde and collagen amine groups. The contact angles for aligned collagen fiber and random collagen fiber were 26° and 11°, respectively. After crosslinking the new contact angles for the collagen fibers were: 41° for aligned and 30° for random. The morphology of the electrospun fiber influence on the water contact angle, for example, larger contact areas on random oriented fibers leads to higher surface spreading and a lower contact angle (higher wettability). The applied crosslinking process with 30% glutaraldehyde vapor for 48 h gave positive results being able to enhance the structural integrity of the electrospun fibers.

Products for wound healing application, first needed to go through crosslinking. Rho et al. [6] has a very similar product and also applied a very similar crosslinking process as Zhong et al. [33]. Nanofibers of collagen type I and 1,1,1,3,3,3-hexafluoro-2-propanol (HFIP) were exposed to less time of reaction 12 h, 0.1M glycine instead of 0.002M glycine and glutaraldehyde solution was 25% rather than 30% [6]. Overall results showed a decrease in porosity from 89% to 71%. Is important to understand that each variation will have a different impact on the mechanical properties of the fibers and because of that, it is crucial when those parameters are being defined. This research group used a very simple yet important test that helps define

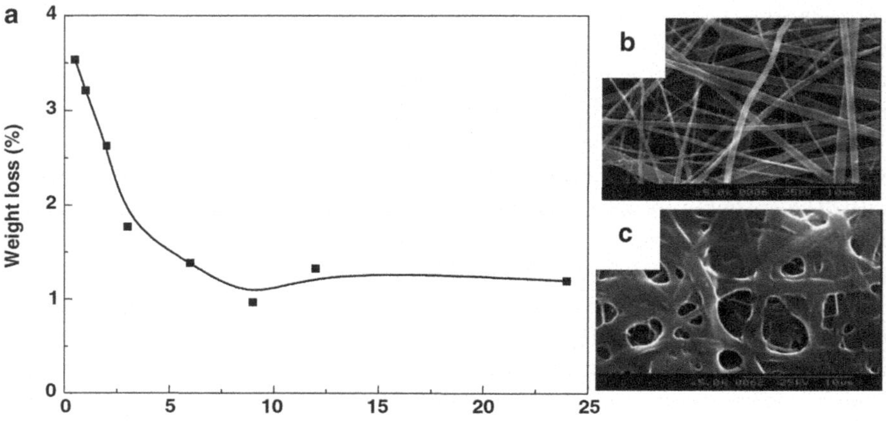

Fig. 8 (**a**) Weight loss of the collagen nanofibrous matrix by cross-linking time. Weight loss of the collagen nanofibrous matrix by cross-linking time. (**b**) Collagen non-crosslinked fibers. (**c**) Collagen crosslinked nanofibers Reprinted with permission from [6]. Copyright 2006 Elsevier

the appropriate time of crosslinking reaction needed to prove its effectiveness. This weight loss (%) of collagen matrix consist of weighting the dried fiber before and after immersion on distilled water for 1 h for different times of crosslinking exposure and example of the plot is shown on Fig. 8a.

The crosslinking time of reaction defines the time of exposure in which the covalent bonds between the glutaraldehyde and amine collagen are formed [66]. This is very useful to determine the time in which the nanofiber is no longer absorbing significant quantities of water, proving the effectiveness of the crosslinking, which was identified to be 12 h. The SEM images on Fig. 8 before crosslinked (b) and after crosslinked (c) represent the reduction in porosity from 89 to 71%, along with the pore volume decreased from 9.328 to 1.566 mL/g, although it is believed that the matrix achieved higher dimensional stability in aqueous medium. The mechanical test of tensile strength matches commercialized products, at dry state, as the collagen matrix (0.2 mm thick) tensile strength was above 10 MPa, although when wet, the tensile strength was 7.40 ± 1.17 MPa, less than commercially product Resolut® LT with 11.72 MPa. Shape and stability were maintained under in vivo procedure and this proved the effectiveness of the crosslinking technique.

As mentioned earlier the assembly for the crosslinking technique remains a mystery but Huang et al. [58] work is one on the few that mentioned several details about the assembly for the crosslinking process. It consisted of a desiccator, Petri dish with 10 mL of 25% glutaraldehyde and a fixed membrane on a glass frame in the upper layer of the desiccator, at room temperature for 2 days. The main purpose was to solve the issue of the collagen-chitosan-TPU scaffold degradation for application as a tubular scaffold to mimic the native artery, which is reported to have a tensile strength of 1.5 MPa. The crosslinking with glutaraldehyde had a positive influence on the tensile strength but a negative influence on the average elongation at break, making the scaffold stiff and brittle.

The thermoplastic TPU is a key component that solved that issue, since it helped to improve elasticity depending upon the ratio, which was set to be higher than 25% [58]. Both tensile strength and elongation at break had significantly different results based on the type of direction being analyzed, parallel or perpendicular. The tensile strength results for aligned scaffold on parallel direction was, 14.93 ± 0.59 and 5.04 ± 0.95 MPa for perpendicular. The elongation at break for parallel direction was 58.92 ± 15.46% and for perpendicular direction was 8.20 ± 0.84%. This marked difference is attributed to the orientation of the fracture. This indicates the importance of fiber orientation combined with crosslinking. The elongation at break for random non-crosslinked fibers was 61.30 ± 3.88 and 9.87 ± 1.77% for crosslinked fibers.

A variety of synthetic and natural crosslinking agents impact the dimensional stability and cytocompatibility of electrospun collagen scaffolds, [68]. The synthetic and natural crosslinking agents that have been used are: genipin, glutaraldehyde, EDC and EDC with *N*-hydroxysulfosuccinimide (EDC-NHS). The structural integrity of the scaffolds was evaluated by an in vitro immersion study for 3 months and cytocompatibility with human mesenchymal stem cell. Two methods were implemented, glutaraldehyde vapor and liquid (immersion) for genipin, EDC, and EDC with EDC-NHS. The composition of the prepared crosslinking agents were: glutaraldehyde (0.5%) in deionized water (DI) for 19 h exposure, 200 mM EDC in ethanol for 18 h, 200 nM EDC with 200 m NHS in ethanol for 4 h, and genipin in (1, 5, and 10%) in ethanol for 72 h at 37 °C. The collagen scaffold retained fiber morphology for 10% genipin and EDC, EDC-NHS for up to 3 months. Genipin concern was that it required extra washing step, which is dependent upon the amount used of genipin, this is important cause it affects the cytotoxicity in culture medium. The degradation of the scaffolds crosslinked with EDC, EDC-NHS is related with the content of free amino groups which was slowly with time. Overall results indicate that EDC and EDC-NHS not only retained fiber structure long-term but also support cell viability.

EDC is a widely used crosslinking agent, and for the first time it was compared with 1,4-butanediol diglycidyl ether (BDDGE) [41]. Is the first time that BDDGE is used as crosslinking agent, especially for collagen electrospun fibers, and is less toxic than glutaraldehyde, because of residual and unstable glutaraldehyde polymers may be retained in the crosslinking procedure [72]. For the two agents, EDC and BDDGE, the solution was 5% w/v in EtOH, the methodology consisted of scaffold fix at CellCrown® 24 plastic rings and immersed in crosslinking solution 37 °C for 7 days. Two types of collagen fibers were developed one using tri-fluoroethanol (TFE) as solvent and other fibers using acetic acid (AcOH) as solvent to evaluate the effect of crosslinking in both. The nanofibers developed with TFE as solvent had 10 wt. % collagen concentration and resulted in fiber diameter 320 ± 80 nm and the ones prepared with dilute acetic acid (AcOH) with 30 wt. % collagen concentration with a diameter of 150 ± 30 nm. SEM characterization, revealed that collagen scaffold crosslinked with BDDGE retained fiber morphology, unlike the collagen scaffold crosslinked with EDC that lost its porous morphology because of fiber swelling. The mechanical properties are dependent upon the crosslinking agents (EDC and

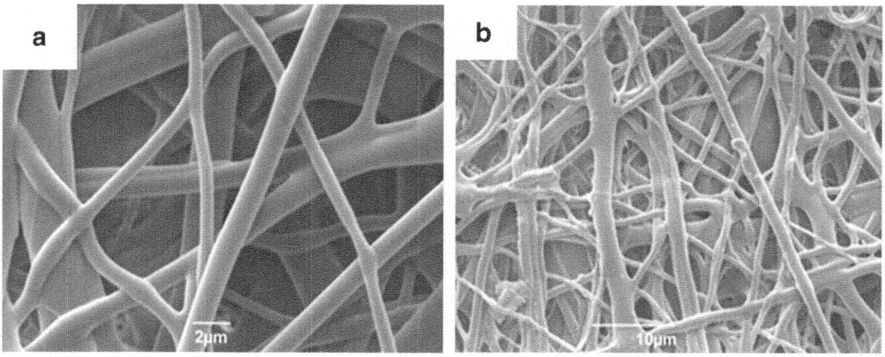

Fig. 9 (**a**) Crosslinked nanofibers, (**b**) crosslinked nanofibers after exposure to water at 37 °C. Reprinted with permission from [15]. Copyright 2016 Wiley-VCH Verlag GmbH & Co. KGaA

BDDGE) and solvents (AcOH and TFA). EDC crosslinking agent resulted with the highest elastic modulus and stress at break when compared with BDDGE agent. The elastic modulus results for EDC crosslinking agent were the following: 870 ± 160 MPa for electrospun fibers with AcOH solvent and 860 ± 80 MPa for TFE solvent. The elastic modulus results for BDDGE crosslinking agent were the following: 590 ± 190 MPa for AcOH solvent and 160 ± 30 MPa for TFE solvent. The stress at break results for EDC agent had no significant difference between solvents with a value of 12.3 ± 0.1 MPa. The stress at break for BDDGE agent had a different value for each of the solvents 6 ± 1 MPa for AcOH and 12 ± 3 MPa for TFE. Both crosslinking agents improved the mechanical properties of non-crosslinked electrospun which had an average elastic modulus of 86 MPa and a stress at break of 1.5 MPa. Clearly crosslinking improved the mechanical properties of the electrospun collagen scaffold by orders of magnitude.

On our most recent work on crosslinking collagen electrospun fibers, we compared the mechanical properties of the electrospun nanofibers with two crosslinking techniques: immersion and vapor exposure both using a 25% glutaraldehyde solution [15, 49]. The assembly consists of a hollow top in which the scaffold is fixed to avoid shrinking. On the bottom of the container the glutaraldehyde solution was placed and the humidity was controlled between 35 and 50%. The time of exposure for the vapor method was 18 h, and 1 min for the immersion method. SEM characterization Fig. 9 confirms the integrity of collagen nanofiber after vapor exposure (a) and after thermal test with 37 °C water (b). The immersion technique has a Young's modulus of 4.1 ± 0.5 MPa and an elongation at break of 29 ± 15%. The vapor method reported a Young's modulus of 2.7 ± 0.7 MPa and an elongation at break of 38 ± 7%. These results help understand how each method can be used to deliver water stable fibers with tunable mechanical properties in order to adapt it to the native tissue.

The crosslinking impact on the electrospun fibers depends upon the components, external parameters, and many other factors [66]. Crosslinking is a very sensitive method, and each researcher should test and prove each technique to evaluate the results of their product. The most common crosslinking agents are glutaraldehyde, by the vapor method, and EDC by immersion. These findings contribute to tissue engineering; it opens the possibility to generate and modify a scaffold that eventually will help regenerate and reestablish tissue functions. The crosslinking method is a crucial step in order to achieve the biomaterial capable to possess the mechanical properties of the native tissue [67].

4 Biomedical Applications

Nanofibrous collagen scaffolds fabricated using the electrospinning process have shown to possess excellent biocompatibility and properties that are necessary for tissue engineering [18], and have been presented throughout this chapter. They have been able to mimic the ECM of native tissue, facilitating cell proliferation and attachment [19]. These scaffolds can, therefore, be applied for the regeneration of nerve [20], bone [21], skin [6], tendon/ligament tissue [22], and vascular grafts [23], among others, by providing promotion of function and tissue repair, both in vivo and in vitro. This section focusses on several biomedical applications of collagen nanofibers that are useful for tissue regeneration and function.

4.1 Nerve

The nervous system has a leading role in the human body, controlling the function of the different organ systems and physiological processes. Injuries to the central and peripheral nervous systems can cause many problems to a patient's everyday life and could therefore benefit from different tissue engineering strategies to facilitate regeneration, such as electrospun collagen nanofibers, providing a good platform for nerve repair [18]. Because nerve regeneration in an injured spinal cord is usually limited, it contributes to the devastating outcome of neurologic impairment below the site of injury and these tissue-engineered scaffolds have evolved as a potential treatment method [18].

Nanofibers mimic the natural extracellular matrix and have therefore been evaluated as potential applications for spinal cord injury both in vivo and in vitro by [18, 38]. These researchers had two main goals, to evaluate topographical effects on astrocyte behavior and to demonstrate the feasibility of using electrospun nanofibers for the treatment of acute spinal cord injury (SCI) using a rat hemi-section model. Type I collagen aligned and random nanofibers were fabricated using collagen extracted from rat-tails with 90% acetic acid, instead of HFIP, to prevent the denaturation of collagen [38]. Primary astrocytes from rat embryos were first seeded in collagen-coated class and randomly oriented or aligned collagen nanofibers [38]. The best response of astrocytes was observed in aligned nanofibers, which exhibited an elongated morphology. Nonspecific orientation on randomly oriented fibers and collagen-coated glass only, was observed [18, 38]. In addition, the elongation factor and cell shape index were significantly higher when using aligned nanofibers than those with random fibers or glass cover slips. Although cell shape index on aligned fibers was significantly higher, they suggested that astrocytes did not truly demonstrate morphological changes and that a higher cell shape index could have been likely caused by the elongated morphology in response to aligned nanofibers instead of enhanced cell activation [38]. Dorsal root ganglia were also seeded onto sterilized random and aligned scaffolds and coated glass. Neurite outgrowth from dorsal root ganglia explants followed the orientation of aligned fibers [38]. They then proceeded to form collagen scaffolds into spiral tubular structures by rolling the wet crosslinked scaffolds into tubes of four to five layers.

These researchers created the lesion on a rat hemi-section model and implanted the tubular structure. At days 10 and 30 post implantation, cellular penetration into the collagen fiber constructs was observed. Scaffolds with aligned fibers appeared more structurally intact at day 30 [18, 38]. They evaluated the lesions for 30 days due to previous research that indicated that 30 days should be enough to observe significant neural regeneration useful for analyses of structural recovery of the lesion [73].

Fluorescent images of the implanted area were taken after 10 and 30 days on both aligned and random nanofibers. Immunofluorescent staining showed a large number of macrophages on day 10 and a drastic increase by day 30. This significant decrease in cell density by day 30 allowed this group of researchers to suggest that there was a reduction in acute inflammatory response and therefore that the collagen scaffolds were biocompatible with the native tissue. Astrocytes with GFAP staining, were found in the borders of the implanted area. This is very significant because it has been demonstrated that astrocytes, like other glial cells, prevent the successful regeneration of the axon after a spinal cord injury [38]. Therefore, it is important for these astrocytes not to appear in the implanted area. Coupled with the ability to support and direct neurite outgrowth, electrospun collagen scaffolds may be more favorable than collagen introduced in 2D format for SCI repair [38]. Finally, Masson's trichrome staining results indicated that the tubular structures were all retained after implantation and cell penetration into the scaffolds was observed. These findings indicated the feasibility of fabricating electrospun collagen nanofiber 3D spiral structures and their potential use for SCI repair [38].

Improving axonal regeneration by combining seeding cells with scaffolds has also been an effective method [18]. Boecker and his research team implanted a predifferentiated MSC-seeded micro-structured collagen nerve guide (Perimaix) in a 20 mm rat sciatic nerve defect [74] and saw that these cells helped in axon regeneration into the Perimaix nerve guide. MSC-seeded Perimaix nerve guide could lead to functional recovery similar to autologous nerve transplantation [18]. Bozkurt et al. have also managed to achieve similar results using a 20 mm rat sciatic nerve bridged with the same collagen-based nerve guide (Perimaix), but using Schwann cells as the seeding cells [75]. These provided a valuable effect on myelination within the scaffold [75].

Additional research findings showed that supplementary physical stimulation such as ultrasound [76], magnetic field [77], electric [78], and laser [79] could enhance functional recovery of peripheral nerve injuries. These physical stimulations combined with electrospun scaffolds could be introduced to promote the repair of injured nerves. Therefore, the use of collagen nanofibers has provided biomimicking signals [38] in addition to showing the topography, especially in aligned nanofibers, for neural tissue engineering and the enhancement of Schwann cell maturation, among other types of cells [38].

Injuries in the nervous system can cause many problems in a patient's life. Therefore, advances like the ones observed here such as astrocyte behavior at the lesion site, macrophage density decrease for spinal cord repair and axonal regeneration by combining cells with scaffolds, could be beneficial to facilitate nerve regeneration [18]. Electrospinning of collagen nanofibers for nerve repair have provided a good platform, due to their biomimicking signals, their topography and their aligned morphology for neural tissue engineering in addition to enhancing cell attachment and proliferation.

4.2 Bone

The main challenge of bone tissue engineering is the fabrication of a bone graft that can mimic the ECM of tissues with the necessary bone mineralization [21]. To develop fibers with these specifications, different biocompatible composite materials, such as nanohydroxyapatite (HA) have been used to fabricate nanofibers, in addition to collagen [80, 81]. Because collagen and hydroxyapatite are two major components of the ECM, these have been highlighted to have advantageous features when used for tissue engineering applications [80]. Prabhakaran et al. fabricated composite nanofibers of PLLA, type I collagen and hydroxyapatite, using the electrospinning technique [18]. They suggested that these membranes could improve the proliferation and mineralization of osteoblasts, which could therefore help in bone regeneration [21, 80]. Figure 10 shows SEM images of PLLA, (a) PLLA/HA (weight ratio: 80:20), (b) and PLLA/collagen/HA (weight ratio: 40:40:20) (c) composites, which displayed HA particles that have been embedded in the nanofibers. The topography of the nanofibers play an important role in cell behavior, for processes such as adhesion and proliferation [38]. In all the blends, HA is at a low ratio in comparison to PLLA and collagen because the purpose is to obtain continuous nanofibers, while having a large amount of HA can produce broken ones [80].

In their study, human fetal osteoblasts were cultured and seeded on PLLA, PLLA/HA, PLLA/collagen/HA and tissue culture polystyrene as control. Cell proliferation was monitored every 5 days for 20 days in total. These showed to adhere and actively grow on all combinations of nanofibers, but the highest proliferation occurred on cells seeded over PLLA/collagen/HA nanofibers [18]. These were found to be 41%, 24% and 11% higher compared to PLLA/HA scaffolds after day 10, 15 and 20, respectively [21]. These showed enhanced mineral deposition of 57% higher than the PLLA/HA nanofibers [18, 80].

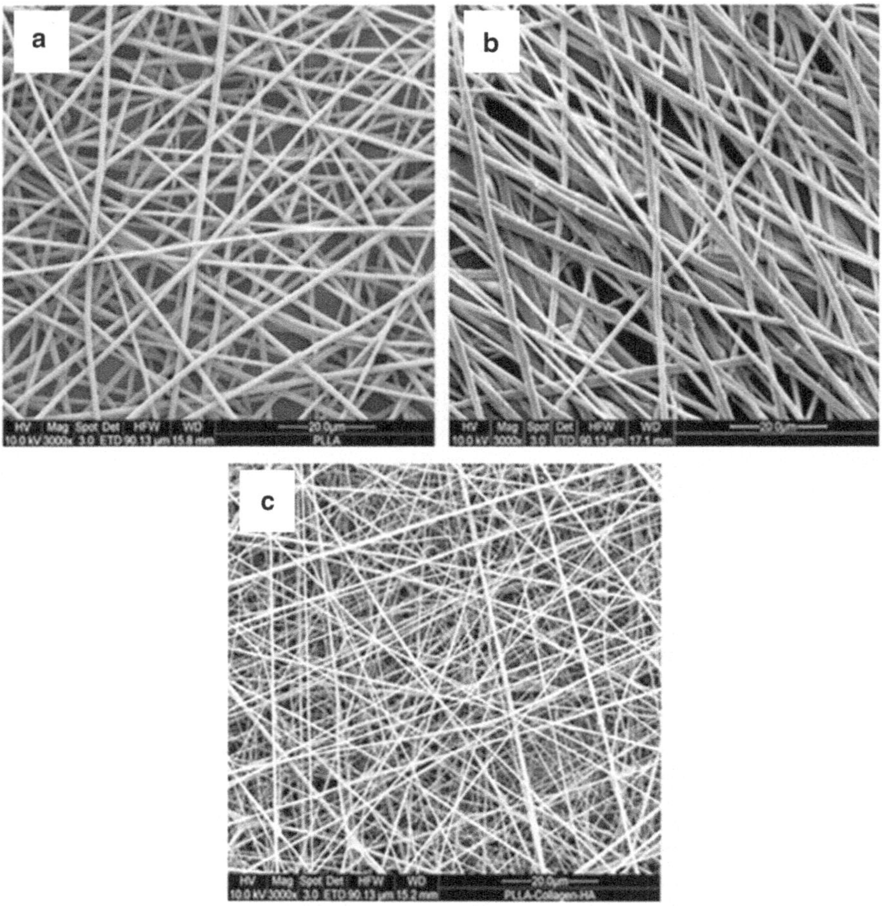

Fig. 10 SEM images. (**a**) PLLA (fiber diameter 860 ± 110 nm) (**b**) PLLA/HA (fiber diameter: 845 ± 140 nm) (**c**) PLLA collagen/HA nanofibers (fiber diameter: 310 ± 125 nm). Reprinted with permission from [21]. Copyright 2009 Elsevier

In another study by Chen et al., an internally structured collagen/HA scaffold was fabricated. They found that the porosity and compressive modulus could be regulated depending on the collagen proportion [18]. The porosity decreased while the compressive modulus increased with an escalation in the collagen proportion [76]. The scaffold with a high collagen proportion also had the best performance of MSCs' viability, proliferation and osteogenic differentiation in contrast to those with low proportion of collagen [18, 76]. Finally, in vivo studies demonstrated a significant effect in cell infiltration, meaning that controlling the internal structure of a collagen/HA scaffold can provide an efficient carrier to repair bone [76].

To provide the necessary characteristics for a good bone graft in orthopedic applications, a combination of structural, biological and mechanical properties are important for cell seeding, proliferation and formation of new tissue. PLLA/collagen/HA scaffolds were fabricated by the electrospinning process and osteoblasts were grown on these scaffolds, showing cell adhesion. In addition, studies have shown that as the collagen proportion increases, MSCs viability, proliferation and osteogenic differentiation increases [18, 76]. This allows for the conclusion that these nanofibrous scaffolds show great potential for cell proliferation, adhesion and mineralization, which could be promising for bone tissue engineering [21].

4.3 Dermal (Skin) and Wound Healing

The skin is the largest organ in the body. Several of the most commonly used skin substitutes, such as allografts and autografts, have shown that they are not able to solve for problems caused by extensive skin loss. Therefore, many researchers have directed their work to tissue engineering in order to promote the regeneration of the skin [18]. Collagen is the most important dermal replacer used in various forms [82]. Rho et al. used the electrospinning of type I collagen to yield nanofibers and observe if they accelerate early wound healing response [6, 18]. They also observed the effects of cell behavior, cell and collagen nanofiber interactions and examined open wound healing in rat [6]. Normal human oral keratinocytes (NHOK) were isolated from epithelial tissue of healthy volunteers undergoing oral surgery, prepared and cultured [6]. Their results showed that scaffolds promoted cell adhesion and the spreading of human keratinocytes in vitro. Because adhesion to nanofibers alone was relatively low, they added several different ECM proteins such as fibronectin, type I collage and laminin to promote adhesion and proliferation. Results of these studies showed that type I collagen and laminin are both functionally active in cell adhesion of normal human keratinocytes [6].

Figure 11 shows the results of open wound healing test where the collagen scaffold accelerated the disappearance of the surface tissue debris and proliferation of fibroblasts in the early stages of wound healing after 4 weeks [6]. As can be observed in the Fig. 11c, d, those containing the collagen nanofibers, show the accelerated decrease in debris when compared to the controls at week 1 and 4, Fig. 11a, b, respectively. In addition, inflammatory cells disappeared and connective tissue was able to form [6].

In another study, dermal injuries were done on adult guinea pigs and treated with scaffold composed of electrospun type I collagen after being vapor crosslinked to varying degrees [83]. They treated each wound with a scaffold and covered it with a piece of silver gauze that was sutured in its place for 5–7 days. To track the healing process, the total wound surface area as a function of time and treatment was measured [83]. Results showed that when crosslinking was increased to 70%, wound surface area was increased dramatically, which indicated increased regeneration.

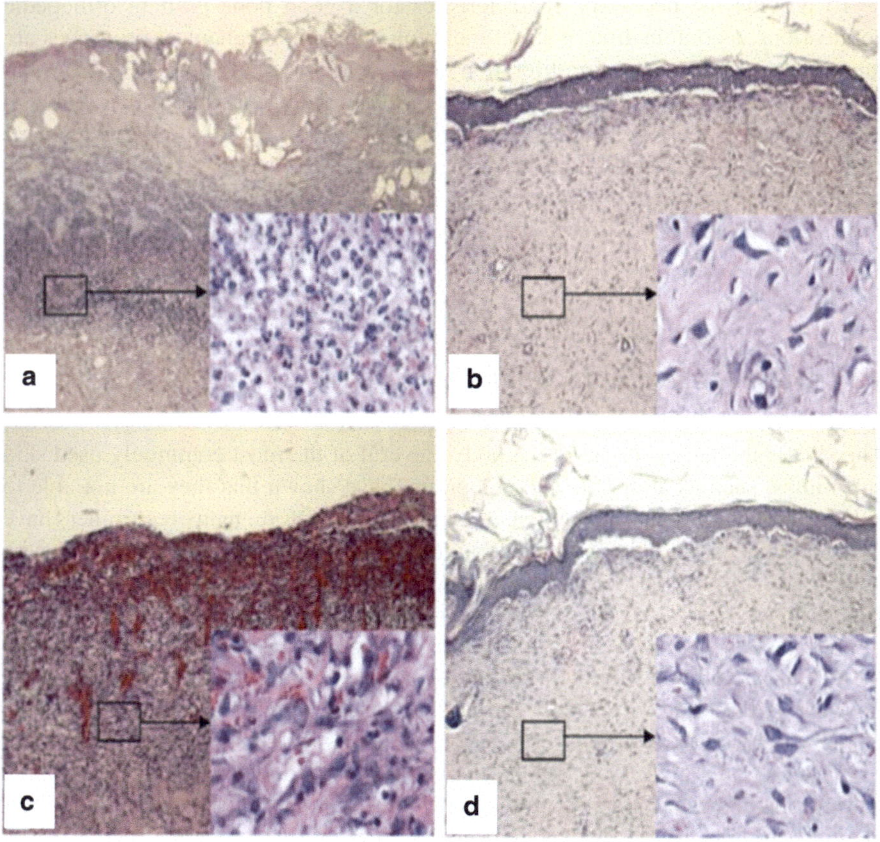

Fig. 11 Photomicrographs of wound healing of rat skin; control group at 1 week (**a**), control group at 4 weeks (**b**), collagen nanofiber group at 1 week (**c**), and collagen nanofiber group at 4 weeks (**d**). Reprinted with permission from [6]. Copyright 2006 Elsevier

This metric has been used to analyze the results because interventions that reduce wound contraction have been associated with less scaring and a more complete regeneration of tissue [84].

Many researchers have directed their work to tissue engineering to help promote the regeneration of skin [18]. Collagen has become one of the most important dermal replacers in various forms. Therefore, a collagen nanofibrous scaffold was produced using the electrospinning process and introduced for applications of wound healing. Results of these different studies have shown that type I collagen and lamina are both active in adhesion of human keratinocytes [6]. In addition, as crosslinking is increased, wound surface area was increased, indicating increased regeneration of tissue. These findings are very useful for advances in dermal repair and regeneration.

4.4 Tendon/Ligament Tissue

Tendons and ligaments are fibrous connective tissue and collagen forms 70–80% of them [18, 85]. They both have weak regenerative properties and do not completely recover from lesions [18]. Therefore, a collagen scaffold can provide a good way to repair or regenerate ligaments and tendons [18], because it is the main load-bearing component of the ECM and contains the mechanical strength to provide the structural required [86]. Gigante et al. used an oriented collagen type I membrane, because fiber orientation provides a pattern for cell growth and alignment that mimics that seen in normal tendons, to study the regenerative properties of tendons by grafting it into the central section of the patellar tendon of a New Zealand white rabbit [22]. Twelve rabbits with a weight of 3 ± 0.5 kg were selected because of their similarity between their knee and the human knee system [87]. The central part of the patellar tendon was explanted from each animal and the collagen membrane was grafted.

Results showed the integration of both the collagen membrane with the native tissue and the tendon with the collagen membrane [18, 22]. Fibroblasts cells were detected on the surface and deeper layers of the membrane after 1 month of implantation. In addition, the orientation and vascularization of collagen in treated tendons and controls were observed after 6 months. Lesions containing the collagen nanofibers showed good orientation and a vessel in a dense connective tissue, with a reduction in cell density. No inflammatory reactions were observed. Meanwhile, control lesions demonstrated the formatted of a rough area, such as scars and vascularization associated with a loose extracellular matrix [18, 22].

This study indicated that a collagen type I membrane can serve as an effective tool for tendon repair and healing without any adverse side effects. This because these fibers can mimic the structure of tendons and ligaments [18], providing the necessary characteristics for promoting cell attachment and proliferation and therefore help in the regeneration of tendon/ligaments.

4.5 Vascular Grafts

Many fabrication techniques have been used through the years to produce vascular scaffolds [88–90]. These scaffolds have been created with the ideal characteristics necessary for vascular grafts to be used for clinical purposes [18]. Lee et al. developed a fibrous scaffold of a blend of PCL and collagen type I from calf skin using the electrospinning technique [74] and crosslinked in 2.5% glutaraldehyde for 6 h to increase strength and stability. To study the tensile properties of the scaffold, they were prepared into tubes of 15 mm length, 5 mm inner diameter and 0.3 mm of thickness and measured using an uniaxial load test machine to achieve results of yield strength, tensile strength, Young's modulus and elongation at break from a stress-strain curve created [23]. On the other hand, to study the burst pressure strength, the pressure was increased within the tubular scaffold until a failure occurred. For biological studies, smooth muscle cells and endothelial cells from bovine carotid artery were seeded on the scaffolds, performing MTS assays at days 1, 3, and 7 to determine cell viability. Biological activity was determined by studying the ability of these types of cells to proliferate in the scaffolds. Finally, to evaluate cell morphology and adhesion, the cells were seeded on the scaffolds and incubated for 4 h to allow the cells to adhere. After 48 h, the cells that adhered were evaluated using SEM and staining [74]. The *b*ECs and *b*SMCs were able to attach to the surface of the PCL/collagen scaffold. The *b*SMCs formed various cell layers on the outer region of the tubular scaffold, while *b*ECs formed a single layer on the inner surface of the scaffold. SEM images also showed a layer of *b*ECs on the lumen while *b*SMCs are seen on the outside of the scaffold.

This membrane's uniaxial tensile properties showed that it initially has an elastic behavior, but is followed by stiffening, which is not unlike the behavior observed in native tissue. The elastic and elongation of this PCL/collagen scaffold was closer to that of native tissues, although ultimate tensile strength and elongation at break were decreased when adding collagen to the PCL blend. This membrane also resisted high degrees of pressure for longer periods of time, with a burst pressure ranging between 4760 and 5070 mmHg, much greater that physiological blood pressure seen in patients, demonstrating that these scaffolds have excellent strength and may be developed to substitute native blood vessels [74]. In addition, this membrane provided a favorable environment for the growth of vascular cells and showed excellent biocompatibility with smooth muscle cells and bovine endothelial cells in comparison to latex, which is what these researchers used as a control.

The fabrication of vascular scaffolds has been done through many different techniques, being the most recent the fabrication of scaffolds with the ideal characteristics necessary for vascular grafts to be used in clinical purposes. The results of the study explained above indicate that these developed PCL/collagen scaffolds may be used with vascular cells to create vessels that would be able to withstand physiological conditions due to their high burst pressure and their excellent strength, among the many other characteristics evaluated [74].

5 Future Studies

Electrospun nanofibers are rapidly moving towards commercialization due to their many applications in a wide range of industrial fields such as biotechnology, pharmaceutics [91], filtration, composite materials and medical [92]. Collagen has many properties that make it very attractive for application in biotechnology, including its biocompatibility, biodegradability, low inflammatory and cytotoxic responses, low antigenicity, high water affinity and availability from a variety of sources [93]. It has become evident that a scaffold that mimics the mechanical structure of the extracellular matrix is not enough to be successful for these applications, they must also promote a natural state of differentiation of the cellular components [93].

The scaling-up process, therefore, has a lot of space to improve, starting with the implementation of methods allowing for an increase of product reproducibility [91]. A high volume of production is especially necessary for biomedical applications, like the ones collagen nanofibers are intended to be used for, including the design of wound healing materials, useful for medical treatment and antimicrobial purposes and drug-delivery for filtration [91]. In addition, scaffolds that could be used in the market must go through rigorous in vivo studies before clinical validations [94]. Most technological advances that have been explored for a high-volume production of scaffolds are mainly based on modifications of the polymer injection system, introducing the introduction of multi-spinneret components, allowing for multiprocessing [91]. These could be useful for the overall set-up, increasing the area of deposition [95].

Industrial-scale electrospinning equipment, although few, are available in the market. For example, two industrial-scale instruments; a Nanospinner416 produced by Inovenso Ltd. and a Nanospider produced by Elmarco have been used [91]. Although these are available, several issues have risen from the use of these instruments such as productivity, a well-established capacity of process monitoring and quality control. In addition, key requirements specific to the equipment are necessary to be able to commercialize, including price reduction, multi-functionality and compactness [91]. Finally, it is important to point out that there is still a long way to go before a collagen scaffold can be scaled-up and used in biomedical applications. Many studies must be done to know the properties of these scaffolds and cellular responses in order to have a better idea of what the best parameters would be to move to the industry. Nevertheless, studies such as the ones presented in this chapter have shown how well these scaffolds can respond to assays in animals in several biomedical applications such as tissue regeneration and drug delivery. These open the doors to more trials to move to an industrial setting.

6 Conclusions

The development of nanofibers to mimic the native tissue is possible through the electrospinning technique and has been experimented with throughout the years. In present day, type I collagen nanofibers by "green" electrospinning have been successfully produced using a non-toxic solvent, preserving the chemical and secondary structure of collagen. The versatility to control the orientation and diameter of collagen electrospun nanofiber can be achieved through tuning of the polymer concentration and via electrospinning parameters such as voltage and polymeric solution flow rate. The orientation which is very important for cell growth direction can be modified mainly through modification of the collector, a rotating collector for aligned nanofiber and a static plate collector for random orientation. Collagen integrity is preserved after the electrospinning process through crosslinking, which serves to tune the mechanical properties of the collagen scaffold and withstand body temperature, crucial for implants and tissue regeneration purposes. Finally, in several biomedical applications it has been possible to use these fabricated nanofibers for nerve, skin, tendon/ligament, vascular grafts and bone regeneration. All of the applications shown have been able to maintain or increase biocompatibility, mechanical properties needed, promotion of function and tissue repair, showing great promise for the field of tissue engineering.

References

1. O'Brien FJ (2011) Biomaterials & scaffolds for tissue engineering. Mater Today 14:88–95. https://doi.org/10.1016/S1369-7021(11)70058-X
2. Ekaputra AK, Prestwich GD, Cool SM, Hutmacher DW (2011) The three-dimensional vascularization of growth factor-releasing hybrid scaffold of poly (e{open}-caprolactone)/collagen fibers and hyaluronic acid hydrogel. Biomaterials 32:8108–8117. https://doi.org/10.1016/j.biomaterials.2011.07.022
3. Kim IL, Khetan S, Baker BM, Chen CS, Burdick JA (2013) Fibrous hyaluronic acid hydrogels that direct MSC chondrogenesis through mechanical and adhesive cues. Biomaterials 34:5571–5580. https://doi.org/10.1016/j.biomaterials.2013.04.004
4. Berthod F, Hayek D, Damour O (1993) Collagen synthesis by fibroblasts cultured within a collagen sponge. Biomaterials 14(10):749–754. https://doi.org/10.1016/0142-9612(93)90039-5
5. Meng S, Liu Z, Shen L, Guo Z, Chou LL, Zhong W, Du Q, Ge J (2009) The effect of a layer-by-layer chitosan-heparin coating on the endothelialization and coagulation properties of a coronary stent system. Biomaterials 30:2276–2283. https://doi.org/10.1016/j.biomaterials.2008.12.075
6. Rho SK, Jeong L, Lee G, Seo B, Jeong Y, Hong S, Roh S, Jin J, Ho W, Min B (2006) Electrospinning of collagen nanofibers: effects on the behavior of normal human keratinocytes and early-stage wound healing. Biomaterials 27:1452–1461. https://doi.org/10.1016/j.biomaterials.2005.08.004
7. Haider S, Park SY (2009) Preparation of the electrospun chitosan nanofibers and their applications to the adsorption of Cu(II) and Pb(II) ions from an aqueous solution. J Membr Sci 328:90–96. https://doi.org/10.1016/j.memsci.2008.11.046
8. Malinen MM, Kanninen LK, Corlu A, Isoniemi HM, Lou Y, Yliperttula ML, Urtti AO (2014) Biomaterials differentiation of liver progenitor cell line to functional organotypic cultures in 3D nano fi brillar cellulose and hyaluronan-gelatin hydrogels. Biomaterials 35:5110–5121. https://doi.org/10.1016/j.biomaterials.2014.03.020
9. Xu F, Weng B, Gilkerson R, Alberto L, Lozano K (2015) Development of tannic acid/chitosan/pullulan composite nanofibers from aqueous solution for potential applications as wound dressing. Carbohydr Polym 115:16–24. https://doi.org/10.1016/j.carbpol.2014.08.081
10. Zuidema JM, Pap MM, Jaroch DB, Morrison FA, Gilbert RJ (2011) Fabrication and characterization of tunable polysaccharide hydrogel blends for neural repair. Acta Biomater 7:1634–1643. https://doi.org/10.1016/j.actbio.2010.11.039
11. Almodovar J, Castilla-Casadiego DA, Ramos-Avilez HV (2015) Polysaccharide-based biomaterials for cell–material interface. CRC Press, Boca Raton, pp 215–244
12. Lu P, Weaver VM, Werb Z (2012) The extracellular matrix: a dynamic niche in cancer progression. J Cell Biol 196:395–406. https://doi.org/10.1083/jcb.201102147
13. Parenteau-Bareil R, Gauvin R, Berthod F (2010) Collagen-based biomaterials for tissue engineering applications. Materials 3:1863–1887. https://doi.org/10.3390/ma3031863
14. Di Lullo GA, Sweeney SM, Körkkö J, Ala-Kokko L, San Antonio JD (2002) Mapping the ligand-binding sites and disease-associated mutations on the most abundant protein in the human, type I collagen. J Biol Chem 277:4223–4231. https://doi.org/10.1074/jbc.M110709200

15. Castilla-Casadiego DA, Ramos-Avilez HV, Herrera-Posada S, Calcagno B, Loyo L, Shipmon J, Acevedo A, Quintana A, Almodovar J (2016) Engineering of a stable collagen nanofibrous scaffold with tunable fiber diameter, alignment, and mechanical properties. Macromol Mater Eng 301(9):1064–1075. https://doi.org/10.1002/mame.201600156
16. Barnes CP, Pemble CW, Brand DD, Simpson DG, Bowlin GL (2007) Cross-linking electrospun type II collagen tissue engineering scaffolds with carbodiimide in ethanol. Tissue Eng 13:1593–1605. https://doi.org/10.1089/ten.2006.0292
17. van der Werf KO, Bennink ML, Yang L, Fitie CFC, Dijkstra PJ, Feijen J (2008) Mechanical properties of single electrospun collagen type I fibers. Biomaterials 29:955–962. https://doi.org/10.1016/j.biomaterials.2007.10.058
18. Dong C, Lv Y (2016) Application of collagen scaffold in tissue engineering: recent advances and new perspectives. Polymers 8:1–20. https://doi.org/10.3390/polym8020042
19. Sahoo S, Ouyang H, Goh JC-H, Tay TE, Toh SL (2006) Characterization of a novel polymeric scaffold for potential application in tendon/ligament tissue engineering. Tissue Eng 12:91–99. https://doi.org/10.1089/ten.2006.12.ft-8
20. Liu T, Houle JD, Xu J, Chan BP, Chew SY (2012) Nanofibrous collagen nerve conduits for spinal cord repair. Tissue Eng Part A 18:1057–1066. https://doi.org/10.1089/ten.TEA.2011.0430
21. Prabhakaran MP, Venugopal J, Ramakrishna S (2009) Electrospun nanostructured scaffolds for bone tissue engineering. Acta Biomater 5:2884–2893. https://doi.org/10.1016/j.actbio.2009.05.007
22. Gigante A, Busilacchi A, Lonzi B, Cecconi S, Manzotti S, Renghini C, Giuliani A, Mattioli-Belmonte M (2013) Purified collagen i oriented membrane for tendon repair: an ex vivo morphological study. J Orthop Res 31:738–745. https://doi.org/10.1002/jor.22270
23. Lee SJ, Liu J, SH O, Soker S, Atala A, Yoo JJ (2008) Development of a composite vascular scaffolding system that withstands physiological vascular conditions. Biomaterials 29:2891–2898. https://doi.org/10.1016/j.biomaterials.2008.03.032
24. Goh Y-F, Shakir I, Hussain R (2013) Electrospun fibers for tissue engineering, drug delivery, and wound dressing. J Mater Sci 48:3027–3054. https://doi.org/10.1007/s10853-013-7145-8
25. Nair LS, Laurencin CT (2007) Biodegradable polymers as biomaterials. Prog Polym Sci 32:762–798. https://doi.org/10.1016/j.progpolymsci.2007.05.017
26. Smith LA, Ma PX (2004) Nano-fibrous scaffolds for tissue engineering. Colloids Surf B Biointerfaces 39:125–131. https://doi.org/10.1016/j.colsurfb.2003.12.004
27. Huang ZM, Zhang YZ, Kotaki M, Ramakrishna S (2003) A review on polymer nanofibers by electrospinning and their applications in nanocomposites. Compos Sci Technol 63:2223–2253. https://doi.org/10.1016/S0266-3538(03)00178-7
28. Pham QP, Sharma U, Mikos AG (2006) Electrospinning of polymeric nanofibers for tissue engineering applications: a review. Tissue Eng 12(5):1197–1211. https://doi.org/10.1089/ten.2006.12.1197
29. Teo WE, Ramakrishna S (2006) A review on electrospinning design and nanofibre assemblies. Nanotechnology 17:R89–R106. https://doi.org/10.1088/0957-4484/17/14/R01
30. Matthews JA, Wnek GE, Simpson DG, Bowlin GL (2002) Electrospinning of collagen nanofibers. Biomacromolecules 3:232–238. https://doi.org/10.1021/bm015533u
31. Kai D, Prabhakaran MP, Jin G, Ramakrishna S (2011) Guided orientation of cardiomyocytes on electrospun aligned nanofibers for cardiac tissue engineering. J Biomed Mater Res B Appl Biomater 98(2):379–386. https://doi.org/10.1002/jbm.b.31862
32. Yang F, Murugan R, Wang S, Ramakrishna S (2005) Electrospinning of nano/micro scale poly(l-lactic acid) aligned fibers and their potential in neural tissue engineering. Biomaterials 26(15):2603–2610. https://doi.org/10.1016/j.biomaterials.2004.06.051
33. Zhong S, Teo WE, Zhu X, Beuerman RW, Ramakrishna S, Yung LYL (2006) An aligned nanofibrous collagen scaffold by electrospinning and its effects on in vitro fibroblast culture. J Biomed Mater Res Part A 79(3):456–463. https://doi.org/10.1002/jbm.a.30870
34. Shields KJ, Beckman MJ, Bowlin GL, Wayne JS (2004) Mechanical properties and cellular proliferation of electrospun collagen type II. Tissue Eng 10:1510–1517. https://doi.org/10.1089/ten.2004.10.1510

35. Ping S, Eong W, Zhu X, Beuerman R (2007) Development of a novel collagen – GAG nanofibrous scaffold via electrospinning. Mater Sci Eng C 27:262–266. https://doi.org/10.1016/j.msec.2006.05.010
36. Zeugolis DI, Khew ST, Yew ESY, Ekaputra AK, Tong YW, Yung LL, Hutmacher DW (2008) Electro-spinning of pure collagen nano-fibers - just an expensive way to make gelatin? Biomaterials 29(15):2293–2305. https://doi.org/10.1016/j.biomaterials.2008.02.009
37. Dong B, Arnoult O, Smith ME, Wnek GE (2009) Electrospinning of collagen nanofiber scaffolds from benign solvents. Macromol Rapid Commun 30:539–542. https://doi.org/10.1002/marc.200800634
38. Liu T, Teng WK, Chan BP, Chew SY (2010) Photochemical crosslinked electrospun collagen nanofibers: synthesis, characterization and neural stem cell interactions. J Biomed Mater Res - Part A 95:276–282. https://doi.org/10.1002/jbm.a.32831
39. Chakrapani VY, Gnanamani A, Giridev VR, Madhusoothanan M, Sekaran G (2012) Electrospinning of type I collagen and PCL nanofibers using acetic acid. J Appl Polym Sci 125(4):3221–3227. https://doi.org/10.1002/app.36504
40. Lin J, Li C, Zhao Y, Hu J, Zhang L (2012) Co-electrospun nanofibrous membranes of collagen and zein for wound healing. ACS Appl Mater Interface 4(2):1050–1057. https://doi.org/10.1021/am201669z
41. Fiorani A, Gualandi C, Panseri S, Montesi M, Marcacci M, Focarete ML, Bigi A (2014) Comparative performance of collagen nanofibers electrospun from different solvents and stabilized by different crosslinkers. J Mater Sci Mater Med 25(10):2313–2321. https://doi.org/10.1007/s10856-014-5196-2
42. Kazanci M (2014) Solvent and temperature effects on folding of electrospun collagen nanofibers. Mater Lett 130:223–226. https://doi.org/10.1016/j.matlet.2014.05.114
43. Elamparithi A, Punnoose AM, Kuruvilla S (2015) Electrospun type 1 collagen matrices preserving native ultrastructure using benign binary solvent for cardiac tissue engineering. Artif Cells Nanomed Biotechnol 44(5):1318–1325. https://doi.org/10.3109/21691401.2015.1029629
44. Alegre-Cebollada J, Martínez del Pozo A, Gavilanes JG, Goormaghtigh E (2007) Infrared spectroscopy study on the conformational changes leading to pore formation of the toxin sticholysin II. Biophys J 93:3191–3201. https://doi.org/10.1529/biophysj.106.102566
45. Belbachir K, Noreen R, Gouspillou G, Petibois C (2009) Collagen types analysis and differentiation by FTIR spectroscopy. Anal Bioanal Chem 395:829–837. https://doi.org/10.1007/s00216-009-3019-y
46. Yao S, Moenner M, Engdahl A, Petibois C (2012) Use of synchrotron-radiation-based FTIR imaging for characterizing changes in cell contents. Anal Bioanal Chem 404:1311–1316. https://doi.org/10.1007/s00216-012-6223-0
47. Agarwal S, Greiner A (2011) On the way to clean and safe electrospinning-green electrospinning: emulsion and suspension electrospinning. Polym Adv Technol 22:372–378. https://doi.org/10.1002/pat.1883
48. Liverani L, Boccaccini A (2016) Versatile production of poly(epsilon-caprolactone) fibers by electrospinning using benign solvents. Nanomaterials 6:75. https://doi.org/10.3390/nano6040075
49. Castilla-Casadiego DA, Maldonado M, Sundaran P, Almodovar J (2016b) "Green" electrospinning of a collagen/hydroxyapatite composite nanofibrous scaffold. MRS Commun 6(4):402–407. https://doi.org/10.1557/mrc.2016.43
50. Jun Z, Hou H, Schaper A, Wendorff JH, Greiner A (2003) Poly-L-lactide nanofibers by electrospinning – Influence of solution viscosity and electrical conductivity on fiber diameter and fiber morphology. Polymers 3(1):102–110. 1–9. https://doi.org/10.1515/epoly.2003.3.1.102
51. Chen R, Qiu L, Ke Q, He C, Mo X (2009) Electrospinning thermoplastic polyurethane-contained collagen nanofibers for tissue-engineering applications. J Biomater Sci Polym Ed 20(11):1513–1536. https://doi.org/10.1163/092050609X12464344958883
52. Xu C, Yang F, Wang S, Ramakrishna S (2004) In vitro study of human vascular endothelial cell function on materials with various surface roughness. J Biomed Mater Res A 71(1):154–161. https://doi.org/10.1002/jbm.a.30143

53. Tuck SJ, Leach MK, Feng ZQ, Corey JM (2012) Critical variables in the alignment of electrospun PLLA nanofibers. Mater Sci Eng C 32(7):1779–1784. https://doi.org/10.1016/j.msec.2012.04.060
54. Gordon MK, Hahn RA (2010) Collagens. Cell Tissue Res 339:247–257. https://doi.org/10.1007/s00441-009-0844-4
55. Chen JP, Chang GY, Chen JK (2008) Electrospun collagen/chitosan nanofibrous membrane as wound dressing. Colloids Surf A Physicochem Eng Asp 313:183–188. https://doi.org/10.1016/j.colsurfa.2007.04.129
56. Zhang Q, Lv S, Lu J, Jiang S, Lin L (2015) Characterization of polycaprolactone/collagen fibrous scaffolds by electrospinning and their bioactivity. Int J Biol Macromol 76:94–101. https://doi.org/10.1016/j.ijbiomac.2015.01.063
57. Jia L, Prabhakaran MP, Qin X, Ramakrishna S (2013) Stem cell differentiation on electrospun nanofibrous substrates for vascular tissue engineering. Mater Sci Eng C 33(8):4640–4650. https://doi.org/10.1016/j.msec.2013.07.021
58. Huang C, Chen R, Ke Q, Morsi Y, Zhang K, Mo X (2011) Electrospun collagen-chitosan-TPU nanofibrous scaffolds for tissue engineered tubular grafts. Colloids Surf B: Biointerfaces 82(2):307–315. https://doi.org/10.1016/j.colsurfb.2010.09.002
59. Theron SA, Zussman E, Yarin AL (2004) Experimental investigation of the governing parameters in the electrospinning of polymer solutions. Polymer 45:2017–2030
60. Tong HW, Wang M (2011) An investigation into the influence of electrospinning parameters on the diameter and alignment of poly(hydroxybutyrate-co-hydroxyvalerate) fibers. J Appl Polym Sci 120:1694–1706. https://doi.org/10.1002/app.33302
61. Meimandi-Parizi A, Oryan A, Moshiri A (2013) Role of tissue engineered collagen based tridimensional implant on the healing response of the experimentally induced large Achilles tendon defect model in rabbits: a long term study with high clinical relevance. J Biomed Sci 20(1):28. https://doi.org/10.1186/1423-0127-20-28
62. Wallace DG, Rosenblatt J (2003) Collagen gel systems for sustained delivery and tissue engineering. Adv Drug Deliv Rev 55(12):1631–1649. https://doi.org/10.1016/j.addr.2003.08.004
63. Prabhakaran MP, Vatankhah E, Ramakrishna S (2013) Electrospun aligned PHBV/collagen nanofibers as substrates for nerve tissue engineering. Biotechnol Bioeng 110(10):2775–2784. https://doi.org/10.1002/bit.24937
64. Ouyang Y, Huang C, Zhu Y, Fan C, Ke Q (2013) Fabrication of seamless electrospun collagen/PLGA conduits whose walls comprise highly longitudinal aligned nanofibers for nerve regeneration. J Biomed Nanotechnol 9(6):931–943. https://doi.org/10.1166/jbn.2013.1605
65. Ashammakhi N, Ndreu A, Nikkola L, Wimpenny I, Yang Y (2008) Advancing tissue engineering by using electrospun nanofibers. Regen Med 3(4):547–574. https://doi.org/10.2217/17460751.3.4.547
66. Zeugolis DI, Paul GR, Attenburrow G (2009) Cross-linking of extruded collagen fibers-A biomimetic three-dimensional scaffold for tissue engineering applications. J Biomed Mater Res A 89(4):895–908. https://doi.org/10.1002/jbm.a.32031
67. Wong SS, Jameson DM (2011) Chemistry of protein and nucleic acid cross-linking and conjugation, 2nd edition. CRC Press, Boca Raton, p 297–315
68. Huang GP, Shanmugasundaram S, Masih P, Pandya D, Amara S, Collins G, Arinzeh TL (2015) An investigation of common crosslinking agents on the stability of electrospun collagen scaffolds. J Biomed Mater Res Part A 103(2):762–771. https://doi.org/10.1002/jbm.a.35222
69. Suwandi JS, Toes REM, Nikolic T, Roep BO (2015) Inducing tissue specific tolerance in autoimmune disease with tolerogenic dendritic cells. Clin Exp Rheumatol 33:97–103. https://doi.org/10.1002/jbm.a
70. Meng L, Arnoult O, Smith M, Wnek GE (2012) Electrospinning of in situ crosslinked collagen nanofibers. J Mater Chem 22(37):19412–19417. https://doi.org/10.1039/c2jm31618h
71. Wu CH, Chen S, Shortreed MR, Kreitinger GM, Yuan Y, Frey BL, Zhang Y, Mirza S, Cirillo LA, Olivier M, Smith LM (2011) Sequence-specific capture of protein-DNA complexes for mass spectrometric protein identification. PLoS One 6(10):e26217. https://doi.org/10.1371/journal.pone.0026217

72. Gough JE, Scotchford CA, Downes S (2002) Cytotoxicity of glutaraldehyde crosslinked collagen/poly(vinyl alcohol) films is by the mechanism of apoptosis. J Biomed Mater Res 61(1):121–130. https://doi.org/10.1002/jbm.10145
73. Stokols S, Tuszynski MH (2006) Freeze-dried agarose scaffolds with uniaxial channels stimulate and guide linear axonal growth following spinal cord injury. Biomaterials 27:443–451. https://doi.org/10.1016/j.biomaterials.2005.06.039
74. Boecker AH, Van Neerven SGA, Scheffel J, Tank J, Altinova H, Seidensticker K, Deumens R, Tolba R, Weis J, Brook GA, Pallua N, Bozkurt A (2016) Pre-differentiation of mesenchymal stromal cells in combination with a microstructured nerve guide supports peripheral nerve regeneration in the rat sciatic nerve model. Eur J Neurosci 43:404–416. https://doi.org/10.1111/ejn.13052
75. Bozkurt A, Boecker A, Tank J, Altinova H, Deumens R, Dabhi C, Tolba R, Weis J, Brook GA, Pallua N, Van Neerven SGA (2016) Efficient bridging of 20 mm rat sciatic nerve lesions with a longitudinally micro-structured collagen scaffold. Biomaterials 75:112–122. https://doi.org/10.1016/j.biomaterials.2015.10.009
76. Lv Y, Nan P, Chen G, Sha Y, Xia B, Yang L (2015) In vivo repair of rat transected sciatic nerve by low-intensity pulsed ultrasound and induced pluripotent stem cells-derived neural crest stem cells. Biotechnol Lett 37:2497–2506. https://doi.org/10.1007/s10529-015-1939-5
77. Kavet R (2015) Dosimetric uncertainties. Health Phys 109:556–565. https://doi.org/10.1097/HP.0000000000000351
78. Koppes AN, Nordberg AL, Paolillo G, Goodsell N, Darwish H, Zhang L, Thompson DM (2013) Electrical stimulation of Schwann cells promotes sustained increases in neurite outgrowth. Tissue Eng Part A 20:130924230853000. https://doi.org/10.1089/ten.TEA.2013.0012
79. Ristic D, Ellrich J (2014) Innocuous peripheral nerve stimulation shifts stimulus-response function of painful laser stimulation in man. Neuromodulation 17:686–694. https://doi.org/10.1111/ner.12133
80. Chen G, Lv Y, Dong C, Yang L (2015) Effect of internal structure of collagen/hydroxyapatite scaffold on the osteogenic differentiation of mesenchymal stem cells. Curr Stem Cell Res Ther 10:99–108. https://doi.org/10.2174/1574888x09666140812112631
81. Cui W, Zhou Y, Chang J (2010) Electrospun nanofibrous materials for tissue engineering and drug delivery. Sci Technol Adv Mater 11:14108. https://doi.org/10.1088/1468-6996/11/1/014108
82. Gu P, Joseph MM, Bs U, Shiji R, Tt S (2015) Biomedical applications of natural polymer based nanofibrous scaffolds. Int J Med Nano Res 2(1):1–9. 10.23937/2378-3664/1410010
83. Simpson DG, Jha BS, Ayres CE, Bowman JR, Telemeco TA, Sell SA, Bowlin GL (2011) Electrospun collagen: a tissue engineering scaffold with unique functional properties in a wide variety of applications. J Nanomater 2011:1–15. https://doi.org/10.1155/2011/348268
84. Yannas IV (2000) Regeneration templates. In: Bronzino JD (ed) The biomedical engineering handbook, 2nd edn. CRC Press, Boca Raton
85. Mienaltowski MJ, Birk D (2014) Structure, physiology, and biochemistry of collagens. Prog Heritable Soft Connect Tissue Dis 802:5–29
86. Buehler MJ (2006) Nature designs tough collagen: explaining the nanostructure of collagen fibrils. Proc Natl Acad Sci U S A 103:12285–12290. https://doi.org/10.1073/pnas.0603216103
87. Gomes-Barrena E, Fernandes-Baillo N (2001) La rodilla en el animal de experimentación, morfología y cinemática comparadas y su applicación a los modelos experimentales de rodilla. Rev Ortop. Trauma 35:100–112
88. Mikos AG, Temenoff JS (2000) Formation of highly porous biodegradable scaffolds for tissue engineering. Electron J Biotechnol 3(2):1995–2000. https://doi.org/10.2225/vol3-issue2-fulltext-5
89. Mironov V, Boland T, Trusk T, Forgacs G, Markwald RR (2003) Organ printing: computer-aided jet-based 3D tissue engineering. Trends Biotechnol 21:157–161. https://doi.org/10.1016/S0167-7799(03)00033-7
90. Yang S, Leong KF, Du Z, Chua CK (2001) The design of scaffolds for use in tissue engineering. Part I. Traditional factors. Tissue Eng 7:679–689. https://doi.org/10.1089/107632701753337645

91. Persano L, Camposeo A, Tekmen C, Pisignano D (2013) Industrial upscaling of electrospinning and applications of polymer nanofibers: a review. Macromol Mater Eng 298:504–520. https://doi.org/10.1002/mame.201200290
92. Mirjalili M, Zohoori S (2016) Review for application of electrospinning and electrospun nanofibers technology in textile industry. J Nanostructure Chem 6:207–213. https://doi.org/10.1007/s40097-016-0189-y
93. Doyle JJ, Choudhari S, Ramakrishna S, Babu RP (2013) Electrospun nanomaterials: biotechnology, food, water, environment, and energy. Conf Pap Mater Sci 2013:1–14. https://doi.org/10.1155/2013/269313
94. Zhong SP, Zhang YZ, Lim CT (2010) Tissue scaffolds for skin wound healing and dermal reconstruction. WIREs Nanomed Nanobiotechnol 2:510–525. https://doi.org/10.1002/wnan.100
95. Kidoaki S, Kwon IK, Matsuda T (2005) Mesoscopic spatial designs of nano- and microfiber meshes for tissue-engineering matrix and scaffold based on newly devised multilayering and mixing electrospinning techniques. Biomaterials 26:37–46. https://doi.org/10.1016/j.biomaterials.2004.01.063

Electrospun Cellulose and Nanocellulose Composites as a Biomaterial

Jaime A. Santillán-Mercado, Yaiel G. Rodríguez-Avilés, Samir A. Bello, José A. González-Feliciano, and Eduardo Nicolau

Abstract The use of electrospun cellulose and nanocellulose composites is a promising novel biomaterial, with particular potential to serve as scaffold for bone tissue regeneration. Nanoscale fibers meet the challenge of mimicking important features of the bone extracellular matrix, and the incorporation of nanofillers based on nanocellulose could enhance mechano-physical properties of these scaffolds. This chapter describes and discusses the properties and applications of cellulose and its derivatives, as well as nanocellulose composites fabricated using the electrospinning technique. Along with a general introduction of cellulose and its main derivatives used in biomedical applications, examples of nanocellulose for mechanical reinforcement and corresponding applications are discussed. Likewise, this review suggests cellulose as a new class of biomaterial that might offer great promise in biomedicine.

Keywords Cellulose • Cellulose acetate • Electrospinning • Nanocomposite • Nanocellulose • Nanofillers • Reinforcement

1 Perspective

Humankind has been always in a quest to find new materials and biomaterials for applications in medicine and other fields. Due to this journey, we now have dental and limb implants that impact the quality of life of people in a positive way. Parallel to all the medical advancements and the development of new materials is also the

J.A. Santillán-Mercado
Department of Physics, University of Puerto Rico, San Juan, PR, USA

Y.G. Rodríguez-Avilés • S.A. Bello
Department of Biology, University of Puerto Rico, San Juan, PR, USA

J.A. González-Feliciano
Molecular Science Research Center, University of Puerto Rico, San Juan, PR, USA

E. Nicolau (✉)
Molecular Science Research Center, University of Puerto Rico, San Juan, PR, USA

Department of Chemistry, University of Puerto Rico, San Juan, PR, USA
e-mail: eduardo.nicolau@upr.edu

J. Almodovar (ed.), *Electrospun Biomaterials and Related Technologies*,
DOI 10.1007/978-3-319-70049-6_3

worry of the environmental impact of exploiting our resources for such applications. For example, old dental implants were typically made of gold, and due to price considerations directly related to its scarcity on the Earth crust, new developments evolved in the area of ceramics and polymers to tackle the situation.

Now, as we look forward to new biomaterials that can serve as platforms for biological applications, cellulose and nanocellulose composites are candidates that must not be overlooked. Recent investigations have shown that cellulosic materials might offer the benefits of a renewable material with the intrinsic characteristics of the material itself such as: high mechanical strength, biocompatibility and bioactivity, among others. In this chapter, we present an extensive literature review on the use of these materials for biological applications such as: bone tissue engineering, biosensing, and wound healing. The possibilities are infinite when we think about cellulose.

2 Introduction

A biomaterial, in agreement to the accepted definition of the National Institutes of Health (NIH), is any matter, surface or construct, either organic or inorganic, intended to interface with biological system to treat, augment or replace any tissue or a function of the body. In addition, such biomaterials must fulfill technical and biological requirements [1], given that they are in contact with bodily fluids and tissues for prolonged periods of time. Indeed, it is clear that in order to achieve these goals, they should have the ability to maintain biofunctionality without producing adverse effects; in other words, good biocompatibility [2, 3]. The special demands of these biomaterials have motivated the interest of researchers to move toward the use of nanotechnology as a potential solution to the current challenges since nanostructured materials exhibit novel and unique properties such as a high surface area and high porosity [4]. Ultimately, these characteristics have direct implications on the design and functionality of scaffolds used within the biomedical field [5, 6].

In recent years, there has been a growing interest in the use of nanotechnology in medicine for a wide range of biomedical applications; developing new opportunities for materials at the nanoscale level to interact with biological systems. Nanoparticles, for example, are used to deliver drugs for cancer treatments and also to stimulate immune responses. On the other hand, nanofibers could be used in wound dressing, vascular grafts, scaffolds for tissue regeneration, etc.

These reasons have revolutionized the design and fabrication of artificial extracellular matrix, referred to as scaffolds, due the current advances on nanotechnology and processing techniques for producing biomaterials. Electrospinning [7], melt blowing [8], template synthesis [9], self-assembly [10] and phase separation [11] are some of the techniques used to produce nanofibers. In particular, electrospinning is a versatile method and has become the most frequently used technique for preparation of porous scaffolds. Electrospun nanofibers also possesses several attractive properties, such as high surface area to volume ratio.

In this extent, the aim of the following chapter is to provide a general overview of cellulose-based nanofibers (both micro and nanostructured form) which has recently recovered a growing interest due its remarkable properties with promising future in the biomedical fields and nanomedicine [12].

3 Electrospinning

Electrospinning is a electrohydrodynamic method used for producing synthetic and natural polymer fibers by electrical force, gathering significant interest due its ability to produce fibers at the micro or nanoscale. From its introduction in 1934 to recent times, electrospinning has been investigated by many researchers to be used in fiber manufacture and recent applications related to nanotechnology, nanomedicine, and material science. Due the innovative uses of electrospun fibers, several articles and reviews have been published highlighting its importance in tissue engineering and the biomedical field [13–16] (Fig. 1). Electrospinning has important tunable working parameters (solution, process and ambient parameters) that can affect the diameter and morphology of the resulting fibers. With control and proper manipulation of these parameters, we can produce electrospun nanofibers with desirable physical properties for advanced applications [18–20].

Due to the great progress in preparation and characterization of nanofibers, many researchers have become interested in studying and exploiting novel properties of nanomaterials exhibited only at such nanoscale level [7]. In such a sense, electrospinning is a versatile, low-cost, useful and efficient technique to produce nanofibers that can form highly porous structures with excellent pore interconnectivity, as well as membranes with large surface to volume ratio than can work as scaffolds for biomedical applications [21, 22].

3.1 Electrospinning Principles

In electrospinning, an intense electrostatic field is created when an electrical potential difference is applied between a needle and a current collector. In consequence, an electrically charged droplet is formed when polymer solution is pumped to flow from the syringe through the needle. This droplet remains at the tip of the needle and acquires a stable meniscus shape as a result of the equilibrium of the forces (Rayleigh instability limit), produced by the surface charge repulsion, and the surface tension of the polymer. At a critical potential, accumulation of charges increase columbic repulsion, causing the elongation of a conical meniscus referred to as the Taylor cone that finally overcomes the Rayleigh limit. In this sense, the equilibrium

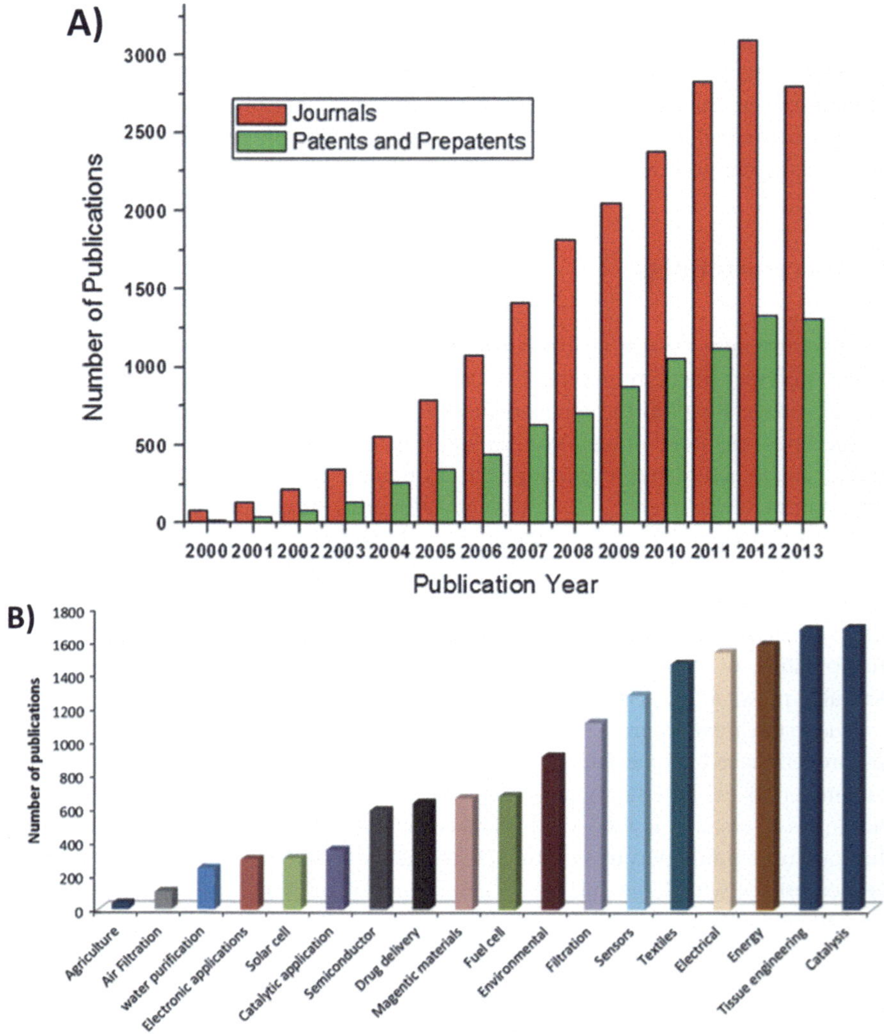

Fig. 1 (**a**) Number of journal and patent publications using electrospun nanofibers. (**b**) Number of publications featuring nanofibers used in various application (reprinted with permission from [17])

is destroyed and the electrostatic forces overcome the surface tension and a charged jet of the polymer solution is expelled from the tip of the cone. First the jet flies in a straight line, between the needle tip and the collector, then its diameter decreases due stretching and solvent evaporation, leaving behind a charged fiber. Repulsive interactions cause the fibers to whip and then to be deposited on the collector. Figure 2 illustrates the electrospinning setup.

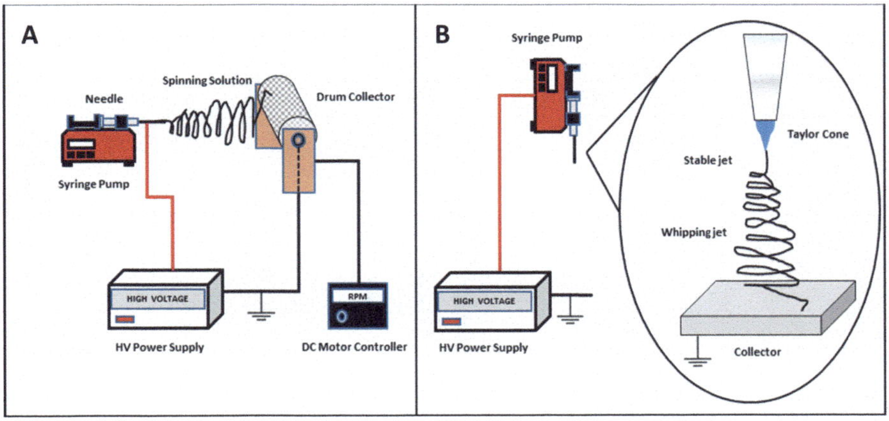

Fig. 2 Schematic diagrams of electrospinning setups. (**a**) Vertical electrospinning process, (**b**) Horizontal rotating electrospinning

3.2 Electrospinning Parameters

As mentioned before, different effects on physical properties of electrospun fibers are exhibited because of variation of the three main working parameters: (1) process parameters such as applied voltage, flow rate, and distance [23]; (2) solution parameters such as concentration, viscosity, polymer molecular weight, surface tension, and conductivity; and (3) ambient parameters such as temperature and humidity.

3.2.1 Process Parameters

Formation of fibers start when a critical potential is achieved and the jet solution is ejected from the Taylor cone and deposited on the collector. Thus, applied voltage has a pivotal role in the initiation of the electrospinning process. There are some discrepancies related about how applied voltage affect the diameters and morphology of electrospun fibers. For example, some researchers concur that fiber length and diameters decreased with increasing voltage, and at the same time continuous and uniform fibers were obtained with increasing voltage at significant levels [24]. It has been reported that these effects are exhibited by the fibers due to the increment of electrostatic repulsive forces on the polymer jet and to the stretching of the polymer solution as result of increasing the electric potential difference between electrodes [25]. However, other authors report the opposite effect, less uniformity, and show bead formation in the collected fibers when voltage is increased beyond a critical value. This critical value of voltage varies according the polymer-solvent system. Beaded electrospun fibers are usually considered as poor quality fibers, and sometimes its formations are attributed to low charge density as well as other working parameters [26]. Figure 3 shows smooth electrospun cellulose acetate nanofibers.

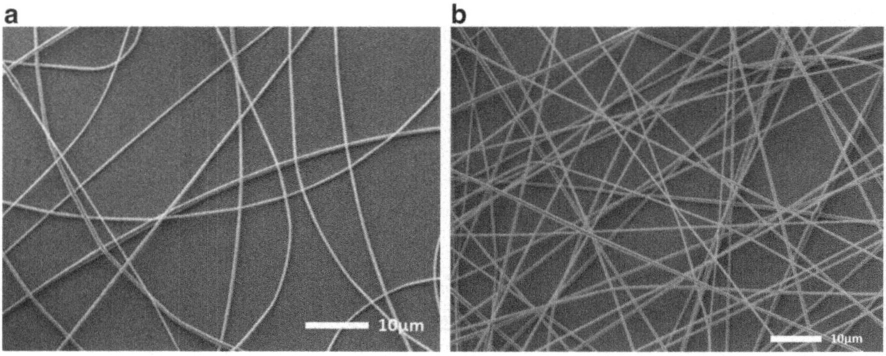

Fig. 3 Electrospun nanofibers at different time of deposition (**a**) $t = 2$ min (**b**) $t = 5$ min

Another parameter that play an important role in the electrospinning process is the flow rate. Recently, a theoretical analysis was conducted to evaluate the effect of flow rate on the diameter of fibers and subsequent experimental studies confirmed that with the increase of the flow rate the diameter of the electrospun nanoporous fibers increases as well as the size of their pores [27]. The use of a proper low flow rate and needle-collector distance allows enough time for evaporation of the solvent and in case of high flow rates, promotes formation of beads due the short time of drying [28].

3.2.2 Solution Parameters

Another important parameter involved on fiber formation and its morphology is the molecular weight. Polymer chain entanglement and consequently viscosity of the solution are determined by molecular weight [29]. It has been reported that by increasing the molecular weight, viscosity will be raised and continuous and uniform fibers are obtained rather than beads [29, 30]. This occur because a higher chain entanglement, derived from the increase of the molecular weight, stabilizes the polymer jet and allow solvent evaporation that finally is deposited on the collector. Studies conducted related to this parameter developed a methodology to predict fiber/bead formation in good solvents, considering only the proper entanglement molecular weight of the polymer. This assessment demonstrates its importance during electrospinning process and establish a correlation between chain entanglement and electrospun fiber formation [31].

Concentration of the polymer solution is also an important parameter considered in the process of fiber formation. At lower concentrations, high surface tension will produce polymeric beads instead nanofibers. Basically, the main difference between electrospinning and electrospraying is related to the critical concentration and the chain entanglement. According to this, Fig. 4 shows a summary of the parameter effects on the morphology of electrospun nanofibers.

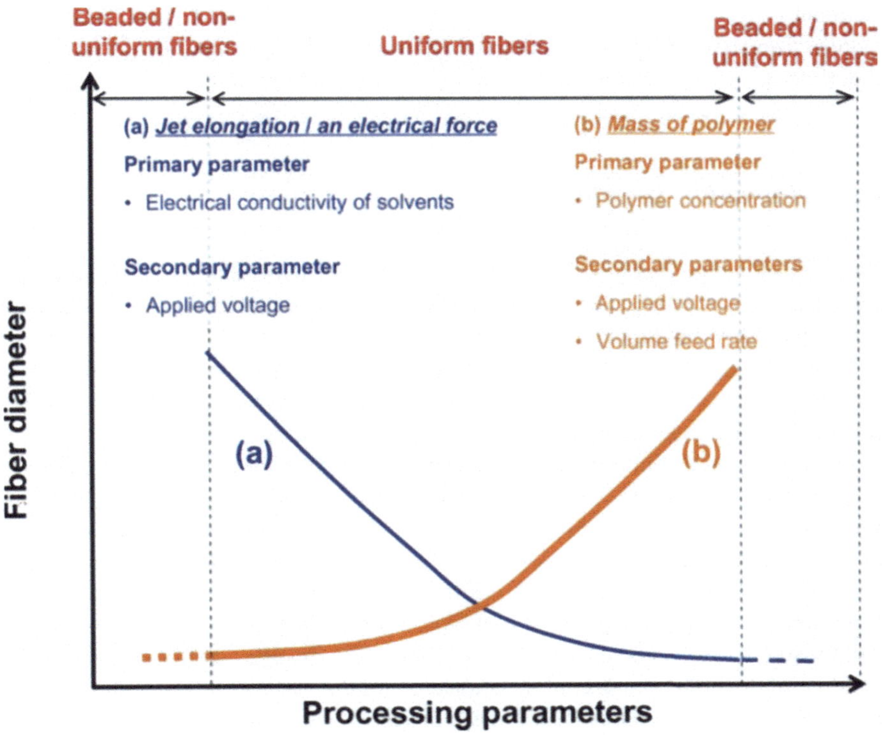

Fig. 4 Fibers diameter as a function of the parameters related to (**a**) forces in solution jet and (**b**) polymer mass (reprinted with permission from [32])

4 Cellulose

Polysaccharide-based polymers have been extensively studied in regards to their innate structure, mechanical properties, functional modifications, preparation techniques and distinctive relevance within various industries, such as those of clothing and agriculture, as well as for pharmaceutical and biomedical applications. Polymers of such interest are chiefly those of chitosan, alginate, glycosaminoglycan and of course, cellulose. Interesting enough, such polymers can be derived from various sources, mainly from plants and trees, but extraction from animals also facilitate other means of obtainment, as well as from microorganisms, such as bacteria and fungi, which have attracted much attention in recent times [33–36].

In regard to what is recognized to be the most abundant polymer on Earth, the properties that characterize cellulose are quite intriguing. This structural polysaccharide is found to be the main constituent of the cell wall of plants and due to its innate hierarchical structure, in combination with other molecular components, gives about intrinsic characteristics and mechanical properties throughout the plant body. Within the following section, the basic molecular and supramolecular structure of this ubiquitous polymer will be reviewed, as well as some of its most pertinent properties. The discussion developed herein will serve as a fundamental source of understanding for the complex nature of this polysaccharide polymer and underline the principals behind its usages as a promising biomaterial.

4.1 Molecular Structure

The intrinsic nature that has characterized celluloses chemical and physical properties (namely its reactivity, geometrical dimensions, crystallinity, high surface area, biocompatibility, biodegradability, among many others) is well attributed to its distinctive molecular organization. In such a sense, it is indispensable to understand with great emphasis the molecular building blocks of this macromolecule and the intrinsic interactions that govern its overall physiognomies.

Its basic arrangement is described as a straight chain or linear syndiotactic homopolymer that lacks ramifications and whose glucose residues adopt a β configuration, through (1 → 4)-glycosidic bonds [37]. Such properties give cellulose its uniqueness, particularly when compared to other closely related homopolymers, such as glycogen and amylose. Both of these saccharides are used for fuel storage in animals and plants, respectively, assume a three-dimensional helical structure and, most noticeably, consist of α(1 → 4) glycosidic linkages. Yet, it is remarkable that this subtle difference in conformation dictates the vast and fundamental properties of these polysaccharides. The essence of such phenomena is related to the steric hindrance upon rotation of the C1–O–C4 linking bonds. To better visualize and understand this concept, one can imagine cellulose as repeating units of its dimer, cellobiose; where its β-D-glucose residue follows a chair conformation with carbon atom 4 in the high position and carbon 1 in the low position (denoted as $^4C_1$) [38]. This in turn causes the hydroxyl groups to be positioned in the equatorial plane, while the hydrogen atoms in the axial plane [37, 39]. The gyration of such bonds, which relate one β-D-glucose residue and the next, are referred as the torsional angles, and are more formally defined in terms of φ_O: O5′–C1′–O4–C4 and ψ_O: C1′–O4′–C4–C5, whose value range within ±180° (Fig. 5) [39, 41]. Thus, it can be reasoned that certain conformational arrangement are much more stable than others. In the case for cellulose, each residue adopts a stable chair conformation that orientates 180° with respect to the adjacent monomer units, and consequently yielding the characteristic elongated chain. This greatly differs from glycogen and amylose, whose α conformation entails a 60° angle between adjacent units, which in turn produces stable, three-dimensional helical structures [42].

This captivating molecular architecture allow for the maximum interaction of weak intermolecular bonds within and between adjacent molecules, which give rise to the supramolecular structure of cellulose; most distinctively, hydrogen bonding [43]. As shown in Fig. 5, intermolecular hydrogen bonding is achieved between the 6′–OH group of one unit and the paralleling 3′–OH of another. Likewise, the same 3′–OH group can form intramolecular hydrogen bonds with the ring oxygen of an adjacent monomer, while, a 6′–OH can link with a 2′–OH. As it can be noted, the 3′ and 6′–OH groups can form both inter- and intramolecular bonds, and thus, each

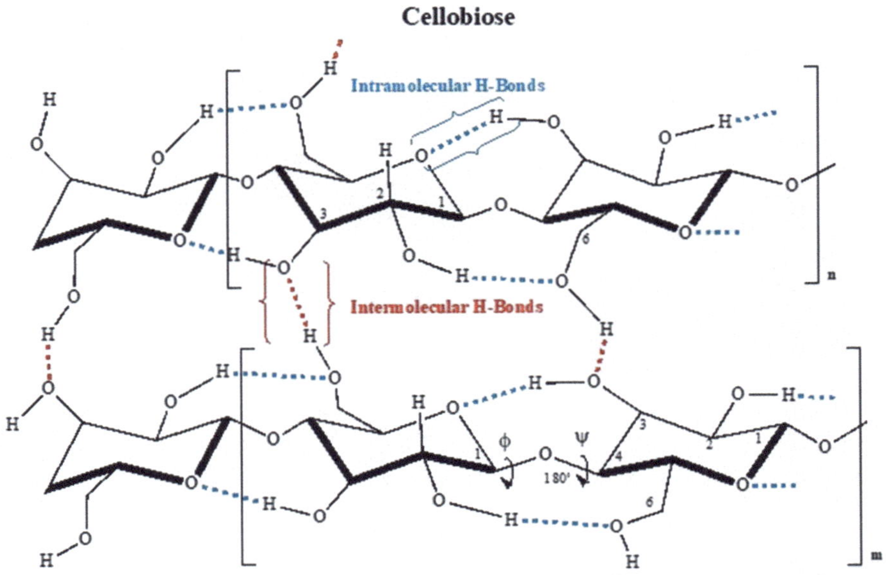

Fig. 5 Inter- and intramolecular hydrogen-bonding network of cellulose, with emphasis on the cellobiose unit (adapted from Credou et al. [40])

residue may form various bonds with other units to give about the natural three-dimensional structure of cellulose. The fact that these interactions maintain a strong coalition is particularly important when discussing the semicrystalline nature of cellulose and, thus, will be further discussed in the subsequent Sect. 4.3.

Based on fundamental concepts of the basic molecular structure of cellulose, we will now turn our discussion to the overall properties of this natural polymer and some of its unique characteristics.

4.2 *Properties of Cellulose*

Cellulose is commonly described to have a hierarchical structure, yet it is well understood that it may exist with different degrees of crystallinity, unit cell dimension and chain orientation, all which depend on the source and mode of extraction. In this sense, several cellulose chains are aligned with one another to form protofilaments (elementary fibrils), which in turn combine with other such structures to form microfibrils, and these in turn come together to form a fibril-matrix structure that is better described as cellulose fibers (Fig. 6). Nevertheless, it is with particular detainment that the microfibril structure must be discussed. This subdivision of the cellulose architecture has been characterized for having regions of high and low crystallinity, and thus confines many of the innate properties of this polysaccharide polymer.

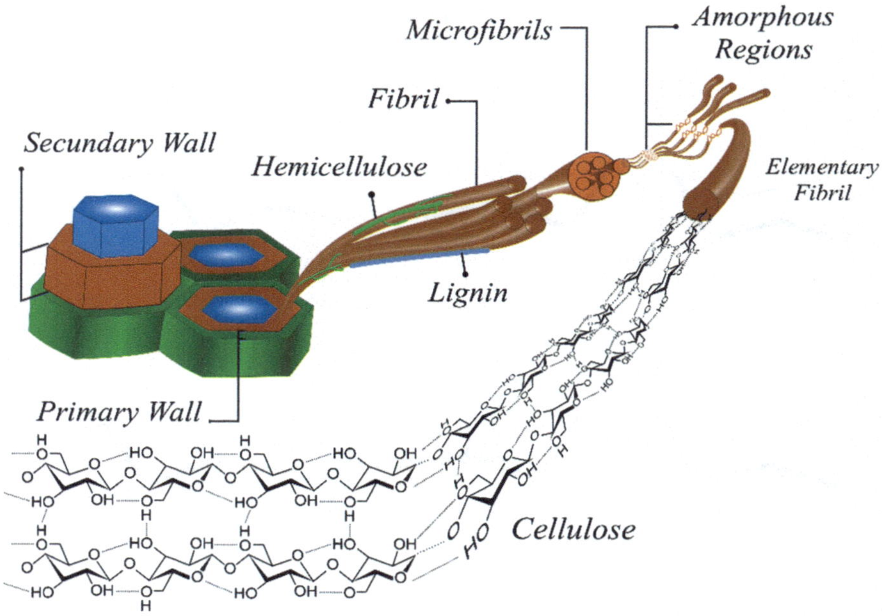

Fig. 6 Hierarchical structure of cellulose. https://www.intechopen.com/copyright-policy.html (reprinted with permission from [26])

4.3 Crystallinity and Supramolecular Structure

Cellulose is considered to possess an alternating crystalline structure given that the microfibrils can be noted to have two regions of distinct crystallinity [44], both of which contribute particular features to the overall structure of the material. In regards to the amorphous (disordered) region, these segments are characterized by a reduce degree of intermolecular bonding's which in turn gives about a certain relaxation that allows for susceptible degradation and for the penetration of water molecules which, altogether, gives flexibility and a higher elastic modulus throughout the supramolecular structure [45–47]. On the other hand, the crystalline (ordered) region is known for its highly compact molecular arrangement, due to the stabilizing network of inter- and intramolecular hydrogen bonds, previously described (Fig. 5). As a consequence of such strong interactions, this portion of the microfibril excludes the penetration of water molecules and thus is highly resistant to degradation. It is this region of the supramolecular structure that is characterized to possess high tensile strength and Young's modulus, as well as being insoluble in water and most organic solvents [40, 48].

Interestingly, this crystalline structure of cellulose is known to exhibit different crystal packing's, due to conformational changes and small structural arrangements of the glycosidic residues, which pertain to a series of polymorphs; namely cellulose I and II (III and IV are also polymorphs of cellulose, but they will not be further discussed) [49]. Native or natural cellulose is classified as cellulose I, and is further categorized as cellulose I_α and I_β. The fundamental difference in such arrangements is due to variations in the unit cell dimensions, where the α form constitutes a triclinic crystal system, with one cellulose chain per unit cell, and the β form has a monoclinic unit cell structure with two chains per unit cell [50]. The prevalence of one structure over the other seems to depend on the source of the cellulose from which it was obtained. In the case of I_α, it is mostly found in algae and bacteria, while I_β is present in cellulose from higher ordered plants [51].

Cellulose II is obtained from cellulose I, be that by a form of alkali treatment (better known as mercerization) or by regenerating cellulose. These structures differ in that they are stacked with opposite polarity; cellulose I runs in parallel, whereas in cellulose II is antiparallel, and as a consequence, is thermodynamically more stable than cellulose I [52]. As a result, the dynamics of the inter- and intra-hydrogen bonding network is changed, which allow for less compaction of the chains. Nevertheless, it should be mentioned that the mode of biological synthesis for cellulose I is what defines the orientation of the microfibrils to be in a parallel fashion, as opposed the entropically favored antiparallel structure found in manufactured cellulose (i.e.; cellulose II) [53].

4.4 Mechanical Properties

Cellulose owes much of its intrinsic mechanical properties to the discussed microfibril structure, particularly that of the crystalline region. Thus, various methods are employed with the aim of extracting these regions for different applications as nanoscale materials. There are two nanostructures commonly extracted from the polysaccharide: cellulose nanocrystals (CNC) and cellulose nanofibrils (CNF) (also termed as microfibrillated cellulose). These nanoscale structures differ in certain properties, such as morphology and elasticity modulus, as well as the mode of extraction or production. For CNC, obtainment is usually due to chemical acid hydrolysis of native cellulose, which results in the removal of the amorphous regions leaving the resulting crystalline structure intact, and is commonly described as "whiskers" or "rod-like". These nanoparticles exhibit a very high Young's modulus, ranging between 100 and 130 GPa, which is quite comparable to that of steel (approx. 200 GPa) [43, 54].

With respect to CNF, the means of production is achieved mainly by mechanical treatment procedures, particularly through high-pressure homogenization, but other techniques are utilized that correspond to different chemical treatments, that is surface modifications. The resulting nanostructure is characterized by the presence of both the crystalline and amorphous domains, in an alternating fashion. When evaluating the elasticity of these nanostructures, it can be assumed that the modulus of nanocellulose should depend greatly on the degree of crystallinity (that is, the ratio of the crystalline to amorphous regions). Thus, in the case of CNF, it should come to no surprise that the Young's modulus be lower than that of CNC, which has been reported to fall within the range of 29−36 GPa [55].

It is quite evident that the molecular and supramolecular structure of cellulose is very well arranged to withstand and in part govern the particular mechanical properties of the plant cell wall. For these reasons, and surely many more, composites of nanocellulose are of great interest for the development of high load-bearing materials, which, in combination with other resources, may produce more novel and complex biomaterial designs.

4.5 *Biocompatibility*

Up to this point, we have described cellulose (more explicitly, nanocellulose) as a sophisticated, rigid structure with regions of high crystallinity and good mechanical tendencies, as well as being insoluble in water. Such quality's found in this one polymeric material marks cellulose as an interesting source for different biomedical applications, particularly as nano-reinforcements or guiding structures and as tissue engineering constructs [38, 48, 56]. Further evaluation on the biomedical applications and relevance of cellulose and its derivatives are further reviewed throughout the succeeding sections, with meticulous discussion in Sects. 5 and 8. Nevertheless, some important aspects shall be mentioned as of now.

Firstly, the toxicity of cellulose has been evaluated by different groups in different types of cells and tissues, which in all, most have shown to conclude that this material has low detrimental effects [57–60]. Some studies have, however, demonstrated that the degree of toxicity is dependent on the mode of processing, particle size and surface chemistry of the cellulose nanoparticles [61]. Bacterial cellulose (BC), on the other hand, has been largely accredited for being highly biocompatible, compared with that of native cellulose, given its signature use as a wound dressing [62]. To such extent, further inquiry is required to better assess the harmful effects that cellulose nano-products may cause as biomaterials.

Another aspect to consider in regard to the cell-material interactions that pertain from cellulose is its lack of biodegradability in humans, which arises from the complex hierarchical order of this polysaccharide and the lack of appropriate degrading enzymes in mammals. Such an issue is relevant given that the ideal biomaterial or scaffolds is sought to be absorbable within the embedded tissue, supporting the initial regeneration stages and eventually leaving behind the fully formed or recovered tissue [63]. One way around this concern is through different chemical modifications, producing functionalized, cellulose derivatives. These modified cellulose products convey in essence different characteristic and properties that highly depend on the mode of treatment; that is: oxidation, esterification, etherification, among others [47]. Some of these cellulose derivatives will be further discussed in the following section.

5 Cellulose Derivatives

As fascinating as cellulose is, its nanoproducts are only the beginning of a magnanimous set of derivatives, whose properties and applications merit distinctive consideration as promising biomaterials. It should be noted, however, that the comprehensive term of "nanocellulose" is used to describe and denote the vast amount of derivatives that pertain from the polysaccharide. The focus of the following section is to introduce some of the most prominent polymeric materials that pertain from cellulose, their mode of synthesis, properties and applications. With the intention of better understanding how such materials arise from cellulose, we will start our discussion once again with the fundamentals. That is, the molecular aspects that governs and allows its modification: the hydroxyl groups and the hydrogen-bonding network.

5.1 *Chemical Reactivity*

As we have previously discussed, the complex nature of cellulose arises from the hierarchical organization of the interacting glucose residues, which eventually form micro- and macrostructures that characterize the three-dimensional architecture of this ubiquitous molecule. In the very epicenter of such complexity lays the intrinsic hydrogen bonding of the adjacent and paralleling glucose units, which stem from the interactions of the hydroxyl groups of C2, C3, and C6. Within the crystalline regions, these groups are tightly packed to the extent that not even water molecules may penetrate the established framework, while in the amorphous segment, this packing is less so. Accordingly, the tendency or degree of reactivity of the anhydroglucose units (AGUs) is very much dependent on the availability of these hydroxyl groups to interact with the reagent of interest. In light of this, one must consider the accessibility of these reactive groups in order to properly bring about the different chemical modifications.

It is well considered that the reactivity of the C6 (primary) hydroxyl group is the most dynamic and available functional moiety within the AGU [47]. Given its attachment to an alkyl group ($HO–CH_2–CHR_2$), this bond allow certain conformations to be assumed, as it pertains dihedral angles that permits multiple possibilities for hydrogen binding, Fig. 7 [40, 41]. The C2 (secondary) hydroxyl group is often considered to be acidic, and its reactivity is greater than that of the C3 (secondary) hydroxyl, but considerably less than the C6 hydroxyl group [52]. Overall, it is evident that a heterogeneous degree of reactivity is found throughout the same AGU molecule, and as consequence distinct measures are employed to increase the likelihood of reaction within these moieties' (i.e.; regioselectivity). Finally, one must consider that the mode of syntheses will influence the degree of substitution (DS), which describes the average number of replaced hydroxyl groups per glucose residue and whose value may range between 0 and 3 (due to the total number of hydroxyl groups present) [64].

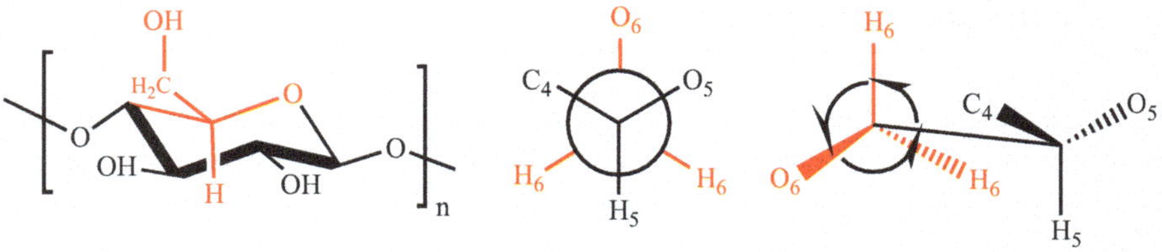

Fig. 7 Conformers depicting the rotation of the C6-OH (hydroxymethyl) group

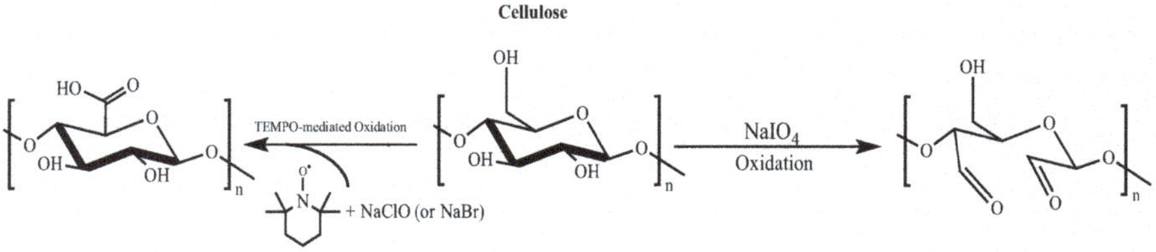

Fig. 8 Different oxidation reactions for cellulose

5.2 *Oxidation*

The addition of carbonyl groups to the cellulose structure imparts a higher degree of reactivity to the polysaccharide but concerns some effects on reducing its crystalline structure by affecting the hydrogen-bonding network [65, 66]. Thus, it is common to employ such chemical treatments for the obtainment of nanofibers (particularly cellulose nanofibrils), in conjunction with subsequent mechanical treatment [54, 66]. One such reaction commonly employed is that of metaperiodate ($NaIO_4$) oxidation, which substitutes the C2 and C3 hydroxyl groups (i.e.; the secondary alcohols) with aldehyde moieties. The resulting reaction causes the cleavage of the C2–C3 bonds, opening the gluco-pyranose ring with the formation of two aldehyde groups Fig. 8, thus producing, 2,3-dialdehyde cellulose (DAC).

As a possible biomaterial, the team lead by Li et al. used the metaperiodate oxidation method to create a 3D nano-network with good degradability in water and PBS but did not test it with cell cultures [67]. On the other hand, Cheng's team designed composites consisting of 2,3-dialdehyde cellulose (from regenerated cellulose) and collagen, forming a hydrogel film structure, and showed low cytotoxicity of the biomaterial with fibroblast in vivo [68].

Another popular method for modifying cellulose is through a 2,2,6,6-tetramethyl -1-piperidyloxy (TEMPO)-mediated oxidation, which requires, in turn, a co-oxidant such as NaClO or NaBrO (Fig. 8). This method is well distinguished for being regioselective for primary hydroxyl groups and, thus, will preferentially react and substitute the C6-OH of the AGUs [69]. The end result of this reaction is the presence of a carboxyl group at C6 and advantageously the gluco-pyranose ring is not severed, maintaining the general structure of cellulose; so named as 6-carboxycellulose.

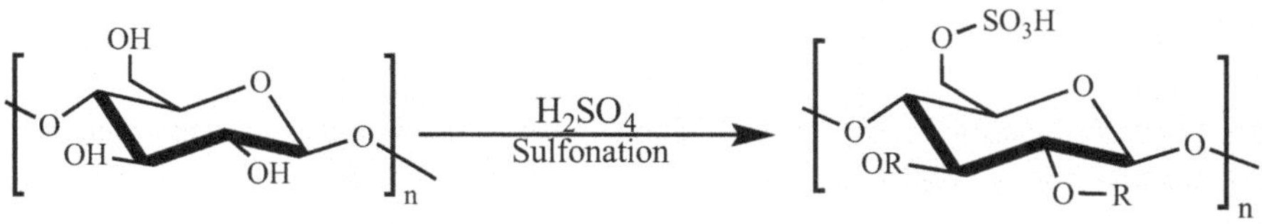

Fig. 9 Sulfonation reaction of cellulose, where R = SO_3H or H

6-Carboxy cellulose has been prepared as an electrospun-composite scaffold with gelatin in different weight ratios, followed by crosslinking of both materials, by Švachová et al. [70]. The group showed that this scaffold biomaterial has a non-toxic nature when cultured with human lung adenocarcinoma cells and exhibited normal morphology. As another application, TEMPO-oxidized nanofibrillated cellulose (TO-NFC) networks (with a DS of 33.2%) were prepared by Weishaupt et al. as a material with good physical adsorbance properties for different bioactive molecules, which could potentially serve distinct medical related applications as biosensors or coatings [71].

5.3 *Esterification*

Esterification of cellulose is of particular importance given that it is a common and efficient way of preparing CNC with different surface chemistry and is of significance within the pharmaceutical industry [54, 72, 73]. Interesting enough, there are many type of modification that can be drawn from this polysaccharide through esterification, among them are: sulfonated, acetylated and nitrated cellulose. Thus, the surface characteristics of the modified polymer will dictate the overall properties and consequently the application or relevance of the prepared material.

5.4 *Sulfonation*

The addition of sulfuric acid allows for acid hydrolysis of the hydroxyl groups present in the cellulose structure. The resulting reaction gives about an anionic surface of sulfate ester groups that allows for the cleavage of the amorphous regions from the crystalline segments and thus may produce cellulose nanocrystals (CNCs), Fig. 9 [54]. It should be noted that other techniques for generating sulfonated cellulose exist and pertain different advantages and purposes [74–76].

5.5 Acetylation

Fischer esterification is a very practical reaction for the synthesis of acetylated cellulose, which may employ the use of acetic acid or acetic anhydride with a strong acid as a catalyst (HCl, H_2SO_4, etc.); Fig. 10. Cellulose acetate (CA) in one noteworthy polymer to consider from such surface modification, as it has been extensively studied as a biomaterial for applications in tissue engineering and regeneration via electrospinning and hydrogel techniques [77–80]. Likewise, it has served within the fields of water remediation [81–83] as membrane and within many other interesting fields [84–87]. A subsequent section will further expand on the properties and uses of electrospun CA polymers.

5.6 Nitration

The main constituent of the nitrated cellulose derivatives is considered to be that of nitrocellulose (also named as cellulose nitrate). The synthesis of this molecule arises mainly from an acidic treatment of cellulose with nitric acid (HNO_3) in the presence of sulfuric acid (H_2SO_4), as showed in Fig. 11. An important remark of this reaction is that the total nitrogen content of the final product influences the its physical-chemical properties and thus its applications. When nitrocellulose has a

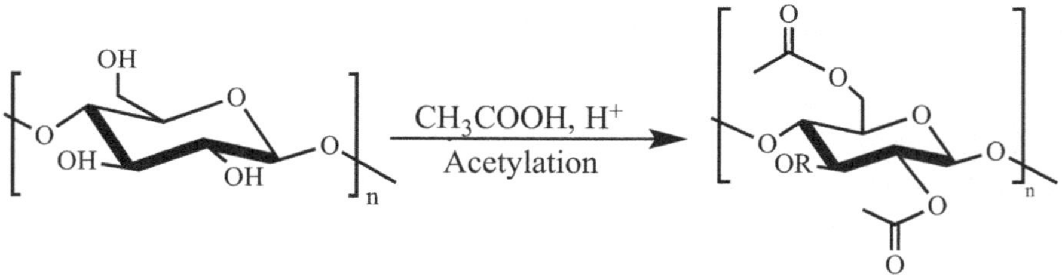

Fig. 10 Acetylation reaction of cellulose to cellulose acetate (CA), where R = CH_3CO or H

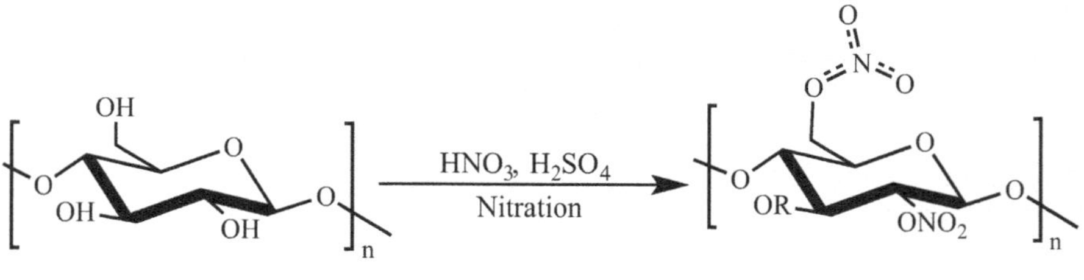

Fig. 11 Nitration reaction of cellulose to nitrocellulose (also named as cellulose nitrate), where R = NO_2 or H. The degree of substitution for this derivative is generally 1 or 2

degree of substitution (DS) below 2, its properties allow for its usage in common products such as cigarettes and nail polish. Yet, above 2, nitrocellulose correlates with explosive tendencies [88, 89].

In regards to biomedical applications, a nano-porous liquid bandage has been recently devised from nitrocellulose and was tested in vitro with mouse fibroblasts and in vivo on mouse models [90]. The team led by Mu et al. showed that their material possesses good anti-infection properties as well as promotes better wound healing than commercial products, based on immunohistochemical and biomechanical testing. Electrospun nitrocellulose fibers have also been achieved by various groups [91–93], which merit consideration as a mean of producing plausible, future biomaterials.

5.7 *Etherification*

The final surface modification that will be discussed is that of cellulose etherification, which may be prepared through the use of alkyl halides (i.e.; O-alkylation), epoxides or by Michael addition. It is common for these reactions to employ the use of sodium hydroxide (NaOH) as an activating agent for the cellulose hydroxyl groups, which in turn causes the cellulose fibers to swell, producing, in effect alkali-cellulose [94]. This process (known as the ballooning effect) thus results in the perturbation of the polysaccharide's supramolecular structure. This intermediate product, whose hydrogen-bonding network has been disrupted, proceeds to react with the etherification agent to form the final cellulose derivative.

One such example is the use of monochloroacetic acid or chloromethane for the synthesis of carboxymethyl cellulose (CMC) and methylcellulose, respectively, both in the presence of sodium hydroxide (Fig. 12).

Cellulose ethers are considered to be of great importance within the pharmaceutical industry and have also been explored for diverse biomedical applications [73, 95–98]. As electrospun materials, hydroxyethyl cellulose/poly(vinyl alcohol) (HEC/PVA) nanofibers were functionalized with calcium phosphate and evaluated in vitro as a possible scaffold for bone tissue engineering (BTE) [99]. On the other hand, silk fibroin and carboxymethyl cellulose based composite scaffolds were also design for BTE by Singh et al. [100]. The team showed that the cross-linked electrospun material had enhanced osteogenic properties when compared to pure silk fibroin nanofibers, improved mechanical properties and better cell attachment and prolif-

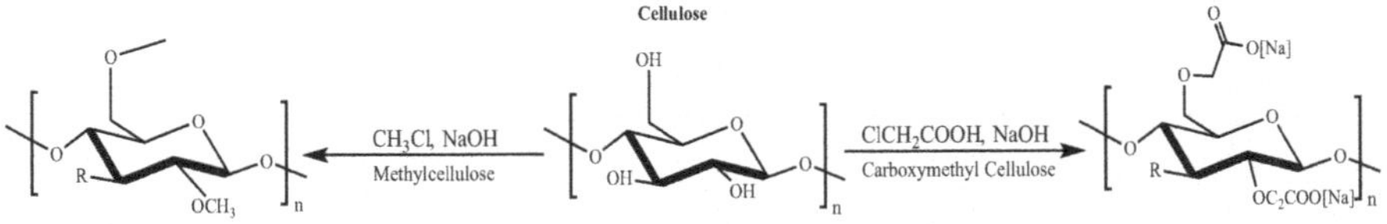

Fig. 12 Etherification reactions of cellulose through the use of alkyl-halides (O-alkylation). R = CH_3, CH_2COOH or H

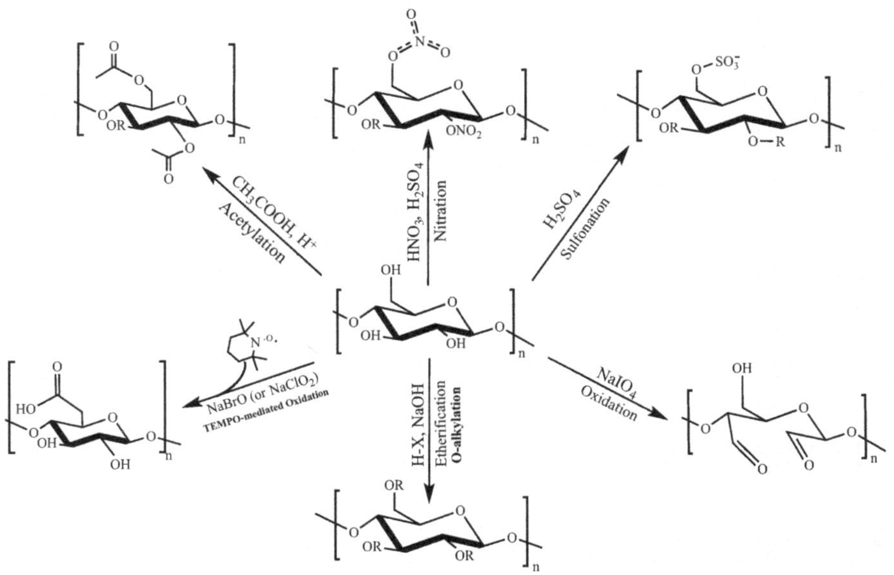

Fig. 13 Summary of the discussed synthesis reactions and their products, where X = alkyl halide

eration, as showed by qPCR and immunocytochemical analysis of human mesenchymal stem cells (hMSCs).

5.8 Section Summary

There are many other cellulose derivatives which merit their own distinctive examination and are of great relevance for the development and enhancement of future biomaterials. Overall, their unique characteristics and modes of synthesis impart different properties that may be modified to the specific needs of the researcher and, thus, their applications may be broad among a diverse range of disciplines. We have discussed until now the relevance of the reactivity of cellulose and its most prominent functional group, the hydroxyl moieties. Likewise, some chemical modifications have been discussed (Fig. 13) in conjunction with recent works that employ the various derivatives mentioned, which pertain from the particular synthesis reaction. The remaining sections of this chapter will expand on some of the properties and relevant aspects of electrospun cellulose and will recapitulate with a focus on the biomedical application of this biopolymer.

6 Electrospun Cellulose Nanofibers

Cellulose is the most abundant organic compound on Earth, so it provides an inexhaustible and sustainable source for the development of new biodegradable materials. Electrospun cellulose nanofibers have outstanding physicochemical properties that make them attractive for applications in tissue engineering, drug delivery, wound dressing, separation membrane, filtration, aerosol purification, coating technology etc. However, processing of cellulose is restricted by its limited solubility in common solvents and its inability to melt because of its numerous intermolecular and intramolecular hydrogen bonding (Fig. 5) and the high crystallinity of its structure [101]. Thus, a high interest of properly selection of non-degrading solvents for cellulose is required. The purpose of this section is to briefly review the suitable solvents for the preparation of electrospinable solutions that directly dissolve cellulose, which include: *N*-methyl-morpholine/*N*-oxide/water (NMMO/H_2O), lithium chloride/dimethylacetamide (LiCl/DMAc), ionic liquids and ethylene diamine/salt [102, 103]. The most important electrospinning parameters to obtain cellulose nanofibers are presented in Table 1.

Table 1 Summary of electrospinning conditions used to produce fine fibers from cellulose solutions (adapted from [18])

Solvent	Cellulose	Coagulant	Fibers	Reference
nMMO/H_2O 85/l 15w/w[a]	DP210, 9 wt. %	Rotating collector into water at 10 °C	750 nm	[104]
nNMMO, 50% H_2O[b]	DP670, 4% w/v	None	90–250 nm	[105]
nNMMO, 50% H_2O[b]	DP800 and DP700, 2 wt. %	Water	200–500 nm	[106]
ED/KSCN[c]	DP140, 8 wt. % DP940, 3 wt. %	None	3 µm 100–1000 nm	[107, 108]
LiCl/DMAc[d]	DP1140, 1–3 wt. %	Heat rotating collector into water	250–750 nm	[104, 109]
LiCl/DMAc[d]	3 wt. %	None	Fibers unstable	[110]
BMIM/CL or EMIM/Ba[e]	10 wt. %	Ethanol	500 nm to 10 µm	[111]

DP degree of polymerization
[a]Weight ratio of *N*-methylmorpholine *N*-oxide to H_2O is 85/15
[b]50% aqueous *N*-methylmorpholine *N*-oxide solution
[c]Ethylene diamine/potassium thiocyanate solution, exact composition not specified
[d]Lithium chloride/dimethylacetamide, exact composition not specified
[e]1-butyl-3-methylimidazolium chloride or 1-ethyl 3-methylimidazolium benzoate

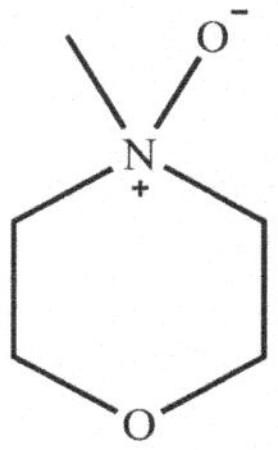

Fig. 14 *N*-methylmorpholine-*N*-oxide (NMMO)

6.1 Fibers Electrospun from nNMMO/H_2O

N-methylmorpholine-*N*-oxide (Fig. 14) is a common solvent for the spinning of cellulosic fibers at the industrial level. This tertiary amine oxide is usually mixed in water or other polar organic solvents that allow for the dissolution of cellulose by disrupting the intermolecular forces that pertain to its hydrogen-bonding network. As a consequence, new bonding interactions may form within the solvent system between the –OH groups of cellulose and the oxide ion ($^{-}$O–) of NMMO [112].

A drawback of utilizing this solvent is its low volatility. Thus, when in the process of electrospinning, the NMMO does not evaporate completely as the spinneret travels to the collector, resulting in ununiformed fibers [113]. Kim et al. redesigned the electrospinning setup for this cellulose/NMMO/H_2O solvent system, such that the temperature of the syringe and needle was kept between 70 and 110 °C and a coagulation system with water was applied to the collector [104]. This in turn allowed for the production of cellulose fibers with diameters within the 250–750 nm range.

6.2 Fibers Electrospun from LiCl/DMAc Solutions

The addition of salts adds a degree of complexity in regards to the molecular interactions or forces experienced by the components of the electrospinning solution. Overall, it is know that the solution conductivity tends to increase by augmenting the electrical charge that is carried within the jet [114–116]. This, in turn, has been observed to decrease the fiber diameter. The LiCl/DMAc system, on the other hand, is a combination of a non-volatile salt with an organic solvent that offers the particular advantage of dissolving cellulose by the work of ion-dipole interactions as well as electrostatic repulsions [117, 118]. However, a study conducted by Frenot et al. revealed that upon electrospinning, the cellulose fibers would orient themselves in an upward, vertical fashion while in the presence of an electric field [110]. In the absence of the electric field, the fibers collapsed and lost their form. Such results may indicate the accumulation of the salts within the fibers and the prevalence of DMAc due to its low volatility. Thus, much consideration must be made in regard to the solution concentration and ratio of the solvent system.

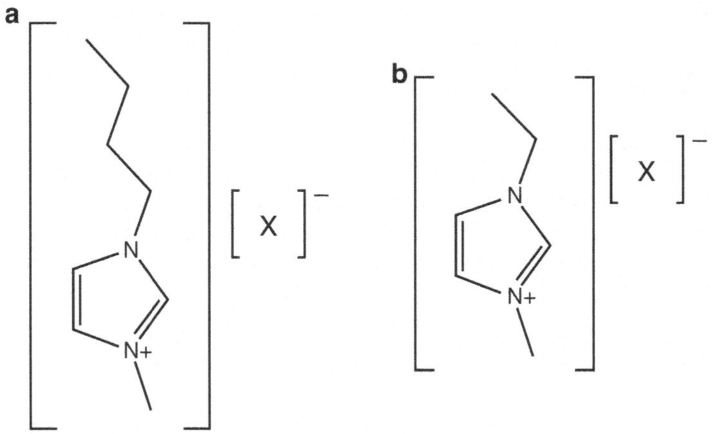

Fig. 15 Structure of the cations (**a**) [BMIM] (1-butyl-3-methyl-imidazolium) and (**b**) [EMIM] (1-ethyl-3-methylimi-dazolium). Where X = Cl, OAc, HCOO

6.3 Electrospinning Cellulose from Ionic Liquids

In a general fashion, ionic liquids (IL) are defined as molten salts whose melting point range up to 100 °C [119]. ILs have gained much recognition in recent times as they are considered to be an environmentally friendly solvent. This appellation is due to their unique properties such as low melting point and, particularly, low or negligible vapor pressures [120]. Other characteristics may also be attributed to these compounds (e.g.; thermal stability, high polarity, non-flammability, biological activity, among others), which in all allow for recycling simplicity [120, 121]. They do pose a problem, however, in regards to there application as solvent systems for the electrospinning of cellulose. Since the evaporation of the solvent is an important factor for proper fiber formation, a coagulation bath is required to remove the IL due to the low volatility as well as for the associated electrical charge retention [122, 123].

Common salts employed for the dissolution of cellulose include cations of 1-butyl-3-methyl-imidazolium ([BMIM]) and 1-ethyl-3-methylimi-dazolium ([EMIM]) accompanied by anions such as acetate, formate and, particularly, chloride (Fig. 15).

Quan et al. have fabricated nonwoven fibers of cellulose dissolved in [BMIM][Cl] at different solution concentrations (from 1.5 to 5% wt/wt) [124]. Their vertical electrospinning setup consisted of a water collector to remove the salts and a constant temperature chamber that enclosed the syringe at 100 °C. The fiber diameter seemed to increase on a dependent fashion to the concentration of the cellulose solution, ranging around 5 μm for the 1.5 wt. % up to 1 μm in the case of the 4 wt. % solution. Interesting enough, they produced cellulose fibers with added DMSO in order to reduce the solution viscosity and generate fine fibers. They found that by adding this component to the 4 wt. % cellulose solution, the produced fibers had a lower diameter mainly in the 500–800 nm range. They also show that the crystallinity of cellulose transform from type I to type II by XRD, which indicated the effective dissolution of the IL by disrupting the hydrogen bonding network of native cellulose and subsequent regeneration (refer to Sect. 4.3).

The use of 1-ethyl-3-methylimidazolium acetate ([EMIM][OAc]; also abbreviated as [C_2MIM][CH_3CO_2]) has also been well employed to dissolve cellulose for electrospinning (Fig. 15b). The group led by Freire's team prepared an 8% cellulose/[C_2MIM][CH_3CO_2] solution with a water coagulation bath and subsequently immersed the produced fibers in ethanol and water to remove the IL. The diameter of such fibers ranged within the 470 ± 110 nm range. They also added a second IL (1-decyl-3-methylimidazolium chloride, [C_{10}MIM][Cl]) to serves as a means to decrease the surface tension and enhance the solvents thermophysical properties. As a result, the average fiber diameter was greatly reduced to 120 ± 55 nm and fiber morphology was stabilized due to changes in the solvent properties. They also show that the crystallinity of cellulose transform from type I to type II by XRD, which indicated the effective dissolution of the IL by disrupting hydrogen bonding network of native cellulose. Likewise, XRD analysis revealed that the crystallinity of cellulose changed from cellulose I (native) to cellulose II from solvent dissolution before electrospinning, as a result of the dissolution and associated regeneration process.

Viswanathan et al. electrosprayed cellulose/heparin composite fibers by combining a 10% cellulose/[BMIM][Cl] solution with a 2% heparin/[EmIm][benzoate] solution, while utilizing an ethanol collector to remove the IL [111]. They observed a rough surface fiber morphology as well as bioactivity from heparin when evaluating clotting kinetics of human whole blood.

7 Electrospinning of Cellulose Acetate Nanofibers

Due the low volatility and toxicity of solvents used to electrospun pure cellulose, additional steps are required for complete removal of solvents from produced nanofibers [18]. To deal with this challenge, the use of cellulose derivatives, as cellulose acetate (CA), has become an alternative approach because it can be dissolved in a wide range of volatile solvents [78], to form films made from cellulose triacetate, and to form membranes or nanofibers made from cellulose 2-2.5-acetate [125]. In our interest to produce electrospun CA nanofibers, available solvent system reported are: single solvent system such as acetone, chloroform, *N,N*-dimethylformamide (DMF), dichloromethane (DCM), formic acid, methanol (MeOH) and pyridine; binary solvent systems such as acetone-dimethylacetamide (DMAc), chloroform-MeOH, and DCM–MeOH; ternary solvent system such as acetone/DMF/trifluoroethanol; mixed solvent system such as acetic acid/water [126]. A summary of solvents used for electrospinning of CA nanofibers and its effect in fiber formation are listed on Table 2. However, several studies considered the development of a facile technique where CA nanofibers were deacetylated to regenerate into cellulose nanofibers. For example, Liu and Hsieh showed that regenerated cellulose

Table 2 Solvents employed in the electrospinning of cellulose acetate (CA)

Solvent	Fiber formation[a]	References
Acetone	Discrepancy	[127–130]
Acetone/water	Smooth fibers	[129, 131]
Acetone/DMAc	Smooth fibers	[127, 128, 132]
Acetic acid	Few fibers	[127]
Acetic acid/acetone	Beaded fibers	[127]
Acetic acid/water	Smooth fibers	[113]
Chloroform	Beads	[128]
Chloroform/methanol	Smooth fibers	[128]
Dichloromethanol	Beads	[128]
Dichloromethanol/methanol	Smooth fibers	[128]
Dimethylformamide (DMF)	Beads	[128]
Acetone/DMF/trifluoroethylene	Smooth fibers	[133, 134]
DMAc	No fibers	[127]
Formic acid	Beads	[128]
Methanol	Beads	[128]
Methylene chloride/ethanol	Smooth fibers	[135]
Pyridine	Beads	[128]

Reprinted with permission from [78]
[a]Solvent is not the only parameter that will impact fiber formation

membranes can be obtained by electrospinning of CA (degree of substitution, DS = 2.45). Moreover, they show evidence that deacetylation process of CA membranes, via alkaline hydrolysis, is more efficient and complete in NaOH/ethanol than in aqueous NaOH, producing DS values between 0.15 and 2.33. Even so, many researchers chose to work with CA because is inexpensive, readily available, and the most stable cellulose derivative.

The solubility of CA is dependent of the degree of esterification or substitution (DS), defined by the number of OH groups substituted by acetate groups. It defines the properties of the CA [136], that include low weathering, heat and chemical resistance, thermal stability, reasonable toughness and dimensional stability. Additional physical properties of CA are listed on Table 3. Furthermore, the electrospinnability of CA and its good mechanical properties made CA a potential candidate for biomedical applications, such as skin tissue engineering [138], drug delivery and therapeutics, wound healing, cardiac cell constructs, bone tissue engineering [139], tissue/organ repair and regeneration, cosmetics and dental applications. Figure 16 demonstrates the cytocompatibility of CA and regenerated cellulose scaffolds, promoting cardiac cells networks, enhancing cell connectivity and addressing the structural complexity of the cardiac muscle [140].

Recently, considerable attention has been paid in the fabrication of scaffolds that can closely mimic the structure of native extracellular matrix (ECM). This approach use electrospinning as one of the preferred methods to fabricate scaffolds, because it provides large surface area for cell attachment, high porosity for cell penetration and colonization, and a possibility to design two-dimensional (2D) and novel 3D scaf-

Table 3 Physical properties of cellulose acetate

	Cellulose acetate
Specific gravity	1.27–1.32
Refractive index (25 °C)	1.47–1.5
Tensile strength (10^3 lbf/in^2) MPa	3.5–11 24–76
Elongation at break (%)	5–55
Izod impact ft lbf in^{-1} notch	0.5–5.0
Rockwell M hardness 70 °F (ASTM D.229-39)	−30 to +75
Flow temperature (°C) (ASTM D.569-40T)	115–165
Heat distortion temperature (°C)	50–100
Volume resistivity (Ωm) 50% RH, 25 °C	~10^{14}
Dielectric constant 60 Hz	3.5–7.5
Power factor 60 Hz	0.01–0.07
% water absorption 2 in × ½ in disc 24 h immersion (ASTM D.570-40T)	1.0–3.0

Reprinted with permission from [137]

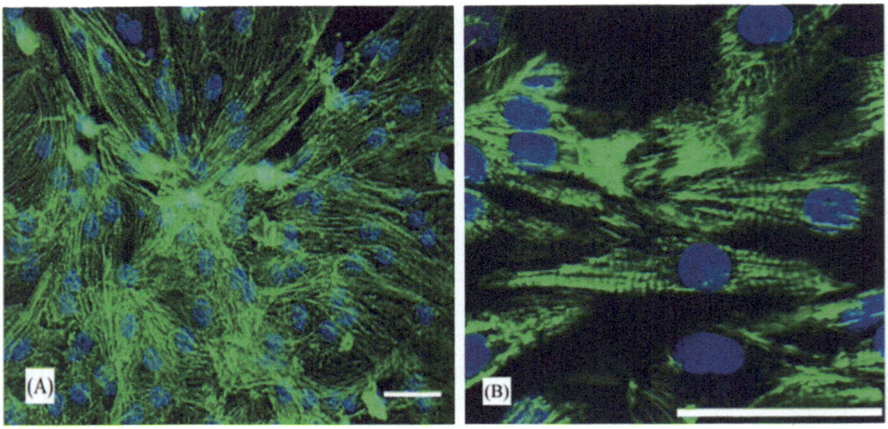

Fig. 16 Cytoskeleton maturity in CA-grown myocytes. (**a**, **b**) shows well-packed myocytes at two different scales. F-actin is fluorescently labeled in green, nuclei—in blue using DAPI (**a**) or TOTO-3 (**b**). Scale bar is 50 mm (reprinted with permission from [140])

folds that resemble a real tissue environment. Electrospun CA nanofibers enhances these features providing good mechanical properties and nanostructure morphology of fibers obtained, mimicking the collagen network of ECM. Biocompatibility of CA scaffolds have been tested *in vitro*, culturing cells on 2D scaffolds and monitoring cell proliferation by BrdU assays as well as cell growth. In vivo studies, conducted on animal models were also used to study biocompatibility of scaffolds in the biological system. According to Unnithan et al. electrospun CA nanofibers play an important role in the fabrication of polyurethane-cellulose acetate-zein (PU–CA–zein) scaffold for application in wound dressing. Authors explains that the presence

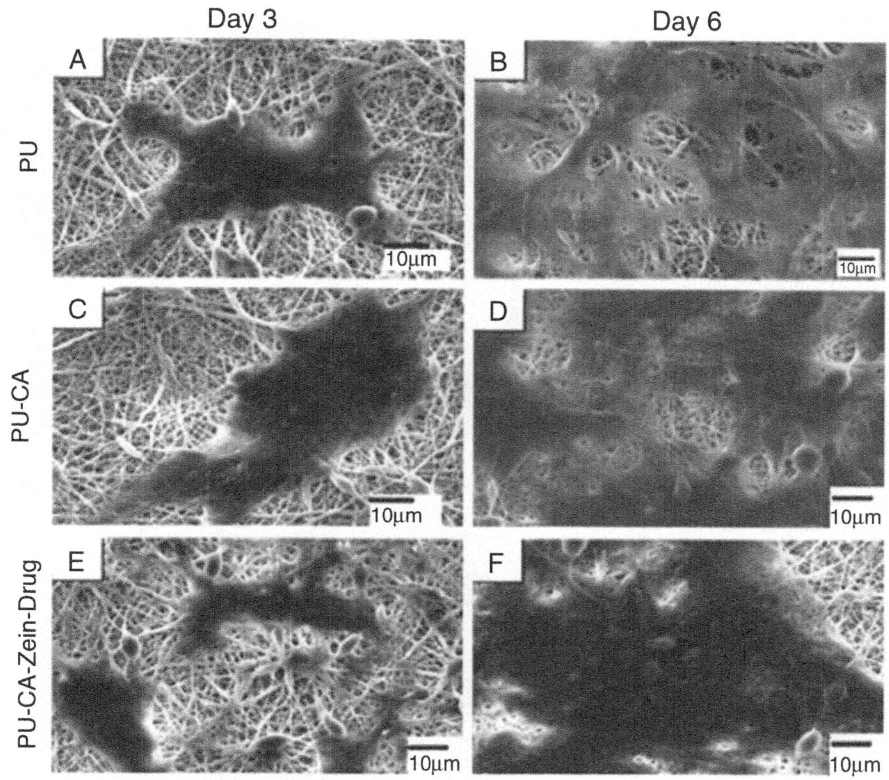

Fig. 17 SEM images showing the cell attachment on PU, PU–CA and PU–CA–zein-drug mats after day 3 (**a**, **c**, **e**) and day 6 (**b**, **d**, **f**), respectively (reprinted with permission from [141])

of CA and zein improves hydrophilicity, bioactivity and also create a moist environment for the wound recovery. To prevent clinical infections, an antimicrobial agent, streptomycin was loaded into the PU–CA–zein scaffolds without changes in the morphology of nanofibers obtained. Therefore, this nanofibrous scaffold achieve the desirable properties required for a wound dressing biomaterial, demonstrating an enhancement in blood clotting ability and platelet activation [141]. Figure 17 shows the cell attachment on the surface of composite scaffold.

Recently, Atila, Keskin and Tezcaner have reported the fabrication of fibrous scaffolds for use in bone tissue engineering applications. Electrospinning of blended pullulan (P) and cellulose acetate (CA) at different ratios: P_{80}/CA_{20}, P_{50}/CA_{50}, and P_{20}/CA_{80} (w/w)% were prepared in order to form 3D scaffold and crosslinked with trisodium trimetaphosphate (SMTP) to improve mechanical properties and to delay fast weight loss. The authors reported that codominance of P and CA resulted in the best blend ratio for adjusting and allowed the proliferation and growth of human osteogenic sarcoma cell line (saos-2). Figure 18 show images of electrospun scaffolds obtained by conventional electrospinning and also SEM images showing saos-2 cells attached to electrospun fibers and proliferated on the surface and inside

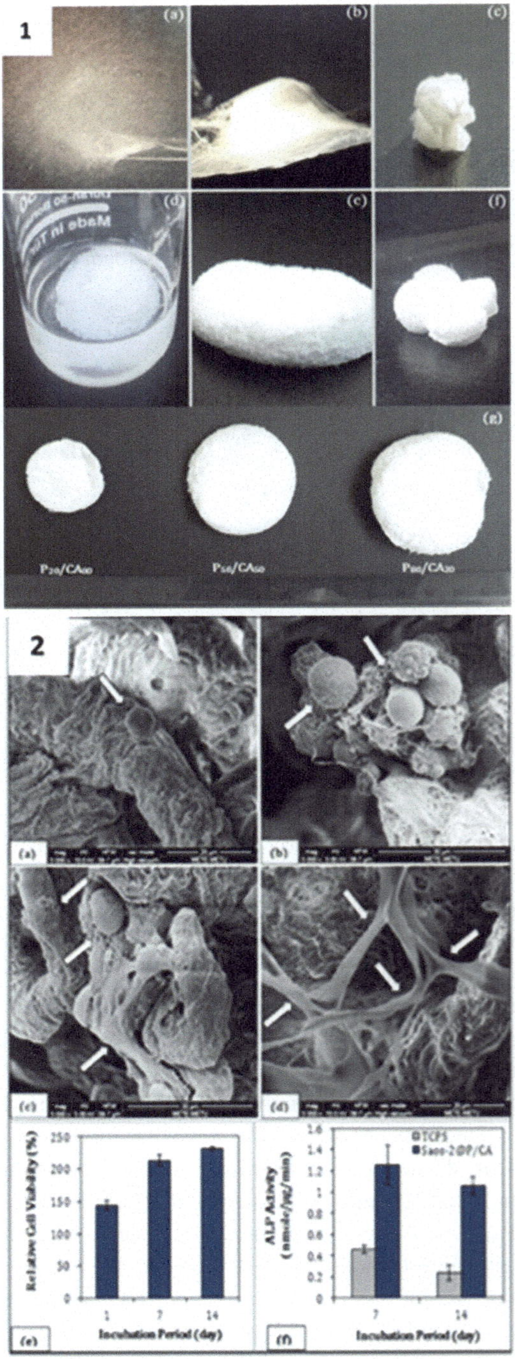

Fig. 18 (**a**) P20/CA80 scaffolds produced by conventional electrospinning (1), cross-sectional view of as-spun form on the collector (2), and punched version (3). P50/CA50 scaffolds produced by wet electrospinning in ethanol bath (4), after freeze-drying (5), and punched forms (6). P20/CA80, P50/CA50, and P80/CA20 scaffolds produced by wet electrospinning (7). (**b**) SEM images of saos-2 cells on P50/CA50 scaffolds at day 1 (scale bar: 20 µm, surface) (1); 4 (scale bar: 30 µm, cross-section) (2); 7 (scale bar: 30 µm, surface) (3); and 14 (scale bar: 20 µm, cross-section) (4). Arrows: cells adhered to the scaffolds. Proliferation (via Alamar Blue test) ($n = 6$) (5) ALP activity (6) of saos-2 cells on crosslinked P50/CA50 scaffolds ($n = 3$). ALP activity of cells seeded on TCPS (control) was statistically lower than the activity of cells seeded on P/CA scaffolds ($p<0.05$) (Reprinted with permission from [139])

the scaffold exhibiting a round and flattened morphology, after 7 and 14 days of culture, respectively.

Other approach found in the literature concerning about the use of cellulose acetate in tissue engineering is presented by Rodriguez et al. that worked with the saponification of electrospun CA mat for the generation of novel scaffold to nucleate bioactive calcium phosphate crystals. The structural characterization of these crystals showed similar diffraction pattern than hydroxyapatite, the mineralized component of bone. In order to improve bioactivity of scaffolds, regenerated cellulose membranes were provided by similar structure, composition and bioactive agents present on ECM, integrated together to promote cell attachment, migration, proliferation and maturation. Actually, nucleation and growth of CaP crystals under physiological conditions is the strategy used to study osteoblast response and to fabricate and ideal scaffold that promotes osseointegration. Figure 19 shows SEM images of the novel mineralized scaffold.

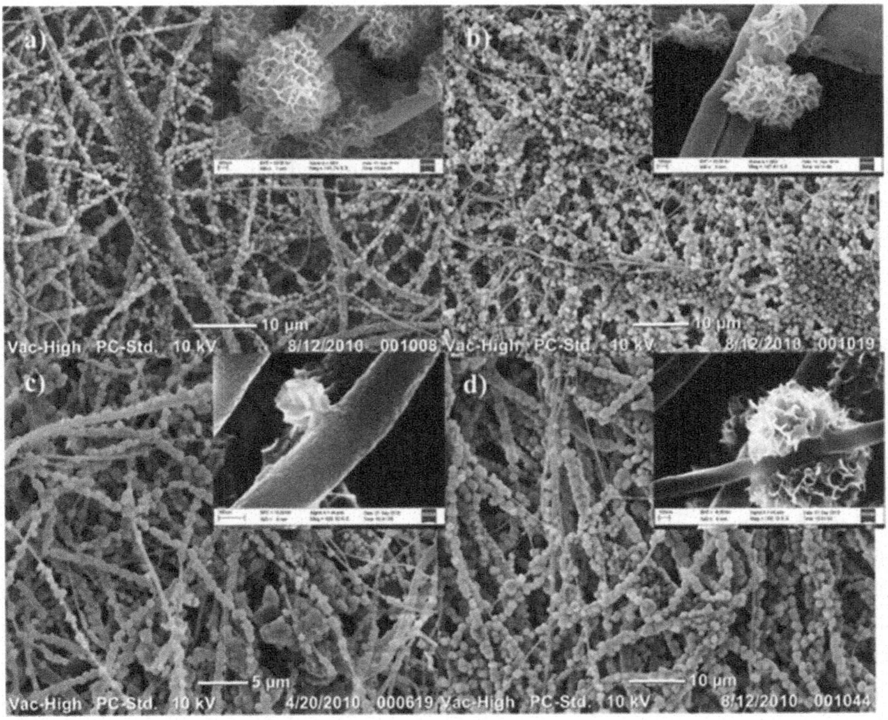

Fig. 19 SEM of mineralized scaffold with CMC treatment (**a**) 45 min RC and (**b**) 24 h RC treated with procedure 1, (**c**) 45 min RC, and (**d**) 24 h RC treated with procedure 2 (reprinted with permission from [80])

8 Biomedical Applications of Nanocellulose

The use of cellulose-based materials in biomedical applications have increased over time due to their biocompatibility, partial biodegradability, hydrophilicity, low cost, and good mechanical properties such as high surface area-to-volume ratio, rigid structure with regions of high crystallinity, tensile strength and high stiffness properties [142]. The complete degradation of nanocellulose in vivo is a desired characteristic in some applications; such as when they are used as temporary scaffolds to induce skin, bone or cartilage regeneration. Conversely, it is expected that medical devices made or reinforced with nanocellulose, such as heart valves, do not degrade in tissues. Usually, nanocellulose is poorly degraded in vivo due to the absence of cellulose degrading enzymes in mammalian tissues. However, this issue has been overcome introducing chemical modifications (oxidation, esterification, etherification) to the nanocellulose producing functionalized nanocellulose derivatives which are more prone to biodegradation in vivo [143].

The biocompatibility of cellulose-based materials has been widely evaluated in vitro and in vivo suggesting that nanocellulose is not cytotoxic and induces a negligible inflammatory response in vitro [144] and after implantation [145–147]. The good biocompatibility and mechanical properties have driven the study of cellulose-based biomaterials in several biomedical applications. In this section, we discuss the use and biocompatibility of cellulose-based materials and specifically, nanocellulose in drug delivery, gene therapy, biosensors/bioassays, wound dressing, and tissue engineering. Figure 20 shows examples of biomedical applications that use cellulose as raw material.

8.1 Drug Delivery/Gene Therapy

The rational basis of using nanocellulose for drug delivery and gene therapy is its high surface area-to-volume ratio and the abundant surface hydroxyl groups that provides sites for chemical modification. Furthermore, the surface negative charge and the possibility of chemical modification provides the nanocellulose the ability to interact with several drugs by ionic (negative/positive compounds) and hydrophobic interactions [149]. In addition to its chemical properties, the notion that nanocellulose can be used as vehicle for drugs and nucleic acid materials is provided by previous studies that showed that nanocellulose can be uptaken by mammalian and insect cells [150, 151]. Previous studies from Mahmoud et al. demonstrated that cellulose nanocrystals (CNCs) conjugated covalently with rhodamine B isothiocyanate by an epoxy-mediated coupling reaction are transported (without affecting the cell viability) inside the human embryonic kidney (HEK 293) and Spodoptera frugiperda (Sf9) cells [150]. However, the fluorescein isothiocyanate (FITC)-conjugated CNCs showed no cellular uptake. Moreover, Liebert et al. showed that conjugated FITC-cellulose nanospheres (produced by dialysis from

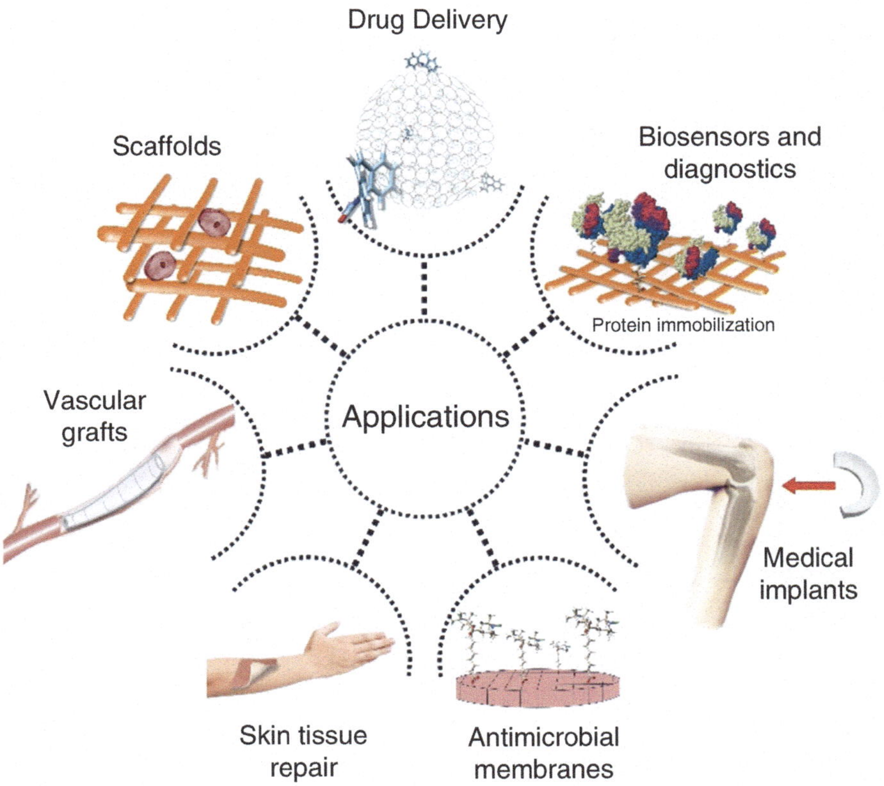

Fig. 20 Biomedical applications. https://www.intechopen.com/copyright-policy.html (reprinted with permission from [148])

trimethylsilylcellulose) are internalized by human fibroblast cells [152]. According to the authors, the use of cellulose nanospheres could be more suitable for drug delivery systems because spheres show a better internalization than rod-like nanoparticles [153]. Therefore, the authors established that particles geometry is important to take in consideration in the design of nanocellulose as drug carriers.

Previous studies mentioned above have demonstrated that nanocellulose material can be endocytosed by eukaryotic cells [150, 151]. However, since several anti-cancer drugs are known to affect normal and tumor cells, the ideal drug carrier is one that can deliver the drug to the malignant cells. It is well known that cancer cells over-expressed the folate receptor (FR) and this characteristic can be used for the specific targeting of tumor cells [154, 155]. Therefore, in order to target cancer cells, Dong et al. designed a folic acid-conjugated CNCs (Fig. 21).

In this study the authors demonstrated that folic acid-conjugated FITC-CNCs (FITC-CNC-FA, FITC labeling is used for measuring the cellular uptake) are internalized by human (DBTRG-05MG, H4) and rat (C6) brain tumor cells. Furthermore, since a decrease in the cellular uptake is observed after the addition of folic acid is

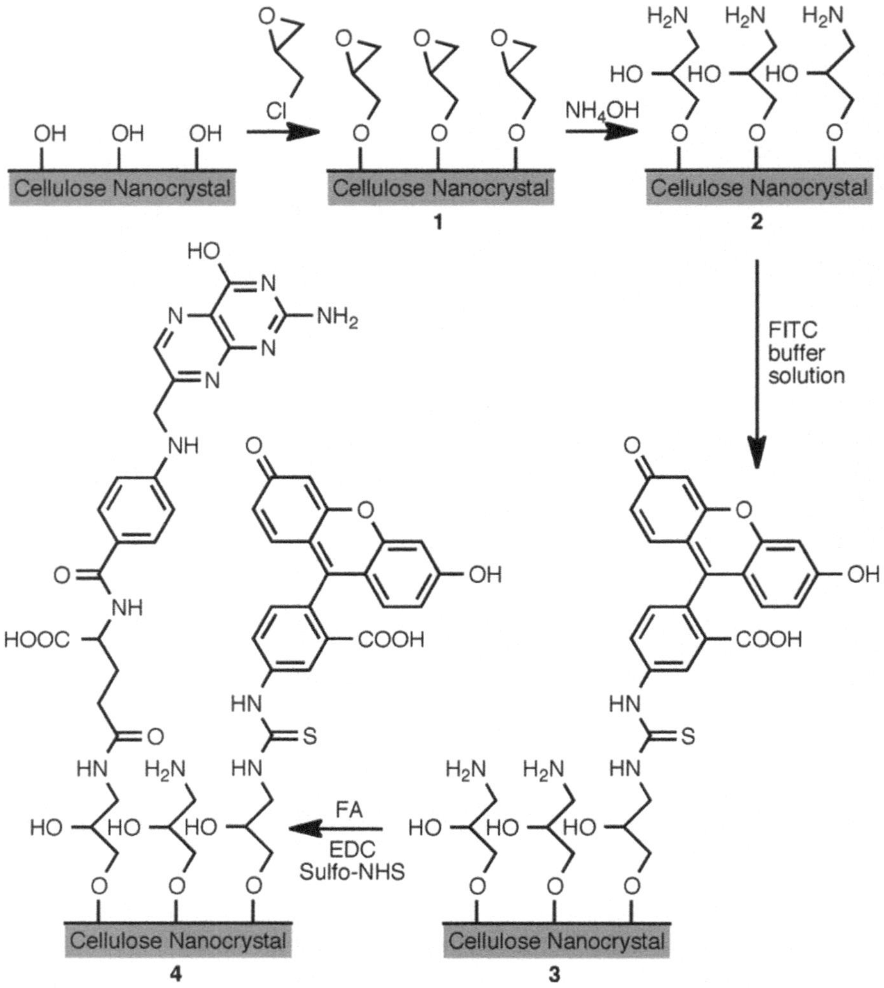

Fig. 21 Scheme of the synthesis of FITC-labeled, folic acid-conjugated CNCs (reprinted with permission from [156])

added to media, the results presented in this study demonstrated that folic acid-conjugated CNCs internalization is dependent of the interaction with the folic acid receptor. In addition, results showed that cells internalized folic acid-conjugated CNCs by using the caveolae (human DBTRG-05MG and rat C6 cells) and clathrin (human H4 cells) endocytic pathway. Moreover, previous results from the same research group have shown that CNC-FA can improve the cytotoxicity induced by the irreversible electroporation (IRE) method, which is a non-thermal focal ablation technique that kill cells by affecting the cell membrane permeability [157]. In this study, the authors showed that the incubation of CNC-FA with human KB and MDA-MB-468 cells induced a higher decrease in cell viability than the IRE alone. More importantly, Colacino et al. showed that CNC-FA potentiate the IRE-mediated decrease in cell viability in folate receptor (FR)-positive cancer cells and not in (FR)- negative cancer cells (A549-human alveolar basal epithelial cells). Collectively, these results demonstrated that CNCs could be a promising agent for drug delivery (for cancer and other diseases) to specific cells in the human body.

In addition, several studies have shown that nanocellulose can bind and release drugs [158–160]. For instance, Jackson et al. demonstrated that nanocrystalline cellulose (NCC) without modifications can bind and release tetracycline (antimicrobial drug) and doxorubicin (anticancer drug) [158]. Nonetheless, the researchers showed that the conjugation of NCC with cetyl trimethylammonium bromide (CTAB) allowed the encapsulation and release of hydrophobic anti-cancer drugs (docetaxel, paclitaxel, etoposide). In addition, this study showed that CTAB-modified NCCs are uptaken by human KU-7 bladder cancer cells. In a more recently study, Cacicedo et al. reported that the modification of bacterial cellulose with alginate (Alg) increase the encapsulation of doxorubicin by threefold [161]. Furthermore, the authors showed that BC-Alg scaffolds loaded with doxorubicin decreased the cell viability of HT-29 human colorectal adenocarcinoma cell line from 95 to 53% after 24 h of incubation. In another study, Gautier et al. demonstrated that nanocellulose composites could be used for the delivery of natural compounds with anti-cancer properties [162]. The results presented by this study showed that cellulose nanocrystals (CNCx) functionalized with β-cyclodextrin (CD) can encapsulate curcumin, which is a natural compound with anti-cancer properties that is extracted from *Curcuma longa*. Results shown in Fig. 22 demonstrate that the presence of CNCx is very important for the internalization of curcumin. More importantly, a decrease in the cell viability is observed after the addition of curcumin-CD/CNCx complexes to human colorectal (HT-29) and prostatic cancer (DU 145 and PC-3) cells. For instance, in vitro studies showed that at 48 h of incubation the cell viability test for the DU 145 cell line showed a IC_{50} (μM) of 18, 20, and 5.5 for the Cur, Cur–CD and Cur–CD/CNCx compounds, respectively. Altogether, these studies demonstrated that nanocellulose compounds can bind and release drugs inside the cells.

Another important properties of nanocellulose composites is that can deliver nucleic acids into cancer cells [163]. Previous results from Anirudhan et al. demonstrated that the delivery of DNA plasmid (pCMVLuc) containing a luciferase reporter into human alveolar epithelial cells (A549) can be performed by using aminated β-cyclodextrin-modified-carboxylated magnetic cobalt/nanocellulose composites (ACDC-Co/NCC). The authors showed that ACDC-Co/NCC has an 88.2% transfection efficiency when 1000 ng of DNA plasmid is used. In addition, this study showed that ACDC-Co/NCC has a small effect in cell viability when is compared with polyethylenimine (PEI), which is a commonly used transfection agent.

Moreover, Hu et al. demonstrated that cellulose nanocrystals (CNCs) functionalized with disulfide bond-linked poly (2-(dimethylamino) ethyl methacrylate (PDMAEMA) brushes can be used as a potential nonviral vector for gene transfection. In this article, the authors showed that CNC-graft-PDMAEMA (CNC-SS-PDs) vectors can deliver a plasmid containing a suicide gene known as a cytosine deaminase (CD) into human cancer cell line (HepG2). The expression of cytosine deaminase (enzyme from the pyrimidine salvage pathway only found in prokaryotes and lower eukaryotes) makes mammalian susceptible to the addition of

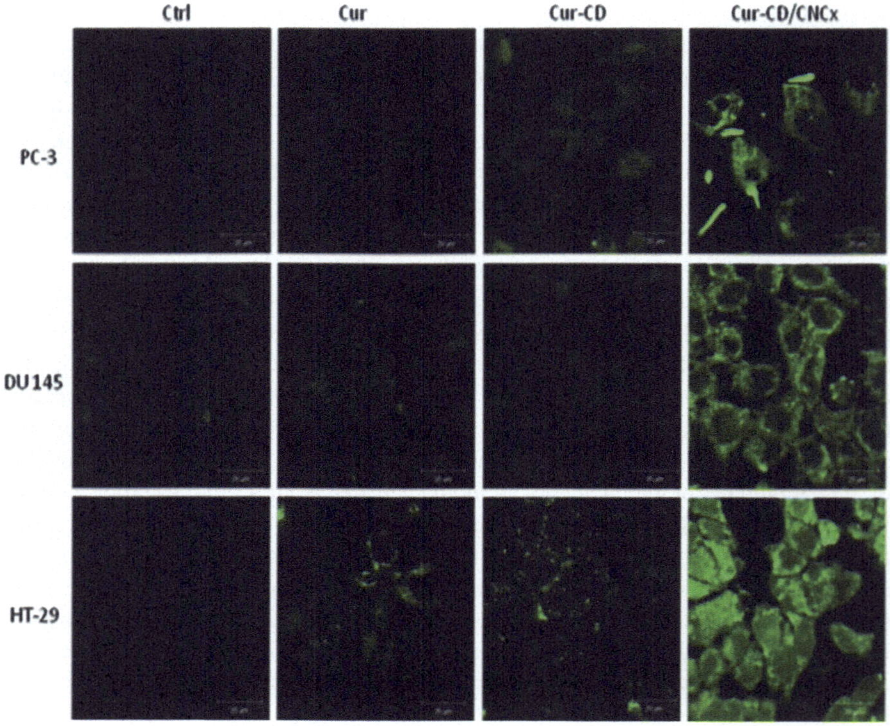

Fig. 22 CNCx increases the accumulation of curcumin I. PC-3, DU 145 and HT-29 cells were incubated with the control (CTL), Cur, Cur–CD, and Cur–CD/CNCx (50 μM) for 4 h. Then, the fluorescence emission from curcumin I was analyzed by confocal microscopy (reprinted with permission from [162])

5-fluorocysteine (5-FC). Cells expressing CD can convert a 5-FC into 5-fluorouracil (5-FU), which is a potent anticancer agent that affects the pyrimidine synthesis. Indeed, the authors noticed a reduction of ~60% of cell viability after the addition of 5-FC to the cells that were transfected with the CNC-SS-PDs carrying the CD gene (pCMV-CD). Furthermore, Hu et al. demonstrated that the treatment with CNC-SS-PDs/pCMV-CD/5-FC can decrease the tumor growth in mice. Results presented in the Fig. 23 shows that there is a decrease in the tumor weight after the intratumoral injection of CNC-SS-PDs/pCMV-CD/5-FC vectors variants.

8.2 *Biosensors/Bioassays*

In the recent years, the use of nanocellulose in biosensors and bioassays for the detection of inflammatory biomarkers and pathogen DNAs have been reported [165–167]. For instance, Edwards et al. have demonstrated that nanocellulose-based biosensors can be employed in the detection of human neutrophil elastase (HNE),

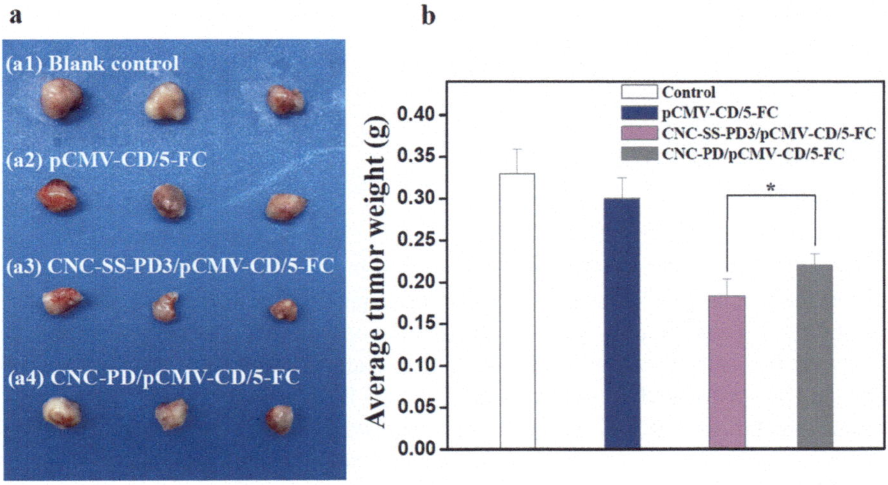

Fig. 23 In vivo antitumor activity of CNC-SS-PDs. Implanted tumors in female balb/c nude mice were injected with pCMV-CD/5-FC, CNC-SS-PD3/pCMV-CD/5-FC, and CNC-PD/pCMV-CD/5-FC for 18 days. After the treatment (**a**) photographs of the tumors were taken and (**b**) the weights of tumors were measured (mean ± SD, $n = 3$, $*p < 0.05$) (reprinted with permission from [164])

which is a serine protease that is a biomarker for several diseases such as psoriasis, pancreatitis and cystic fibrosis [167]. In this study, the authors linked a fluorescent tetrapeptide (substrate of HNE) to a cotton cellulose nanocrystals and incubate it with different concentrations of HNE. Results demonstrated that this nanocellulose-based sensor can emit fluorescence after the addition of HNE. Indeed, in a more recent publication from the same research group was shown that the lowest detection range for this type of protease biosensors is from 0.125 to 0.050 U/mL [168]. Moreover, a previous study from Saikrishnan et al. has demonstrated that nanocellulose can be used in the detection of *Mycobacterium tuberculosis* [165]. To detect this pathogen bacteria, the authors immobilized a DNA sequence *IS*6110 of *M. tuberculosis* into a tosylated cellulose. Then, the DNA-cellulose conjugate was incubated with the biotinylated complementary DNA target from the *M. tuberculosis* (oligonucleotide or PCR amplified) and the colorimetric signal was obtained by the addition of a streptavidin-HRP conjugate. The detection limit for this cellulose-based bioassay is up to a concentration of 0.05 μM.

In a recent study, Weishaupt et al. demonstrated that a protein-nanocellulose biosensor can be used in the detection of copper ions at the nanomolar level [169]. For the detection of copper ions, the authors use cyanobacterial red-fluorescent phycobiliprotein C-phycocyanin (CPC) and functionalized this sensing biomolecule into a TEMPO-oxidized nanofibrillated cellulose matrix (TONCF). Importantly, this research group demonstrated that this biosensor (called by the authors Cysense films) can detect cupper from human serum filtrate spiked with $CuCl_2$. Altogether, the research evidence presented here showed that nanocellulose could be an important component in the development of future biosensors and bioassays focused in the detection of human diseases and pathogens.

8.3 Wound Dressing

Cellulose-based nanomaterials have been widely tested as skin wound dressings [170–172]. These materials have high water-holding capacity to maintain the moisture at the wound site, high mechanical strength and great conformability, high permeability, and low toxicity which are desirable properties to perform the function as temporal biologic dressings. Cellulose-based scaffolds tested in pre-clinical and clinical trials as wound dressings for skin lesions have been produced mainly with bacterial cellulose (BC) [173]. However, some studies have also studied the suitability of nanocellulose derived from plants as wound dressings [174].

BC pellicles have been modified to improve their performance as temporary skin substitutes. One of the modifications introduced to BC scaffolds is the degree of porosity. Preclinical studies of nanocellulose scaffolds with high porosity showed that they accelerate the healing of skin lesions in rats compared to low porosity scaffolds and gauze [171]. They also found that the inflammatory response was less intense in animals where the skin wounds were covered with BC scaffolds independent of the porosity compared to controls (wounds covered with gauze). Li et al. [170] also evaluated BC nanoscaffolds where the glycosaminoglycan hyaluronic acid (HA) was incorporated at different concentrations to improve their mechanical properties. The scaffolds composed by nanocellulose and 0.1% HA displayed the best performance as wound dressings in rats. The lesions covered by this type of composite closed more rapidly and resembled accurately the original skin characteristics after healing compared to controls. The authors claim that the loosely structured nanofibers in scaffolds with high porosity facilitate the infiltration of cells and cytokines from tissues to scaffolds and that the presence of HA increases the water retention resulting in a moist environment that increases the tissue remodeling and wound healing.

The utility of bacterial cellulose (BC) pellicles as dressings in skin wounds in rats was evaluated by [175]. They coated induced full-thickness skin lesions with BC pellicles for up to 28 days. The wounds were dressed every 3 days. As controls, they dressed skin wounds with Vaseline gauze or Algisite®, a commercial dressing composed of alginate. They observed that the wounds dressed with the BC pellicles closed more rapidly than both control lesions (dressed with gauze or Algisite®). Additionally, they found that the fibroblast infiltration, collagen deposition, and the neovascularisation were higher in the wounds dressed with BC films and Algisite® compared to the lesions dressed with gauze.

Some of the BC scaffolds that have been tested in clinical trials as wound dressings are already commercially available and are known as Biofill® [176], Bioprocess®, XCell®, and Gengiflex®, among others [173]. Biofill® and Bioprocess® were proposed to treat burns and ulcers. Gengiflex® has been used for the treatment of peri-

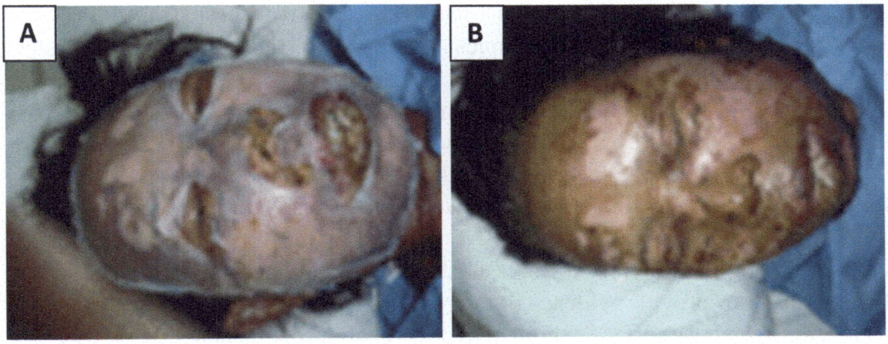

Fig. 24 Bacterial cellulose scaffold used as biological dressing in facial skin burns. (**a**) Photograph showing the application of the BC scaffold on the injured facial skin and (**b**) complete epithelialization two weeks after use of BC as temporary skin substitute (Reprinted with permission from [177])

odontal diseases and XCell® has been commercialized for chronic wound care [173]. A BC scaffold named Nanocell® was tested as a biological dressing to accelerate the skin healing in a patient who suffered second degree burns on face, both upper arms and trunk [177]. The BC dressing adhered to the wound sites and its elasticity allowed its molding to facial contours (Fig 24a). After two weeks the facial wounds displayed complete epithelialization (Fig. 24b). Remarkably, the patient did not show irritation, allergic reaction or infection during the healing process suggesting the suitability of BC dressings to treat skin burns. Nanofibrillar cellulose (NFC) derived from wood was also tested as biological coverage of split-thickness skin wounds originated after skin removal which was utilized as autografts on burn lesions [174].

8.4 Biocompatibility of Nanocellulose Scaffolds Designed for Applications in Tissue Engineering In Vitro Studies

A fund They seeded hDFCs for 3 days on the cellulose-nanoHA scaffolds amental property of a novel biomaterial is its biocompatibility. This property may be defined as the ability of a foreign material implanted in the body to perform its function without elicit a deleterious host response, such as foreign body response. In vitro, the biocompatibility is defined as the ability of a material to promote the adhesion, growth and differentiation of cells. The biocompatibility depends on various factors. Among them are the chemical properties of the material, the biodegradability rate and the products released to the biological environment after degradation, the surface functionalization, and the material topography. The biocompatibility of nanocellulose biomaterials in the form of hydro- gels, composites, electrospun nanofibers, sponges, and membranes from different sources and exhibiting different chemical and structural modifications has been extensively tested in vitro [142]. Here we present some relevant examples of the in vitro studies performed to determine the nanocellulose biocompatibility.

The biocompatibility of mineralized macroporous bacterial nanocellulose (BNC) scaffolds was evaluated by [178]. Mesenchymal stem cells (MSCs) derived from human fetal livers were seeded on mineralized and non-mineralized BNC scaffolds for up to 21 days. They found that MCS were able to attach, proliferate, spread, and differentiate into osteoblasts on both BNC scaffolds. However, MSC formed confluent monolayers in mineralized BNC substrates compared to non-mineralized scaffolds. Thus, they provide evidence of the biocompatibility of macroporous bacterial nanocellulose whose surface exhibited hydroxyapatite coating and propose them for bone tissue engineering.

The suitability of scaffolds made with non-woven cellulose coated with calcium phosphate (CaP) as cell culture substrates for primary bovine cartilage cells (chondrocytes) was evaluated by [179]. After 6 weeks of culture, the authors observed an increased cell adhesion and cell spreading, and higher expression of components of the cartilage extracellular matrix, specifically of collagens type I and II and proteoglycans, in non-woven cellulose coated with CaP compared to uncoated scaffolds. Based on these results, the authors concluded that non-woven cellulose scaffolds were moderately biocompatible, a property that was enhanced by CaP nanocrystals deposition on their surfaces. In addition, they proposed that CaP-coated cellulose scaffolds might be useful for cartilage tissue engineering applications.

The orientation of nanofibers in CA-based scaffolds appears to be a key factor to stimulate cell ingrowth, cell spreading and viability. He et al. observed that primary human dental follicle cells (hDFCs) seeded on scaffolds made of randomly oriented nanofibers were able to attach to scaffolds but in a scattered manner. hDFCs failed to spread and exhibited mainly a rounded morphology. Conversely, when the hDFCs were seeded on scaffolds made of parallel or uniaxially aligned cellulose fibers they attached to the fibers and additionally were able to spread and acquire an elongated morphology. They suggest that the aligned nanofibers resemble more accurately the extracellular matrix (ECM) morphology and in this manner constitute a more natural environment for cell growth [180].

The biocompatibility of scaffolds made from cotton cellulose and different concentrations (3, 5, or 10%) of nano-hydroxyapatite (nano-HA) was evaluated by [181]. They seeded hDFCs for 3 days on the cellulose-nanoHA scaffolds; then they evaluated the hDFCs viability and adhesion. Furthermore, Ao et al. found that a higher number of hDFCs adhered on the scaffolds with 10% of nano-HA compared to the other concentrations. Additionally, cells were evenly distributed in the scaffolds containing 10% nano-HA and displayed higher viability compared to the cells seeded on the cellulose scaffolds containing 3 or 5% nano-HA. From these results, the authors concluded that the addition of nano-HA to the cellulose improved the tensile strength and Young's modulus as well as the biocompatibility of the scaffolds.

The surface texture of each nanofiber also appears to be an important factor that modulates the behavior of cells on cellulose acetate (CA) butyrate scaffolds. Huang et al. compared the growth of Schwann cells (a type of non-neuronal cells forming part of the peripheral nervous system) on either aligned or randomly oriented CA-butyrate scaffolds with smooth or parallel line surface texture. They found that both the nanofiber alignment and the parallel lined texture improve the Schwann cell attachment, growth and elongation. A greater number of cells were attached on CA scaffolds made with aligned nanofibers whose surface exhibited a parallel lined texture compared to scaffolds made with randomly oriented nanofibers and smooth surface. Additionally, Schwann cells exhibited typical cell morphology with elongated cell bodies since early culture times on aligned nanofibers with surface texture compared to controls. They claim that CA scaffolds with aligned and textured nanofibers may be tested for neural tissue engineering [182].

Primary rat astrocytes (a non-neuronal type of cells forming part of the central nervous system) were seeded on electrospun cellulose acetate (CA) nanofibrous scaffolds that had been heat-treated to increase the scaffold's stiffness. They used thermal treatment of CA nanofibers because they were interested in improving their stiffness without inducing changes in their chemical properties. They evaluated the astrocyte's viability and attachment on thermally-treated CA scaffolds compared to non-heat treated CA scaffolds or conventional culture plates (polystyrene surfaces) [183]. They found that astrocyte's viability and adhesion increased on heat-treated CA scaffolds compared to controls. They also determined that cells growing on heat-treated CA scaffolds expressed higher level of glial fibrillary acidic protein (GFAP), which was used as an indicator of the astrocyte activity, compared to non-thermally treated CA scaffolds. Their results suggest that the increase in CA stiffness improves astrocyte viability and adhesion.

CA-based nanoscaffolds were used to coat metallic devices employed to replace heart valves. Bioactive groups, mainly the synthetic peptides RGD (Arg–Gly–Asp) and YIGSRG (tyrosine–isoleucine–glycine–serine–arginine–glycine), were chemically bound to the CA nanoscaffolds surface to promote cell adhesion. RGD and YIGSRG are motifs found in the ECM proteins fibronectin and laminin, respectively. The ability of CA-based constructs to support the adhesion and growth of mouse fibroblasts (L29 cell line) was assessed. They found that the biofunctionalized CA scaffolds improved the cell growth compared to controls, suggesting their applicability for future preclinical studies of valve tissue engineering [184].

Overall, the biocompatibility in vitro of nanocellulose, nanofibrillar cellulose, and bacterial cellulose materials has been evaluated using cells derived from different sources showing cell adhesion, survival, and proliferation abilities on contact to the cellulose-based materials. The results show that the degree of toxicity and biocompatibility are dependent on the mode of processing, particle size and surface chemistry of the cellulose nanoparticles.

8.5 Tissue Engineering

Tissue engineering is a scientific field that applies the principles of engineering and biological sciences to develop tissue and organ analogs to replace or improve lost or damaged human body parts [185, 186]. Cellulose-based nanoscaffolds and hydrogels from different sources (plants, bacteria, and algae) have been explored as novel biomaterials to elaborate tissue analogs. Specifically, electrospun cellulose acetate nanoscaffolds have been proposed for skin, bone and cartilage tissue engineering because of its biocompatibility and mechanical strength. However, most studies have only addressed the in vitro biocompatibility of nanocellulose scaffolds to date as explained in the previous section. Here we present some examples of CA scaffolds that have been tested as tissue analogs in vivo. Surprisingly, data about the use of nanocellulose as implant material in preclinical and clinical studies are scarce indicating that this is an incipient area of research.

The in vivo biocompatibility of scaffolds made of BC nanofibrils was evaluated in rats [187]. They implanted the BC nanoscaffolds subcutaneously in rats for 1, 4, and 12 weeks. They found a mild number of inflammatory cells without signs of foreign body reaction around the BC implants at the different time points evaluated. Additionally, they observed fibroblast infiltration in the BC scaffolds suggesting their integration in the subcutaneous tissue. Collagen production and angiogenesis were observed in the implanted BC scaffolds over time.

Shi et al. studied the biocompatibility of bacterial cellulose (BC) nanoscaffolds in vitro and in vivo [188]. The viability of C2C12 cells seeded on the nanoscaffolds was evaluated in vitro at different time points. They found no significant differences between the viability and proliferation of cells on BC scaffolds and conventional cell culture plates, suggesting the BC biocompatibility. More importantly, the authors evaluated the biocompatibility of BC scaffolds alone or carrying two different concentrations of bone morphogenetic protein 2 (BMP-2) after subcutaneous implantation in rats. Results showed a cell ingrowths in BC scaffolds carrying the both doses of BMP-2 whereas no cells were observed in BC scaffolds without BMP-2 two weeks of post-implantation. Additionally, ectopic bone tissue formation was observed around the BC scaffolds loaded with BMP-2 after 4 weeks of implantation. Conversely, BC scaffolds without BMP-2 showed some cells that have grown in the scaffold at 4 weeks post-implantation but no signs of bone formation were observed, suggesting that BC scaffolds without BMP-2 are biocompatible but are not osteoinductive [188]. A shortcoming of this study was the lack of information regarding the potential presence of inflammatory infiltration or foreign body reaction around the BC scaffolds to support the BC biocompatibility in vivo.

Plant-derived cellulose scaffold biocompatibility was tested after subcutaneous implantation in mice for 1, 4, or 8 weeks [189]. Results from this report found histological evidence of acute inflammation and foreign body response to the scaffold 1 week post-implantation. Nevertheless, these immune responses have disappeared at week 8 post-implantations, suggesting that the cellulose scaffold reduces its immunogenicity over time. Remarkably, cellulose scaffolds retained most of their original structure and shape but exhibited a significant reduction in size (48%) after 8 weeks, suggesting slow biodegradation in vivo. However, this slow degradation rate in vivo appears not to be correlated with immunogenicity. Additionally, fibroblast ingrowth, vascularization and collagen deposition were also observed into the cellulose scaffold at week 8 after implantation.

An ideal polymeric scaffold designed to induce bone, cartilage, or skin tissue regeneration after grafting must act as a temporary environment to induce host cell ingrowth, proliferation, neovascularization and extracellular matrix deposition concomitantly with scaffold degradation over time. As was mentioned previously, the poor biodegradability of nanocellulose in vivo is a concern when used as implant. However, various mechanisms to modulate the biodegradability of the CA based scaffolds have been studied. Czaja et al. evaluated the biodegradability in vitro and in vivo of bacterial cellulose (BC) scaffolds with varying degrees of oxidation (50–94%) obtained through treatment with sodium periodate. They implanted γ-irradiated and oxidized or native BC scaffolds subcutaneously in rabbits for up to 26 weeks. When they evaluated the site 26 weeks' post-implantation they found that the γ-irradiated scaffolds with the different degrees of oxidation have almost completely degraded and only some remnants were observed. Conversely, the native BC scaffolds remained unchanged. At the histological level, they found an immune response around all the implanted scaffolds. The immune response consisted of a foreign body response characterized by variable numbers of macrophages and giant cells with mild numbers of neutrophils around implants. However, the intensity of the response varied among treatments. It was scored as slightly irritant for most of the oxidized BC scaffolds whereas it was scored as irritant for native BC scaffolds. Additionally, around native BC scaffolds was observed a defined fibrous capsule, an indicator of the strong foreign body response elicited by native cellulose. Overall, their results suggest that γ-irradiation and oxidation of cellulose improves the degradation rate of scaffolds in vivo, a phenomenon that is correlated with a reduction in the intensity of the immune response to this type of implants [143].

Cellulose–hydroxyapatite (BC–HA) nanocomposite membranes were implanted in noncritical bone lesions in rat tibiae for 1, 4, and 16 weeks. At week 1 osteoid and newly formed bone was observed at the lesion defects implanted with the BC–HA. Conversely, fibrotic tissue was observed in the control lesions. Since week 4 new bone tissue was observed in the bone defects that were implanted with the BC–HA scaffolds. The BC–HA membranes were observed during the entire period of study suggesting poor biodegradation. However, no inflammatory reaction was observed around the BC–HA membranes at the different time points [190].

Tubes made with bacterial nanocellulose (Fig. 25a) were grafted in lesions of peripheral nerves in rats to induce nerve regeneration as depicted in Fig. 25b [191]. The spontaneous healing of peripheral nerves frequently led to partial reconnection of damaged nerves. This process in accompanied by the formation of scars and neuromas (connective tissue overgrowths) at the sites where the nerves were damaged (Fig. 25c, white arrow). This spontaneous healing process results in partial recovery of motor activity. However, when the nanocellulose tubes served as guidance channels for the damaged nerves a significant lesser number neuromas were

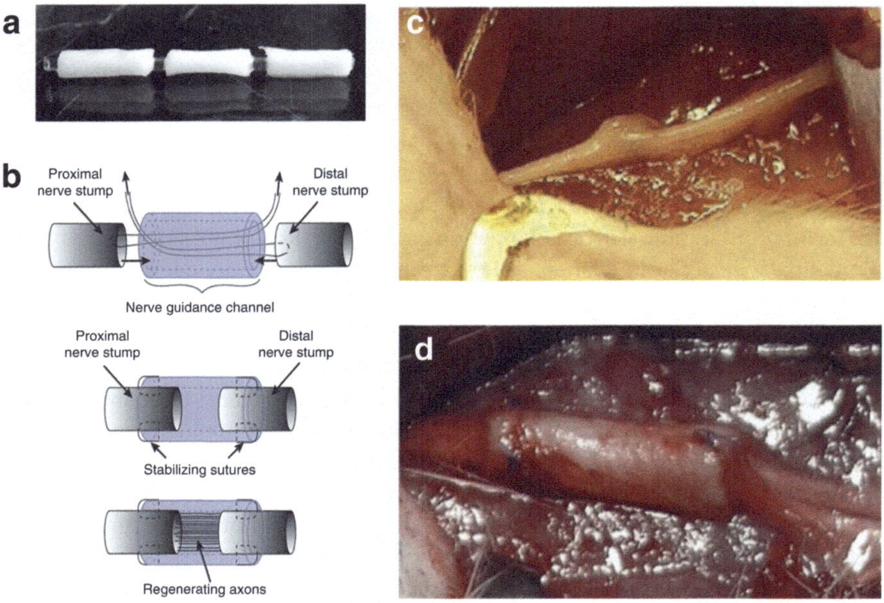

Fig. 25 Bacterial nanocellulose tubes used for nerve tissue engineering. (**a**) Macroscopically appearance of the nanocellulose tubes used as guidance channels for nerve regeneration. (**b**) Schematic diagram of the surgical procedure for tube implantation in rats. (**c**) Clinical aspect of the regenerated nerves after 30 days in the control group (regenerated without nanocellulose tube implantation) showing the formation of a neuroma (white arrow); and (**d**) Clinical aspect of a regenerated nerve grafted with a nanocellulose tube after 90 days without the formation of neuromas. https://www.termedia.pl/Journal/Archives_of_Medical_Science-19/Info (reprinted with permission from [191])

observed at the implantation site 90 days post-grafting (Fig. 25d) and complete recovery of motor activity was also observed. They suggested that the nanocellulose tubes prevented the growth of connective tissue around the regenerating nerves thus accelerating peripheral nerve regeneration. The biocompatibility of the nanocellulose tubes was demonstrated by the slight inflammatory response found around the implants tubes after 90 days of grafting.

9 Conclusions

The body of work reviewed in this chapter certainly provides extensive evidence of the advantages of using cellulose-based materials for tissue regeneration. Specially, the advantages that offer the use of cellulose nanofibers for tissue regeneration is remarkable and more research in this area are foreseen. Particularly the use of electrospinning to generate such fibers opens many possibilities for the modification and fabrication of the fibers using this technique. Electrospinning is a simple and cost effective technique that yields large amounts of nanofibers in short periods of time and with minimum effort. It is anticipated that more studies with these materials would occur particularly in the area of composites and colloids with other polymers and block copolymers.

References

1. Sternberg K (2009) Current requirements for polymeric biomaterials in otolaryngology. GMS Curr Top Otorhinolaryngol Head Neck Surg 8:11. https://doi.org/10.3205/cto000063
2. Doremus RH (1992) Bioceramics. J Mater Sci 27(2):285–297. https://doi.org/10.1007/bf00543915s
3. Vert M, Doi Y, Hellwich K-H, Hess M, Hodge P, Kubisa P, Rinaudo M, Schué F (2012) Terminology for biorelated polymers and applications (IUPAC recommendations 2012). Pure Appl Chem 84(2). https://doi.org/10.1351/pac-rec-10-12-04
4. Ramakrishna S, Fujihara K, Teo W-E, Yong T, Ma Z, Ramaseshan R (2006) Electrospun nanofibers: solving global issues. Mater Today 9(3):40–50. https://doi.org/10.1016/s1369-7021(06)71389-x
5. Loh QL, Choong C (2013) Three-dimensional scaffolds for tissue engineering applications: role of porosity and pore size. Tissue Eng Part B Rev 19(6):485–502. https://doi.org/10.1089/ten.TEB.2012.0437
6. Zhou XH, Wei DX, Ye HM, Zhang X, Meng X, Zhou Q (2016) Development of poly(vinyl alcohol) porous scaffold with high strength and well ciprofloxacin release efficiency. Mater Sci Eng C 67:326–335. https://doi.org/10.1016/j.msec.2016.05.030
7. Bhardwaj N, Kundu SC (2010) Electrospinning: a fascinating fiber fabrication technique. Biotechnol Adv 28(3):325–347. https://doi.org/10.1016/j.biotechadv.2010.01.004
8. Feng J (2017) Preparation and properties of poly(lactic acid) fiber melt blown non-woven disordered mats. Mater Lett 189:180–183. https://doi.org/10.1016/j.matlet.2016.12.013
9. Yuan R, Fu X, Wang X, Liu P, Wu L, Xu Y, Wang X, Wang Z (2006) Template synthesis of hollow metal oxide fibers with hierarchical architecture. Chem Mater 18(19):4700–4705. https://doi.org/10.1021/cm060991l
10. Eimura H, Umeta Y, Tokoro H, Yoshio M, Ohkoshi S, Kato T (2016) Self-assembled fibers containing stable organic radical moieties: alignment and magnetic properties in liquid crystals. Chemistry 22(26):8872–8878. https://doi.org/10.1002/chem.201505213
11. Rezabeigi E, Sta M, Swain M, McDonald J, Demarquette NR, Drew RAL, Wood-Adams PM (2017) Electrospinning of porous polylactic acid fibers during nonsolvent induced phase separation. J Appl Polym Sci 134(20). https://doi.org/10.1002/app.44862
12. Liao IC, Chew SY, Leong KW (2006) Aligned core-shell nanofibers delivering bioactive proteins. Nanomedicine 1(4):465–471. https://doi.org/10.2217/17435889.1.4.465
13. Zafar M, Najeeb S, Khurshid Z, Vazirzadeh M, Zohaib S, Najeeb B, Sefat F (2016) Potential of electrospun nanofibers for biomedical and dental applications. Materials 9(2):73. https://doi.org/10.3390/ma9020073
14. Liu H, Ding X, Zhou G, Li P, Wei X, Fan Y (2013) Electrospinning of nanofibers for tissue engineering applications. J Nanomater 2013:1–11. https://doi.org/10.1155/2013/495708
15. Kim PH, Cho JY (2016) Myocardial tissue engineering using electrospun nanofiber composites. BMB Rep 49(1):26–36. https://doi.org/10.5483/BMBRep.2016.49.1.165
16. Vashisth P, Pruthi V (2016) Synthesis and characterization of crosslinked gellan/PVA nanofibers for tissue engineering application. Mater Sci Eng C 67:304–312. https://doi.org/10.1016/j.msec.2016.05.049

17. Kaur S, Sundarrajan S, Rana D, Sridhar R, Gopal R, Matsuura T, Ramakrishna S (2014) Review: the characterization of electrospun nanofibrous liquid filtration membranes. J Mater Sci 49(18):6143–6159. https://doi.org/10.1007/s10853-014-8308-y
18. Frey MW (2008) Electrospinning cellulose and cellulose derivatives. Polym Rev 48(2):378–391. https://doi.org/10.1080/15583720802022281
19. Merkle VM, Zeng L, Slepian MJ, Wu X (2014) Core-shell nanofibers: integrating the bioactivity of gelatin and the mechanical property of polyvinyl alcohol. Biopolymers 101(4):336–346. https://doi.org/10.1002/bip.22367
20. Williams GR, Chatterton NP, Nazir T, D-G Y, Zhu L-M, Branford-White CJ (2012) Electrospun nanofibers in drug delivery: recent developments and perspectives. Ther Deliv 3(4):515–533. https://doi.org/10.4155/tde.12.17
21. Fang J, Wang X, Li T (2011) Functional applications of electrospun nanofibers. doi:https://doi.org/10.5772/24998
22. Biazar E (2016) Application of polymeric nanofibers in medical designs, part IV: drug and biological materials delivery. Int J Polym Mater Polym Biomater 66(2):53–60. https://doi.org/10.1080/00914037.2016.1180621
23. Shahabadi SMS, Kheradmand A, Montazeri V, Ziaee H (2015) Effects of process and ambient parameters on diameter and morphology of electrospun polyacrylonitrile nanofibers. Polym Sci Ser A 57(2):155–167. https://doi.org/10.1134/s0965545x15020157
24. Beachley V, Wen X (2009) Effect of electrospinning parameters on the nanofiber diameter and length. Mater Sci Eng C 29(3):663–668. https://doi.org/10.1016/j.msec.2008.10.037
25. Sill TJ, von Recum HA (2008) Electrospinning: applications in drug delivery and tissue engineering. Biomaterials 29(13):1989–2006. https://doi.org/10.1016/j.biomaterials.2008.01.011
26. Fong H, Chun I, Reneker DH (1999) Beaded nanofibers formed during electrospinning. Polymer 40(16):4585–4592. https://doi.org/10.1016/s0032-3861(99)00068-3
27. Tang X-P, Na S, Lan X, Liu H-Y (2014) Effect of flow rate on diameter of electrospun nanoporous fibers. Therm Sci 18:1447–1449. https://doi.org/10.2298/TSCI1405447T
28. AK S, Sanpui P, Chatterjee K (2014) Fabrication of poly(Caprolactone) nanofibers by electrospinning. J Polym Biopolym Phys Chem 2(4):62–66
29. Abdul Khalil HPS, Davoudpour Y, Bhat AH, Rosamah E, Tahir PM (2015) Electrospun cellulose composite nanofibers. In: Handbook of polymer nanocomposites. processing, performance and application. Springer, Berlin, pp 191–227. https://doi.org/10.1007/978-3-642-45232-1_61
30. Jacobs V, Anandjiwala RD, Maaza M (2010) The influence of electrospinning parameters on the structural morphology and diameter of electrospun nanofibers. J Appl Polym Sci 115(5):3130–3136. https://doi.org/10.1002/app.31396
31. Shenoy SL, Bates WD, Frisch HL, Wnek GE (2005) Role of chain entanglements on fiber formation during electrospinning of polymer solutions: good solvent, non-specific polymer–polymer interaction limit. Polymer 46(10):3372–3384. https://doi.org/10.1016/j.polymer.2005.03.011
32. Tan SH, Inai R, Kotaki M, Ramakrishna S (2005) Systematic parameter study for ultrafine fiber fabrication via electrospinning process. Polymer 46(16):6128–6134. https://doi.org/10.1016/j.polymer.2005.05.068
33. Florea M, Hagemann H, Santosa G, Abbott J, Micklem CN, Spencer-Milnes X, de Arroyo Garcia L, Paschou D, Lazenbatt C, Kong D, Chughtai H, Jensen K, Freemont PS, Kitney R, Reeve B, Ellis T (2016) Engineering control of bacterial cellulose production using a genetic toolkit and a new cellulose-producing strain. Proc Natl Acad Sci U S A 113(24):E3431–E3440. https://doi.org/10.1073/pnas.1522985113
34. Romling U, Galperin MY (2015) Bacterial cellulose biosynthesis: diversity of operons, subunits, products, and functions. Trends Microbiol 23(9):545–557. https://doi.org/10.1016/j.tim.2015.05.005
35. Ghormade V, Pathan EK, Deshpande MV (2017) Can fungi compete with marine sources for chitosan production? Int J Biol Macromol. https://doi.org/10.1016/j.ijbiomac.2017.01.112

36. Lenardon MD, Munro CA, Gow NA (2010) Chitin synthesis and fungal pathogenesis. Curr Opin Microbiol 13(4):416–423. https://doi.org/10.1016/j.mib.2010.05.002
37. Klemm D, Philipp B, Heinze T, Heinze U, Wagenknecht W (1998) Comprehensive cellulose chemistry, vol 1. Wiley, Weinheim. https://doi.org/10.1002/3527601929
38. Brown RM, Saxena IM (2007) Cellulose: molecular and structural biology. Springer, Dordrecht. https://doi.org/10.1007/978-1-4020-5380-1
39. Cocinero EJ, Gamblin DP, Davis BG, Simons JP (2009) The building blocks of cellulose: the intrinsic conformational structures of cellobiose, its epimer, lactose, and their singly hydrated complexes. J Am Chem Soc 131(31):11117–11123. https://doi.org/10.1021/ja903322w
40. Credou J, Berthelot T (2014) Cellulose: from biocompatible to bioactive material. J Mater Chem B 2(30):4767. https://doi.org/10.1039/c4tb00431k
41. French AD (2012) Combining computational chemistry and crystallography for a better understanding of the structure of cellulose. Adv Carbohydr Chem Biochem 67:19–93. https://doi.org/10.1016/B978-0-12-396527-1.00002-4
42. Imberty A, Perez S (1988) A revisit to the three-dimensional structure of B-type starch. Biopolymers 27(8):1205–1221. https://doi.org/10.1002/bip.360270803
43. Habibi Y, Lucia LA, Rojas OJ (2010) Cellulose nanocrystals: chemistry, self-assembly, and applications. Chem Rev 110(6):3479–3500. https://doi.org/10.1021/cr900339w
44. Desvaux M (2005) Clostridium cellulolyticum: model organism of mesophilic cellulolytic clostridia. FEMS Microbiol Rev 29(4):741–764. https://doi.org/10.1016/j.femsre.2004.11.003
45. Ponni R, Kontturi E, Vuorinen T (2013) Accessibility of cellulose: structural changes and their reversibility in aqueous media. Carbohydr Polym 93(2):424–429. https://doi.org/10.1016/j.carbpol.2012.12.025
46. Müller M, Czihak C, Schober H, Nishiyama Y, Vogl G (2000) All disordered regions of native cellulose show common low-frequency dynamics. Macromolecules 33(5):1834–1840. https://doi.org/10.1021/ma991227l
47. Gümüskaya E, Usta M, Kirci H (2003) The effects of various pulping conditions on crystalline structure of cellulose in cotton linters. Polym Degrad Stab 81(3):559–564. https://doi.org/10.1016/s0141-3910(03)00157-5
48. Gross AS, Bell AT, Chu JW (2011) Thermodynamics of cellulose solvation in water and the ionic liquid 1-butyl-3-methylimidazolim chloride. J Phys Chem B 115(46):13433–13440. https://doi.org/10.1021/jp202415v
49. Kovalenko VI (2010) Crystalline cellulose: structure and hydrogen bonds. Russ Chem Rev 79(3):231–241. https://doi.org/10.1070/RC2010v079n03ABEH004065
50. Nishiyama Y, Langan P, Chanzy H (2002) Crystal structure and hydrogen-bonding system in cellulose iβ from synchrotron x-ray and neutron fiber diffraction. J Am Chem Soc 124(31):9074–9082. https://doi.org/10.1021/ja025731 9
51. Wang T, Yang H, Kubicki JD, Hong M (2016) Cellulose structural polymorphism in plant primary cell walls investigated by high-field 2D solid-state NMR spectroscopy and density functional theory calculations. Biomacromolecules 17(6):2210–2222. https://doi.org/10.1021/acs.biomac.6b00441
52. Klemm D, Heublein B, Fink HP, Bohn A (2005) Cellulose: fascinating biopolymer and sustainable raw material. Angew Chem Int Ed Engl 44(22):3358–3393. https://doi.org/10.1002/anie.200460587
53. Thomas LH, Forsyth VT, Sturcova A, Kennedy CJ, May RP, Altaner CM, Apperley DC, Wess TJ, Jarvis MC (2013) Structure of cellulose microfibrils in primary cell walls from collenchyma. Plant Physiol 161(1):465–476. https://doi.org/10.1104/pp.112.206359
54. Dufresne A (2013) Nanocellulose: a new ageless bionanomaterial. Mater Today 16(6):220–227. https://doi.org/10.1016/j.mattod.2013.06.004
55. Tanpichai S, Quero F, Nogi M, Yano H, Young RJ, Lindstrom T, Sampson WW, Eichhorn SJ (2012) Effective Young's modulus of bacterial and microfibrillated cellulose fibrils in fibrous networks. Biomacromolecules 13(5):1340–1349. https://doi.org/10.1021/bm300042t

56. Herdocia-Lluberes CS, Laboy-López S, Morales S, Gonzalez-Robles TJ, González-Feliciano JA, Nicolau E (2015) Evaluation of synthesized nanohydroxyapatite-nanocellulose composites as biocompatible scaffolds for applications in bone tissue engineering. J Nanomater 2015:9. https://doi.org/10.1155/2015/310935
57. Hanif Z, Ahmed FR, Shin SW, Kim YK, Um SH (2014) Size- and dose-dependent toxicity of cellulose nanocrystals (CNC) on human fibroblasts and colon adenocarcinoma. Colloids Surf B Biointerfaces 119:162–165. https://doi.org/10.1016/j.colsurfb.2014.04.018
58. Roman M (2015) Toxicity of cellulose nanocrystals: a review. Ind Biotechnol 11(1):25–33. https://doi.org/10.1089/ind.2014.0024
59. Alexandrescu L, Syverud K, Gatti A, Chinga-Carrasco G (2013) Cytotoxicity tests of cellulose nanofibril-based structures. Cellulose 20(4):1765–1775. https://doi.org/10.1007/s10570-013-9948-9
60. Vartiainen J, Pöhler T, Sirola K, Pylkkänen L, Alenius H, Hokkinen J, Tapper U, Lahtinen P, Kapanen A, Putkisto K, Hiekkataipale P, Eronen P, Ruokolainen J, Laukkanen A (2011) Health and environmental safety aspects of friction grinding and spray drying of microfibrillated cellulose. Cellulose 18(3):775–786. https://doi.org/10.1007/s10570-011-9501-7
61. Yanamala N, Farcas MT, Hatfield MK, Kisin ER, Kagan VE, Geraci CL, Shvedova AA (2014) In vivo evaluation of the pulmonary toxicity of cellulose nanocrystals: a renewable and sustainable nanomaterial of the future. ACS Sustain Chem Eng 2(7):1691–1698. https://doi.org/10.1021/sc500153k
62. Czaja WK, Young DJ, Kawecki M, Brown RM Jr (2007) The future prospects of microbial cellulose in biomedical applications. Biomacromolecules 8(1):1–12. https://doi.org/10.1021/bm060620d
63. Bohner M (2010) Resorbable biomaterials as bone graft substitutes. Mater Today 13(1–2):24–30. https://doi.org/10.1016/s1369-7021(10)70014-6
64. Degree of Substitution (2011). doi:https://doi.org/10.1002/0471440264.pst445
65. Eyley S, Thielemans W (2014) Surface modification of cellulose nanocrystals. Nanoscale 6(14):7764–7779. https://doi.org/10.1039/c4nr01756k
66. Saito T, Kimura S, Nishiyama Y, Isogai A (2007) Cellulose nanofibers prepared by TEMPO-mediated oxidation of native cellulose. Biomacromolecules 8(8):2485–2491. https://doi.org/10.1021/bm0703970
67. Li J, Wan Y, Li L, Liang H, Wang J (2009) Preparation and characterization of 2,3-dialdehyde bacterial cellulose for potential biodegradable tissue engineering scaffolds. Mater Sci Eng C 29(5):1635–1642. https://doi.org/10.1016/j.msec.2009.01.006
68. Cheng Y, Lu J, Liu S, Zhao P, Lu G, Chen J (2014) The preparation, characterization and evaluation of regenerated cellulose/collagen composite hydrogel films. Carbohydr Polym 107:57–64. https://doi.org/10.1016/j.carbpol.2014.02.034
69. Habibi Y (2014) Key advances in the chemical modification of nanocelluloses. Chem Soc Rev 43(5):1519–1542. https://doi.org/10.1039/c3cs60204d
70. Svachova V, Vojtova L, Pavlinak D, Vojtek L, Sedlakova V, Hyrsl P, Alberti M, Jaros J, Hampl A, Jancar J (2016) Novel electrospun gelatin/oxycellulose nanofibers as a suitable platform for lung disease modeling. Mater Sci Eng C 67:493–501. https://doi.org/10.1016/j.msec.2016.05.059
71. Weishaupt R, Siqueira G, Schubert M, Tingaut P, Maniura-Weber K, Zimmermann T, Thony-Meyer L, Faccio G, Ihssen J (2015) TEMPO-oxidized nanofibrillated cellulose as a high density carrier for bioactive molecules. Biomacromolecules 16(11):3640–3650. https://doi.org/10.1021/acs.biomac.5b01100
72. Moon RJ, Martini A, Nairn J, Simonsen J, Youngblood J (2011) Cellulose nanomaterials review: structure, properties and nanocomposites. Chem Soc Rev 40(7):3941–3994. https://doi.org/10.1039/c0cs00108b
73. Shokri J, Adibki K (2013) Application of cellulose and cellulose derivatives in pharmaceutical industries. In: Cellulose - medical, pharmaceutical and electronic applications. InTech. doi:https://doi.org/10.5772/55178

74. Liimatainen H, Visanko M, Sirviö J, Hormi O, Niinimäki J (2013) Sulfonated cellulose nanofibrils obtained from wood pulp through regioselective oxidative bisulfite pre-treatment. Cellulose 20(2):741–749. https://doi.org/10.1007/s10570-013-9865-y
75. Shaabani A, Rahmati A, Badri Z (2008) Sulfonated cellulose and starch: new biodegradable and renewable solid acid catalysts for efficient synthesis of quinolines. Catal Commun 9(1):13–16. https://doi.org/10.1016/j.catcom.2007.05.021
76. Bhatt N, Gupta PK, Naithani S (2008) Preparation of cellulose sulfate from α-cellulose isolated from Lantana camara by the direct esterification method. J Appl Polym Sci 108(5):2895–2901. https://doi.org/10.1002/app.27773
77. Vatankhah E, Prabhakaran MP, Jin G, Mobarakeh LG, Ramakrishna S (2014) Development of nanofibrous cellulose acetate/gelatin skin substitutes for variety wound treatment applications. J Biomater Appl 28(6):909–921. https://doi.org/10.1177/0885328213486527
78. Rodríguez K, Gatenholm P, Renneckar S (2012) Electrospinning cellulosic nanofibers for biomedical applications: structure and in vitro biocompatibility. Cellulose 19(5):1583–1598. https://doi.org/10.1007/s10570-012-9734-0
79. Huang R, Li W, Lv X, Lei Z, Bian Y, Deng H, Wang H, Li J, Li X (2015) Biomimetic LBL structured nanofibrous matrices assembled by chitosan/collagen for promoting wound healing. Biomaterials 53:58–75. https://doi.org/10.1016/j.biomaterials.2015.02.076
80. Rodriguez K, Renneckar S, Gatenholm P (2011) Biomimetic calcium phosphate crystal mineralization on electrospun cellulose-based scaffolds. ACS Appl Mater Interfaces 3(3):681–689. https://doi.org/10.1021/am100972r
81. Huang Y, Miao Y-E, Liu T (2014) Electrospun fibrous membranes for efficient heavy metal removal. J Appl Polym Sci 131(19):n/a–n/a. https://doi.org/10.1002/app.40864
82. Wang S-D, Ma Q, Liu H, Wang K, Ling L-Z, Zhang K-Q (2015) Robust electrospinning cellulose acetate@TiO2ultrafine fibers for dyeing water treatment by photocatalytic reactions. RSC Adv 5(51):40521–40530. https://doi.org/10.1039/c5ra03797b
83. Arslan O, Aytac Z, Uyar T (2016) Superhydrophobic, hybrid, electrospun cellulose acetate nanofibrous mats for oil/water separation by tailored surface modification. ACS Appl Mater Interfaces 8(30):19747–19754. https://doi.org/10.1021/acsami.6b05429
84. Su J, Raghuwanshi VS, Raverty W, Garvey CJ, Holden PJ, Gillon M, Holt SA, Tabor R, Batchelor W, Garnier G (2016) Smooth deuterated cellulose films for the visualisation of adsorbed bio-macromolecules. Sci Rep 6:36119. https://doi.org/10.1038/srep36119
85. Volpi N, Maccari F (2006) Electrophoretic approaches to the analysis of complex polysaccharides. J Chromatogr B 834(1–2):1–13. https://doi.org/10.1016/j.jchromb.2006.02.049
86. Demirci S, Celebioglu A, Uyar T (2014) Surface modification of electrospun cellulose acetate nanofibers via RAFT polymerization for DNA adsorption. Carbohydr Polym 113:200–207. https://doi.org/10.1016/j.carbpol.2014.06.086
87. Schulze P, Gericke M, Scholz F, Wondraczek H, Miethe P, Heinze T (2016) Incorporation of hydrophobic dyes within cellulose acetate and acetate phthalate based nanoparticles. Macromol Chem Phys 217(16):1823–1833. https://doi.org/10.1002/macp.201600160
88. Alinat E, Delaunay N, Archer X, Mallet JM, Gareil P (2015) A new method for the determination of the nitrogen content of nitrocellulose based on the molar ratio of nitrite-to-nitrate ions released after alkaline hydrolysis. J Hazard Mater 286:92–99. https://doi.org/10.1016/j.jhazmat.2014.12.032
89. Fernandez de la Ossa MA, Ortega-Ojeda F, Garcia-Ruiz C (2013) Discrimination of non-explosive and explosive samples through nitrocellulose fingerprints obtained by capillary electrophoresis. J Chromatogr A 1302:197–204. https://doi.org/10.1016/j.chroma.2013.06.034
90. Mu X, Yu H, Zhang C, Chen X, Cheng Z, Bai R, Wu X, Yu Q, Wu C, Diao Y (2016) Nano-porous nitrocellulose liquid bandage modulates cell and cytokine response and accelerates cutaneous wound healing in a mouse model. Carbohydr Polym 136:618–629. https://doi.org/10.1016/j.carbpol.2015.08.070
91. Yan S, Jian G, Zachariah MR (2012) Electrospun nanofiber-based thermite textiles and their reactive properties. ACS Appl Mater Interfaces 4(12):6432–6435. https://doi.org/10.1021/am3021125

92. Nartker S, Drzal LT (2010) Electrospun cellulose nitrate nanofibers. J Nanosci Nanotechnol 10(9):5810–5813. https://doi.org/10.1166/jnn.2010.2447
93. Sovizi MR, Hajimirsadeghi SS, Naderizadeh B (2009) Effect of particle size on thermal decomposition of nitrocellulose. J Hazard Mater 168(2–3):1134–1139. https://doi.org/10.1016/j.jhazmat.2009.02.146
94. Budtova T, Navard P (2015) Cellulose in NaOH–water based solvents: a review. Cellulose 23(1):5–55. https://doi.org/10.1007/s10570-015-0779-8
95. Pakulska MM, Vulic K, Shoichet MS (2013) Affinity-based release of chondroitinase ABC from a modified methylcellulose hydrogel. J Control Release 171(1):11–16. https://doi.org/10.1016/j.jconrel.2013.06.029
96. Pakulska MM, Vulic K, Tam RY, Shoichet MS (2015) Hybrid crosslinked methylcellulose hydrogel: a predictable and tunable platform for local drug delivery. Adv Mater 27(34):5002–5008. https://doi.org/10.1002/adma.201502767
97. Park CH, Jeong L, Cho D, Kwon OH, Park WH (2013) Effect of methylcellulose on the formation and drug release behavior of silk fibroin hydrogel. Carbohydr Polym 98(1):1179–1185. https://doi.org/10.1016/j.carbpol.2013.07.028
98. Huang CC, Liao ZX, Chen DY, Hsiao CW, Chang Y, Sung HW (2014) Injectable cell constructs fabricated via culture on a thermoresponsive methylcellulose hydrogel system for the treatment of ischemic diseases. Adv Healthc Mater 3(8):1133–1148. https://doi.org/10.1002/adhm.201300605
99. Chahal S, Jahir Hussain FS, Kumar A, Yusoff MM, Bahari Abdull Rasad MS (2015) Electrospun hydroxyethyl cellulose nanofibers functionalized with calcium phosphate coating for bone tissue engineering. RSC Adv 5(37):29497–29504. https://doi.org/10.1039/c4ra17087c
100. Singh BN, Panda NN, Mund R, Pramanik K (2016) Carboxymethyl cellulose enables silk fibroin nanofibrous scaffold with enhanced biomimetic potential for bone tissue engineering application. Carbohydr Polym 151:335–347. https://doi.org/10.1016/j.carbpol.2016.05.088
101. Prasanth R, Nageswaran S, Thakur VK, Ahn J-H (2014) Electrospinning of cellulose: process and applications. In: Nanocellulose polymer nanocomposites. Wiley, Hoboken, pp 311–340. https://doi.org/10.1002/9781118872246.ch12
102. Xu S, Zhang J, He A, Li J, Zhang H, Han CC (2008) Electrospinning of native cellulose from nonvolatile solvent system. Polymer 49(12):2911–2917. https://doi.org/10.1016/j.polymer.2008.04.046
103. Lee KY, Jeong L, Kang YO, Lee SJ, Park WH (2009) Electrospinning of polysaccharides for regenerative medicine. Adv Drug Deliv Rev 61(12):1020–1032. https://doi.org/10.1016/j.addr.2009.07.006
104. Kim C-W, Kim D-S, Kang S-Y, Marquez M, Joo YL (2006) Structural studies of electrospun cellulose nanofibers. Polymer 47(14):5097–5107. https://doi.org/10.1016/j.polymer.2006.05.033
105. Khil MS, Kim HY, Kang YS, Bang HJ, Lee DR, Doo JK (2013) Preparation of electrospun oxidized cellulose mats and theirin vitro degradation behavior. Macromol Res 13(1):62–67. https://doi.org/10.1007/bf03219016
106. Kulpinski P (2005) Cellulose nanofibers prepared by the N-methylmorpholine-N-oxide method. J Appl Polym Sci 98(4):1855–1859. https://doi.org/10.1002/app.22123
107. Frey M, Joo Y, Kim C-W (2003) New solvents for cellulose electrospinning and preliminary nanofiber spinning results. Polym Prepr-Am 44(2):168–169
108. Frey M, Song H (2003) Cellulose fibers formed by electrospinning from solution. In: Abstracts of papers of the American Chemical Society. ACS, Washington, DC, pp U288–U288
109. Kim C-W, Frey MW, Marquez M, Joo YL (2005) Preparation of submicron-scale, electrospun cellulose fibers via direct dissolution. J Polym Sci B Polym Phys 43(13):1673–1683. https://doi.org/10.1002/polb.20475
110. Frenot A, Henriksson MW, Walkenström P (2007) Electrospinning of cellulose-based nanofibers. J Appl Polym Sci 103(3):1473–1482. https://doi.org/10.1002/app.24912

111. Viswanathan G, Murugesan S, Pushparaj V, Nalamasu O, Ajayan PM, Linhardt RJ (2006) Preparation of biopolymer fibers by electrospinning from room temperature ionic liquids. Biomacromolecules 7(2):415–418. https://doi.org/10.1021/bm050837s
112. Chanzy H, Nawrot S, Peguy A, Smith P, Chevalier J (1982) Phase behavior of the quasiternary system N-methylmorpholine-N-oxide, water, and cellulose. J Polym Sci Polym Phys Ed 20(10):1909–1924. https://doi.org/10.1002/pol.1982.180201014
113. Han SO, Youk JH, Min KD, Kang YO, Park WH (2008) Electrospinning of cellulose acetate nanofibers using a mixed solvent of acetic acid/water: Effects of solvent composition on the fiber diameter. Mater Lett 62(4-5):759–762. https://doi.org/10.1016/j.matlet.2007.06.059
114. Sun B, Long YZ, Zhang HD, Li MM, Duvail JL, Jiang XY, Yin HL (2014) Advances in three-dimensional nanofibrous macrostructures via electrospinning. Prog Polym Sci 39(5):862–890. https://doi.org/10.1016/j.progpolymsci.2013.06.002
115. Zong X, Kim K, Fang D, Ran S, Hsiao BS, Chu B (2002) Structure and process relationship of electrospun bioabsorbable nanofiber membranes. Polymer 43(16):4403–4412. https://doi.org/10.1016/s0032-3861(02)00275-6
116. Haider A, Haider S, Kang I-K (2015) A comprehensive review summarizing the effect of electrospinning parameters and potential applications of nanofibers in biomedical and biotechnology. Arabian J Chem. doi:https://doi.org/10.1016/j.arabjc.2015.11.015
117. Zhang C, Liu R, Xiang J, Kang H, Liu Z, Huang Y (2014) Dissolution mechanism of cellulose in N,N-dimethylacetamide/lithium chloride: revisiting through molecular interactions. J Phys Chem B 118(31):9507–9514. https://doi.org/10.1021/jp506013c
118. El Seoud OA, Heinze T (2005) Organic esters of cellulose: new perspectives for old polymers. In: Advances in polymer science, vol 186. Springer, Berlin, pp 103–149. https://doi.org/10.1007/b136818
119. Hayes R, Warr GG, Atkin R (2015) Structure and nanostructure in ionic liquids. Chem Rev 115(13):6357–6426. https://doi.org/10.1021/cr500411q
120. Egorova KS, Gordeev EG, Ananikov VP (2017) Biological activity of ionic liquids and their application in pharmaceutics and medicine. Chem Rev 117(10):7132–7189. https://doi.org/10.1021/acs.chemrev.6b00562
121. Pinkert A, Marsh KN, Pang S, Staiger MP (2009) Ionic liquids and their interaction with cellulose. Chem Rev 109(12):6712–6728. https://doi.org/10.1021/cr9001947
122. Freire MG, Teles ARR, Ferreira RAS, Carlos LD, Lopes-da-Silva JA, Coutinho JAP (2011) Electrospun nanosized cellulose fibers using ionic liquids at room temperature. Green Chem 13(11):3173. https://doi.org/10.1039/c1gc15930e
123. Meli L, Miao J, Dordick JS, Linhardt RJ (2010) Electrospinning from room temperature ionic liquids for biopolymer fiber formation. Green Chem 12(11):1883. https://doi.org/10.1039/c0gc00283f
124. Quan S-L, Kang S-G, Chin I-J (2009) Characterization of cellulose fibers electrospun using ionic liquid. Cellulose 17(2):223–230. https://doi.org/10.1007/s10570-009-9386-x
125. Fischer S, Thümmler K, Volkert B, Hettrich K, Schmidt I, Fischer K (2008) Properties and applications of cellulose acetate. Macromol Symp 262(1):89–96. https://doi.org/10.1002/masy.200850210
126. Konwarh R, Karak N, Misra M (2013) Electrospun cellulose acetate nanofibers: the present status and gamut of biotechnological applications. Biotechnol Adv 31(4):421–437. https://doi.org/10.1016/j.biotechadv.2013.01.002
127. Liu H, Hsieh Y-L (2002) Ultrafine fibrous cellulose membranes from electrospinning of cellulose acetate. J Polym Sci B Polym Phys 40(18):2119–2129. https://doi.org/10.1002/polb.10261
128. Tungprapa S, Puangparn T, Weerasombut M, Jangchud I, Fakum P, Semongkhol S, Meechaisue C, Supaphol P (2007) Electrospun cellulose acetate fibers: effect of solvent system on morphology and fiber diameter. Cellulose 14(6):563–575. https://doi.org/10.1007/s10570-007-9113-4

129. Son WK, Youk JH, Lee TS, Park WH (2004) Electrospinning of ultrafine cellulose acetate fibers: studies of a new solvent system and deacetylation of ultrafine cellulose acetate fibers. J Polym Sci B Polym Phys 42(1):5–11. https://doi.org/10.1002/polb.10668
130. Han D, Gouma PI (2006) Electrospun bioscaffolds that mimic the topology of extracellular matrix. Nanomedicine 2(1):37–41. https://doi.org/10.1016/j.nano.2006.01.002
131. Song J, Birbach NL, Hinestroza JP (2012) Deposition of silver nanoparticles on cellulosic fibers via stabilization of carboxymethyl groups. Cellulose 19(2):411–424. https://doi.org/10.1007/s10570-011-9647-3
132. Liu H, Tang C (2006) Electrospinning of cellulose acetate in solvent mixture N,N-dimethylacetamide (DMAc)/acetone. Polym J 39(1):65–72. https://doi.org/10.1295/polymj.PJ2006117
133. Ma Z, Kotaki M, Ramakrishna S (2005) Electrospun cellulose nanofiber as affinity membrane. J Membr Sci 265(1-2):115–123. https://doi.org/10.1016/j.memsci.2005.04.044
134. Ma Z, Ramakrishna S (2008) Electrospun regenerated cellulose nanofiber affinity membrane functionalized with protein A/G for IgG purification. J Membr Sci 319(1-2):23–28. https://doi.org/10.1016/j.memsci.2008.03.045
135. Han SO, Son WK, Youk JH, Lee TS, Park WH (2005) Ultrafine porous fibers electrospun from cellulose triacetate. Mater Lett 59(24-25):2998–3001. https://doi.org/10.1016/j.matlet.2005.05.003
136. Cao X, Sun S, Peng X, Zhong L, Sun R, Jiang D (2013) Rapid synthesis of cellulose esters by transesterification of cellulose with vinyl esters under the catalysis of NaOH or KOH in DMSO. J Agric Food Chem 61(10):2489–2495. https://doi.org/10.1021/jf3055104
137. Brydson J (1999) Plastic materials. Butterworth Heinemann, Oxford
138. Atila D, Keskin D, Tezcaner A (2015) Cellulose acetate based 3-dimensional electrospun scaffolds for skin tissue engineering applications. Carbohydr Polym 133:251–261. https://doi.org/10.1016/j.carbpol.2015.06.109
139. Atila D, Keskin D, Tezcaner A (2016) Crosslinked pullulan/cellulose acetate fibrous scaffolds for bone tissue engineering. Mater Sci Eng C 69:1103–1115. https://doi.org/10.1016/j.msec.2016.08.015
140. Entcheva E, Bien H, Yin L, Chung CY, Farrell M, Kostov Y (2004) Functional cardiac cell constructs on cellulose-based scaffolding. Biomaterials 25(26):5753–5762. https://doi.org/10.1016/j.biomaterials.2004.01.024
141. Unnithan AR, Gnanasekaran G, Sathishkumar Y, Lee YS, Kim CS (2014) Electrospun antibacterial polyurethane-cellulose acetate-zein composite mats for wound dressing. Carbohydr Polym 102:884–892. https://doi.org/10.1016/j.carbpol.2013.10.070
142. Lin N, Dufresne A (2014) Nanocellulose in biomedicine: current status and future prospect. Eur Polym J 59:302–325. https://doi.org/10.1016/j.eurpolymj.2014.07.025
143. Czaja W, Kyryliouk D, DePaula CA, Buechter DD (2014) Oxidation of γ-irradiated microbial cellulose results in bioresorbable, highly conformable biomaterial. J Appl Polym Sci 131(6):n/a–n/a. https://doi.org/10.1002/app.39995
144. Lopes VR, Sanchez-Martinez C, Stromme M, Ferraz N (2017) In vitro biological responses to nanofibrillated cellulose by human dermal, lung and immune cells: surface chemistry aspect. Part Fibre Toxicol 14(1):1. https://doi.org/10.1186/s12989-016-0182-0
145. Roman M, Dong S, Hirani A, Lee YW (2009) Cellulose nanocrystals for drug delivery. In: Polysaccharide materials: performance by design, ACS symposium series, vol 1017. American Chemical Society, Washington DC, pp 81–91. https://doi.org/10.1021/bk-2009-1017.ch004
146. Dong S, Hirani AA, Colacino KR, Lee YW, Roman M (2012) Cytotoxicity and cellular uptake of cellulose nanocrystals. Nano Life 2(3):1241006. https://doi.org/10.1142/s1793984412410061
147. Herdocia-Lluberes CS, Laboy-López S, Morales S, Gonzalez-Robles TJ, González-Feliciano JA, Nicolau E (2015) Evaluation of synthesized nanohydroxyapatite-nanocellulose com-

posites as biocompatible scaffolds for applications in bone tissue engineering. J Nanomater 2015:9. https://doi.org/10.1155/2015/310935
148. Rojas J, Bedoya M, Ciro Y (2015) Current trends in the production of cellulose nanoparticles and nanocomposites for biomedical applications. In: Cellulose - fundamental aspects and current trends. InTech, Rijeka. https://doi.org/10.5772/61334
149. Jackson JK, Letchford K, Wasserman BZ, Ye L, Hamad WY, Burt HM (2011) The use of nanocrystalline cellulose for the binding and controlled release of drugs. Int J Nanomedicine 1(6):321–330. https://doi.org/10.2147/IJN.S16749
150. Mahmoud KA, Mena JA, Male KB, Hrapovic S, Kamen A, Luong JH (2010) Effect of surface charge on the cellular uptake and cytotoxicity of fluorescent labeled cellulose nanocrystals. ACS Appl Mater Interfaces 2(10):2924–2932. https://doi.org/10.1021/am1006222
151. Dong S, Roman M (2007) Fluorescently labeled cellulose nanocrystals for bioimaging applications. J Am Chem Soc 129(45):13810–13811. https://doi.org/10.1021/ja076196l
152. Liebert T, Kostag M, Wotschadlo J, Heinze T (2011) Stable cellulose nanospheres for cellular uptake. Macromol Biosci 11:1387–1392. https://doi.org/10.1002/mabi.201100113
153. Chithrani BD, Ghazani AA, Chan WCW (2006) Determining the size and shape dependence of gold nanoparticle uptake into mammalian cells. Nano Lett 6(4):662–668. https://doi.org/10.1021/nl052396o
154. Zwicke GL, Mansoori GA, Jeffery CJ (2012) Utilizing the folate receptor for active targeting of cancer nanotherapeutics. Nano Rev 3:18496. https://doi.org/10.3402/nano.v3i0.18496
155. Ledermann JA, Canevari S, Thigpen T (2015) Targeting the folate receptor: diagnostic and therapeutic approaches to personalize cancer treatments. Ann Oncol 26(10):2034–2043. https://doi.org/10.1093/annonc/mdv250
156. Dong S, Cho HJ, Lee YW, Roman M (2014) Synthesis and cellular uptake of folic acid-conjugated cellulose nanocrystals for cancer targeting. Biomacromolecules 15(5):1560–1567. https://doi.org/10.1021/bm401593n
157. Colacino KR, Arena CB, Dong S, Roman M, Davalos RV, Lee YW (2015) Folate conjugated cellulose nanocrystals potentiate irreversible electroporation-induced cytotoxicity for the selective treatment of cancer cells. Technol Cancer Res Treat 14(6):757–766. https://doi.org/10.7785/tcrt.2012.500428
158. Jackson JK, Letchford K, Wasserman BZ, Ye L, Hamad WY, Burt HM (2011) The use of nanocrystalline cellulose for the binding and controlled release of drugs. Int J Nanomedicine 6:321–330. https://doi.org/10.2147/IJN.S16749
159. Lin N, Huang J, Chang PR, Feng L, Yu J (2011) Effect of polysaccharide nanocrystals on structure, properties, and drug release kinetics of alginate-based microspheres. Colloids Surf B Biointerfaces 85(2):270–279. https://doi.org/10.1016/j.colsurfb.2011.02.039
160. Akhlaghi SP, Berry RC, Tam KC (2013) Surface modification of cellulose nanocrystal with chitosan oligosaccharide for drug delivery applications. Cellulose 20(4):1747–1764. https://doi.org/10.1007/s10570-013-9954-y
161. Cacicedo ML, Leon IE, Gonzalez JS, Porto LM, Alvarez VA, Castro GR (2016) Modified bacterial cellulose scaffolds for localized doxorubicin release in human colorectal HT-29 cells. Colloids Surf B Biointerfaces 140:421–429. https://doi.org/10.1016/j.colsurfb.2016.01.007
162. Ndong Ntoutoume GM, Granet R, Mbakidi JP, Bregier F, Leger DY, Fidanzi-Dugas C, Lequart V, Joly N, Liagre B, Chaleix V, Sol V (2016) Development of curcumin-cyclodextrin/cellulose nanocrystals complexes: new anticancer drug delivery systems. Bioorg Med Chem Lett 26(3):941–945. https://doi.org/10.1016/j.bmcl.2015.12.060
163. Anirudhan TS, Rejeena SR (2014) Aminatedβ-cyclodextrin-modified-carboxylated magnetic cobalt/nanocellulose composite for tumor-targeted gene delivery. J Appl Chem 2014:10. https://doi.org/10.1155/2014/184153
164. Hu H, Yuan W, Liu FS, Cheng G, FJ X, Ma J (2015) Redox-responsive polycation-functionalized cotton cellulose nanocrystals for effective cancer treatment. ACS Appl Mater Interfaces 7(16):8942–8951. https://doi.org/10.1021/acsami.5b02432

165. Saikrishnan D, Goyal M, Rossiter S, Kukol A (2014) A cellulose-based bioassay for the colorimetric detection of pathogen DNA. Anal Bioanal Chem 406(30):7887–7898. https://doi.org/10.1007/s00216-014-8257-y
166. Fontenot KR, Edwards JV, Haldane D, Graves E, Citron MS, Prevost NT, French AD, Condon BD (2016) Human neutrophil elastase detection with fluorescent peptide sensors conjugated to cellulosic and nanocellulosic materials: part II, structure/function analysis. Cellulose 23(2):1297–1309. https://doi.org/10.1007/s10570-016-0873-6
167. Vincent Edwards J, Prevost N, French A, Concha M, DeLucca A, Wu Q (2013) Nanocellulose-based biosensors: design, preparation, and activity of peptide-linked cotton cellulose nanocrystals having fluorimetric and colorimetric elastase detection sensitivity. Sci Res 5:20–28. https://doi.org/10.4236/eng.2013.59A003
168. Vincent Edwards J, Fontenot KR, Prevost NT, Haldane D, Pircher N, Liebner F, French A, Condon BD (2016) Protease biosensors based on peptide-nanocellulose conjugates: from molecular design to dressing interface. Int J Med Nano Res 3(2). 10.23937/2378-3664/1410018
169. Weishaupt R, Siqueira G, Schubert M, Kämpf MM, Zimmermann T, Maniura-Weber K, Faccio G (2017) A protein-nanocellulose paper for sensing copper ions at the nano- to micromolar level. Adv Funct Mater 27(4):1604291. https://doi.org/10.1002/adfm.201604291
170. Li Y, Jiang H, Zheng W, Gong N, Chen L, Jiang X, Yang G (2015) Bacterial cellulose–hyaluronan nanocomposite biomaterials as wound dressings for severe skin injury repair. J Mater Chem B 3(17):3498–3507. https://doi.org/10.1039/c4tb01819b
171. Li Y, Wang S, Huang R, Huang Z, Hu B, Zheng W, Yang G, Jiang X (2015) Evaluation of the effect of the structure of bacterial cellulose on full thickness skin wound repair on a microfluidic chip. Biomacromolecules 16(3):780–789. https://doi.org/10.1021/bm501680s
172. Fu L, Zhou P, Zhang S, Yang G (2013) Evaluation of bacterial nanocellulose-based uniform wound dressing for large area skin transplantation. Mater Sci Eng 33(5):2995–3000. https://doi.org/10.1016/j.msec.2013.03.026
173. Czaja W, Krystynowicz A, Bielecki S, Brown RM Jr (2006) Microbial cellulose–the natural power to heal wounds. Biomaterials 27(2):145–151. https://doi.org/10.1016/j.biomaterials.2005.07.035
174. Hakkarainen T, Koivuniemi R, Kosonen M, Escobedo-Lucea C, Sanz-Garcia A, Vuola J, Valtonen J, Tammela P, Makitie A, Luukko K, Yliperttula M, Kavola H (2016) Nanofibrillar cellulose wound dressing in skin graft donor site treatment. J Control Release 244(Pt B):292–301. https://doi.org/10.1016/j.jconrel.2016.07.053
175. Park SU, Lee BK, Kim MS, Park KK, Sung WJ, Kim HY, Han DG, Shim JS, Lee YJ, Kim SH, Kim IH, Park DH (2014) The possibility of microbial cellulose for dressing and scaffold materials. Int Wound J 11(1):35–43. https://doi.org/10.1111/j.1742-481X.2012.01035.x
176. Fontana JD, De Souza AM, Fontana CK, Torriani IL, Moreschi JC, Gallotti BJ, De Souza SJ, Narcisco GP, Bichara JA, Farah LFX (1990) Acetobacter cellulose pellicle as a temporary skin substitute. Appl Biochem Biotechnol 24–25(1):253–264. https://doi.org/10.1007/bf02920250
177. Muangman P, Opasanon S, Suwanchot S, Thangthed O (2011) Efficiency of microbial cellulose dressing in partial-thickness burn wounds. J Am Col Certif Wound Spec 3(1):16–19. https://doi.org/10.1016/j.jcws.2011.04.001
178. Sundberg J, Gotherstrom C, Gatenholm P (2015) Biosynthesis and in vitro evaluation of macroporous mineralized bacterial nanocellulose scaffolds for bone tissue engineering. Biomed Mater Eng 25(1):39–52. https://doi.org/10.3233/BME-141245
179. Muller FA, Muller L, Hofmann I, Greil P, Wenzel MM, Staudenmaier R (2006) Cellulose-based scaffold materials for cartilage tissue engineering. Biomaterials 27(21):3955–3963. https://doi.org/10.1016/j.biomaterials.2006.02.031
180. He X, Cheng L, Zhang X, Xiao Q, Zhang W, Lu C (2015) Tissue engineering scaffolds electrospun from cotton cellulose. Carbohydr Polym 115:485–493. https://doi.org/10.1016/j.carbpol.2014.08.114
181. Ao C, Niu Y, Zhang X, He X, Zhang W, Lu C (2017) Fabrication and characterization of electrospun cellulose/nano-hydroxyapatite nanofibers for bone tissue engineering. Int J Biol Macromol 97:568–573. https://doi.org/10.1016/j.ijbiomac.2016.12.091

182. Huang C, Tang Y, Liu X, Sutti A, Ke Q, Mo X, Wang X, Morsi Y, Lin T (2011) Electrospinning of nanofibres with parallel line surface texture for improvement of nerve cell growth. Soft Matter 7(22):10812. https://doi.org/10.1039/c1sm06430d
183. Min SK, Jung SM, JH J, Kwon YS, Yoon GH, Shin HS (2015) Regulation of astrocyte activity via control over stiffness of cellulose acetate electrospun nanofiber. In Vitro Cell Dev Biol Anim 51(9):933–940. https://doi.org/10.1007/s11626-015-9925-8
184. Chainoglou E, Karagkiozaki V, Choli-Papadopoulou T, Mavromanolis C, Laskarakis A, Logothetidis S (2016) Development of biofunctionalized cellulose acetate nanoscaffolds for heart valve tissue engineering. World J Nano Sci Eng 6(4):129–152. https://doi.org/10.4236/wjnse.2016.64013
185. Mohan T, Hribernik S, Kargl R, Stana-Kleinschek K (2015) Nanocellulosic materials in tissue engineering applications. In: Cellulose - fundamental aspects and current trends. InTech, Rijeka. https://doi.org/10.5772/61344
186. O'Brien FJ (2011) Biomaterials & scaffolds for tissue engineering. Mater Today 14(3):88–95. https://doi.org/10.1016/s1369-7021(11)70058-x
187. Helenius G, Backdahl H, Bodin A, Nannmark U, Gatenholm P, Risberg B (2006) In vivo biocompatibility of bacterial cellulose. J Biomed Mater Res A 76(2):431–438. https://doi.org/10.1002/jbm.a.30570
188. Shi Q, Li Y, Sun J, Zhang H, Chen L, Chen B, Yang H, Wang Z (2012) The osteogenesis of bacterial cellulose scaffold loaded with bone morphogenetic protein-2. Biomaterials 33(28):6644–6649. https://doi.org/10.1016/j.biomaterials.2012.05.071
189. Modulevsky DJ, Cuerrier CM, Pelling AE (2016) Biocompatibility of subcutaneously implanted plant-derived cellulose biomaterials. PLoS One 11(6):e0157894. https://doi.org/10.1371/journal.pone.0157894
190. Saska S, Barud HS, Gaspar AM, Marchetto R, Ribeiro SJ, Messaddeq Y (2011) Bacterial cellulose-hydroxyapatite nanocomposites for bone regeneration. Int J Biomater 2011:175362. https://doi.org/10.1155/2011/175362
191. Kowalska-Ludwicka K, Cala J, Grobelski B, Sygut D, Jesionek-Kupnicka D, Kolodziejczyk M, Bielecki S, Pasieka Z (2013) Modified bacterial cellulose tubes for regeneration of damaged peripheral nerves. AMS 9(3):527–534. https://doi.org/10.5114/aoms.2013.33433

Biopolymers Nanofibers for Biomedical Applications and Environmental Applications

Ana L. Vega-Avila, Oscar Perales-Perez, and Ricky Valentín Rullan

Abstract Electrospinning is a manufacturing technique of growing interest by its versatility in the generation of membranes constituted by fibers. Its relevance has been evidenced by the growing number of researches conducted in the recent years in the field of biotechnology and nanotechnology. In the field of biotechnology, the research interest has been focused in the development of tridimensional assemblies from biocompatible polymers and biopolymers which are able to mimics biological structures and functions for use in tissue regeneration, wound healing, drug delivery, among other applications. In the environmental protection field, the use of electrospun biopolymers has been favored by the feasibility to functionalize them to improve its behavior as adsorbent material, filter or catalyst in the decomposition of toxic compounds while keeping its biodegradability.

This review will provide an overview on the electrospinning parameters, focusing in the processing and the stabilization conditions for nanofibers synthesized from biopolymers such as chitosan, collagen, alginate and gelatin. The results of some promising researches in biomedical and environmental applications using these biopolymers will be also summarized here.

Keywords Electrospinning • Biopolymers • Nanofibers • Nanomaterials • Composites • Nanoscale manufacturing • Membranes

A.L. Vega-Avila
Chemistry Department, University of Puerto Rico, Mayagüez, Puerto Rico

O. Perales-Perez (✉)
Department of Engineering Science and Materials, University of Puerto Rico, Mayagüez, Puerto Rico
e-mail: oscarjuan.perales@upr.edu

R. Valentín Rullan
Mechanical Engineering Department, University of Puerto Rico, Mayagüez, Puerto Rico

J. Almodovar (ed.), *Electrospun Biomaterials and Related Technologies*,
DOI 10.1007/978-3-319-70049-6_4

Abbreviations

BDDGE	1,4-butanediol diglycidyl ether
BSA	Bovine serum albumin
CLSM	Confocal laser scanning microscopy
CNCs	Cellulose nanocrystals
DMAc	Dimethylacetamide
DMEM	Dulbecco's modified Eagle's medium
DMF	Dimethylformamide
DMSO	Dimethyl sulfoxide
E. coli	*Escherichia coli*
EC	Endothelial cells
EDC	1-ethyl-3-(3-dimethylaminopropyl) carbodiimide
FTIR	Fourier-transform infrared spectroscopy
GRGDSP	Glycine–arginine–glycine–aspartic acid–serine–proline
HAP	Hydroxyapatite
HDFCs	Human dental follicle cells
HFIP	Hexafluoroisopropanol
HFP	1,1,1,3,3,3-hexafluoro-2-propanol
hMSCs	Human mesenchymal stem cells
HTCC	*N*-[(2-hydroxyl-3-trimethylammonium) propyl] chitosan
ICP–AES	Inductively Coupled Plasma Atomic Emission Spectroscopy
MIC	Minimal inhibitory concentration
NMR	Nuclear magnetic resonance
P. aureginosa	*Pseudomona aureginosa*
PBS	Phosphate buffer solution
PCL	Poly(caprolactone)
PEO	Poly(ethylene oxide)
PLA	Poly(lactic acid)
PPV	Porcine parvovirus
PVA	Poly(vinyl alcohol)
S-aureus	*Staphylococcus aureus*
SDS	Sodium dodecyl sulphate
SHP	Sodium hypophosphite monohydrate
SMC	Smooth muscle cells
TEM	Transmission electron microscopy
TFA	Trifluoroacetic acid
TFE	Trifluoroethanol
THF	Tetrahydrofuran
WVTR	Water vapor transmission rate

1 Introduction

Electrospinning is a manufacturing method used to generate tri-dimensional structures constituted by fibers whit a high surface area to volume ratio, high porosity and good mechanical strength [1, 2]. This technique is characterized by its feasibility to processing multiple material including polymers, composites, ceramics and metals and generate membranes with tunable size and morphology [3–6].

This versatility, has allowed the use of electrospun membranes in multiple applications such as micro and nano-filtration [7, 8], tissue engineering [9, 10], drug delivery [11], wound healing dressings [12], environmental remediation [13, 14], antibacterial treatment [14], electrodes manufacturing [15], protective clothing [16], sensors manufacturing, hydrogen storage [17], among others [18].

Electrospun fibers are generated by the application of high voltage to a precursor solution flowing at a controlled rate. The applied voltage generates repulsive electrostatic forces at the solution surface. When the electrostatic force overcomes the surface tension of the solution a jet is ejected from the solution towards an opposite charged surface or collector. During this process the polymer chains are stretched and the solvent is evaporated generating in this way nanofibers [6].

Several processing parameters influence nanofibers properties including the applied voltage, the flow rate, the distance between the collector and the needle tip, the tip gauge, environmental parameters such as relative humidity, solution properties such as viscosity, conductivity, surface tension, solution composition, as same as collector geometry, among others [19, 20].

The achievable high specific surface area is one of the most prominent attributes of electrospun fibers, a feature that can be enhanced by the generation of fibers exhibiting diameters at the submicron o nano-size level. Thinner nanofibers with narrower size distribution are usually obtained by increasing the applied voltage [21, 22]. However, increasing this diameter beyond a critical value (which is specific for each polymer) could give place to loss of diameter uniformity because of the formation of beaded nanofibers [18]. On the other hand, some researchers claim that the applied voltage does not influence the fibers diameter, but improves its morphology and homogeneity by decreasing the fibers junctions [23].

Thinner nanofibers can be also obtained by increasing the solution conductivity [24], however, the research conducted by Heikkilä and co-workers [25] revealed that the addition of $ZnCl_2$ to PAN solutions during electrospinning leads to the formation of fibers with larger diameter. This was attributed to an increasing in the mass flow generated by the higher conductivity resulting of the salt addition.

Other electrospinning parameters that influenced the diameter and size distribution of the fibers are the flow rate [26, 27] and the viscosity of the precursor solution [25]. An increasing in the flow rate generates fibers with higher diameter and could lead to the formation of beaded fibers due to an incomplete solvent evaporation [28].

On the other hand, the increasing in the viscosity of the solution leads to the formation of larger nanofibers. This effect was evident in the electrospinning of

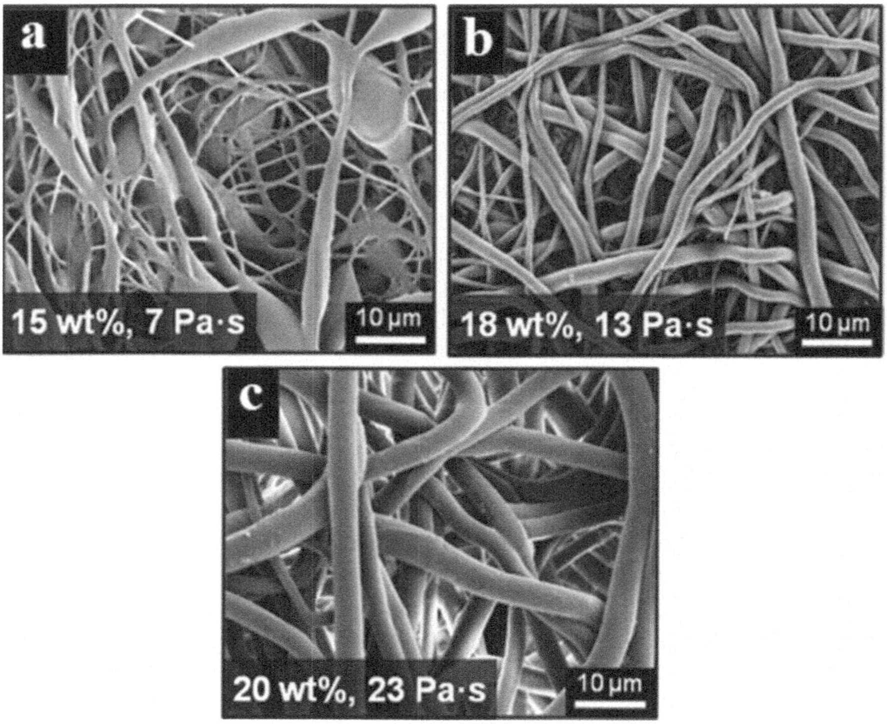

Fig. 1 SEM micrographs of poly-carbonate urethane electrospun fibers synthesized from solutions of different viscosity: (**a**) low viscosity, (**b**) medium viscosity, (**c**) high viscosity. Reproduced from [29]

poly-carbonate urethane fibers from solutions with low (7.2 ± 1.7 Pa s), medium (10.1 ± 0.5 Pa s) and large viscosity (22.5 ± 1.4 Pa s). Beaded fibers were obtained from the low viscosity solution, while uniform fibers from the solution with intermediate viscosity. The diameter of fibers rises in approximately 192% when the solution of high viscosity was used as fibers precursor [29], as can be seen in Fig. 1.

Fibers formation requires a threshold viscosity that allows the generation of a stable solution jet during electrospinning [30]. Therefore, solutions with low viscosity lacks of the viscoelastic forces required to suppress the effect of the surface tension leading to the formation of charged droplets instead of a continuous jet [29]; which leads to electrospraying or to the formation of beaded fibers.

Some biopolymers exhibit high molecular weight and electrical charges, which makes its solutions highly viscous and difficult to process via electrospinning. For this reason, understanding of the processing conditions becomes relevant to successfully synthesize fibers from these materials.

2 Nanofibers from Biopolymers

Biopolymers are biodegradable, biocompatible and exhibit low toxicity. These are also obtained from sustainable and renewable sources. Some naturally occurring biopolymers such as collagen, keratin, fibroin fibrils exhibit complex hierarchical morphology which are inspiration for the synthesis of materials that mimic these structures at a reduced scale while keeping its desired properties.

Properties such as the high surface area and high porosity of the electrospun nanofibers allow its use in the generation of tri-dimensional structures that simulate the extracellular matrix. However, the poor electrospinnability exhibited by most biopolymers [31] and the low stability in aqueous environments remains as a drawback for practical applications.

Several approaches have been developed to improve the electrospinnability of biopolymers including co-electrospinning with assistant polymers [32], addition of salts to increase the conductivity of its solutions, addition of surfactants to decrease the surface tension [33], among other approaches. In the same way, several stabilization methods such as chemical and photochemical crosslinking and plasma oxidation have been developed to increase the resistance of electrospun membranes in the presence of water.

The general properties of some relevant biopolymers such as chitosan, alginate, collagen and cellulose. As same as a detailed description of its electrospinning conditions will be discussed below.

2.1 *Chitosan Nanofibers*

Chitosan is a polysaccharide obtained by deacetylation of chitin, which is the second most abundant biopolymer [34, 35]. Because of its biocompatibility, biodegradability, anti-bacterial and anti-fungal properties, and wound-healing properties [35, 36], chitosan has great potential for multiple applications including biomedical and pharmaceutical, environmental applications [13, 37], membrane filtration [38], among others.

Chitosan contains reactive amino groups and hydroxyl groups in its structure which allows it to act as chelating agent for metal ions to bind to biomolecules, mammalian and microbial cells [34]. The presence of amine groups is also responsible for the solubility of chitosan in acidic media, which allows it to be processed by electrospinning.

The high molecular weight of chitosan makes necessary the use of acids as solvents to prepare electrospinning solutions. Chitosan electrospun nanofibers were prepared from 4 to 6 wt.% chitosan solutions using trifluoroacetic acid (TFA) as solvent [39], these nanofibers were treated with glutaraldehyde in order to increase their chemical resistance in water.

Xinying and coworkers [40], synthesized chitosan nanofibers with average diameter of 130 nm, using acetic acid as solvent. The effects of the solvent concentration, the molecular weight of chitosan and the voltage, in the morphology of the nanofibers were assessed in this work. Chitosan solutions were prepared by dissolving chitosan of three different molecular weight and deacetylation degrees (30,000 g/mol and 56%; 106,000 g/mol and 54%; 398,000 g/mol and 65%) in acetic acid solutions of several concentrations (10–90%) at room temperature for 12 h. The electrospinning parameters used to prepare the nanofibers were: a flow rate of 20 μL/min, the voltage was varied from 1 to 5 kV/cm. Nanofibers were only obtained from solutions with acetic acid concentration higher than 30%. Which was explained by a decreasing in the surface tension of the solution and an increasing in its charge density observed when the concentration of acetic acid is increased. Regarding the effect of the molecular weight, only chitosan of 106,000 g/mol produced bead-free chitosan nanofibers, while the use of chitosan of low molecular weight (30,000 g/mol) and high molecular weight (398,000 g/mol) leads to defective nanofibers. Fragile nanofibers containing large sized beads were obtained from low molecular weight chitosan while non- uniform nanofibers with some beads were obtained from high molecular weight chitosan solutions (2.5–3 wt.%). In general, the average diameter and size distribution of the fibers decreased with increasing of the applied electric field. Summarizing, free beads nanofibers were obtained from 7% chitosan solution of 106,000 g/mol dissolved in 90% acetic acid under an applied electric field of 4 kV/cm.

Chitosan is usually blended with other polymers such as Poly(vinyl alcohol) (PVA) [41] and Poly(ethylene oxide) PEO [13, 42, 43] to decrease the high surface tension of chitosan solutions and allows the use of less aggressive solvents such as aqueous solutions of acetic acid [40, 42, 44] in the preparation of electrospinning solutions from this compound.

Martinova and co-workers [42], prepared electrospun chitosan: PEO nanofibers with a 95:5 weight ratio and average diameter under 500 nm. The researchers assessed the effects of the molecular weight of chitosan and PEO, salt addition, surfactant addition, solvent type and composition in the morphology of Chitosan—PEO electrospun nanofibers.

Electrospinning solutions were prepared by blending separately prepared solutions of chitosan and PEO. Chitosan solutions were prepared from three different grades of chitosan as follows: 10 wt.% of chitosan A (low molecular weight and 78% deacetylation degree), 2% Chitosan B (medium molecular weight and 75–85% deacetylation degree) and 1% Chitosan C (high molecular weight and over 75% deacetylation degree) dissolved in 10, 30, or 60 wt.% acetic acid, and in 30 wt.% lactic acid.

Alternatively, 10 wt.% citric acid was used to obtain an 8 wt.% Chitosan A solution. The citric acid solution was used to enable crosslinking through esterification reaction between hydroxyl groups of chitosan and carboxylic groups of the acid.

In turn 5% PEO (400 kD, 900 kD) aqueous solutions containing NaCl in concentrations ranging from 0 to 2.4 mol/L were combined with chitosan solutions to obtain solution blends with a chitosan: PEO ratio ranging from 70:30 to 95:5.

Triton X-100 was added to these solutions to adjust the surface tension to 51.0–32.0 mN/m.

An alternative electrospinning process that does not needs nozzles (nanospider) was used to prepare the nanofibers. A potential difference of 65 kV was applied between the rotating polymer solution reservoir and the collector, while the solution flowing from reservoir by the action of the centrifugal force is collected in the form of nanofibers.

Water-resistant electrospun membranes were obtained after heat treatment at 145 °C for 15 min. Citric acid was used as crosslinking agent. Heat treated membranes exhibited water resistance after immersion for 24 h. The molecular weight of PEO affected the stability of the membranes, those containing PEO with molecular weight of 400 kD experienced PEO leaching from the nanofibers. The FTIR spectra analysis demonstrated that heat treatment did not lead to chemical crosslinking despite it increases the stability of nanofibers in water.

NaCl addition in a concentration of 0.24–0.66 mol/L leaded to fibers with improved morphology, which was explained by the shrinkage of the polymer chains generated by the interaction of the salt ions with the chitosan polyelectrolyte.

The effect of the addition of metal ions in the morphology of chitosan- PEO electrospun nanofibers was also assessed by Su and coworkers [45]. The electrospinning solutions were blended in adequate proportions of inorganic salt solutions (16 wt.% aqueous solution of NaCl, KCl, $CaCl_2$, $MgCl_2$, $SrCl_2$, $ZnCl_2$, and $FeCl_3$), 3.0 wt.% chitosan dissolved in 0.5 wt.% acetic acid and 5.0 wt.% PEO aqueous solution. The electrospinning parameters used to prepare the nanofibers were flow rate of 0.2 mL/h, 12 kV and a distance from collector to needle tip of 8 cm.

The addition of monovalent cations to the electrospinning solutions (0.4–1.6 wt.% NaCl or KCl) leads to the formation of nanocrystals of inorganic salts in the fibers, while the addition of divalent and trivalent cations (0.8 wt.% $CaCl_2$ or $FeCl_3$) generated defect-free nanofibers. The addition of Ca^{2+} and Fe^{3+} ion also leads to smaller fibers, fibers with average diameter of 200 nm were obtained with additions of 0.4–1% $CaCl_2$. However, branched fibers were obtained when the Fe^{3+} concentration was increased from 0.8 to 1.2 wt.%. Fourier-Transform Infrared Spectroscopy (FTIR) and rheological analysis results suggested that the addition of metal cations disrupt the intermolecular and intramolecular hydrogen bonds in chitosan chains decreasing in this way the viscosity of the electrospinning solutions, which leads to smaller nanofibers.

Chitosan-PEO electrospun nanofibers were assessed by Li and coworkers for tissue engineering applications [46]. The electrospinning solutions were prepared by dissolving chitosan-PEO blends with mass ratios of 85:15, 90:10, 95:5, and 97:3 in aqueous solution containing 3 wt.% acetic acid and DMSO (10:1, w/w). The nanofibers were prepared using a voltage of 18–30 kV, a flow rate of 2–4 mL/h and a distance from needle tip to collector of 10–12 cm as electrospinning parameters. The electrospun fibers were crosslinked with genipin. After crosslinking the membranes were insoluble in PBS after 72 h of immersion at 37 °C, however these still exhibited swelling.

Improvement in the young's modulus of the membranes was observed after crosslinking with genipin, the membranes crosslinked with 0.5% genipin exhibited young's modulus of 670 MPa, which was 59.6% superior to the exhibited by the un-crosslinked membranes (476 MPa). However, its tensile strength (4.87 MPa) was inferior in 300% compared to the showed by the non-crosslinked membranes (14.63 MPa).

Cytocompatibility studies conducted culturing L929 fibroblast onto the genipin crosslinked nanofibers suggested that these scaffolds promoted the adhesion and growth of the cells, being the crosslinked with 0.5% genipin the more biocompatible.

2.2 *Cellulose Nanofibers*

Cellulose is the most abundant natural biopolymer, it can be obtained from wood, leaves and cotton. Cellulose has been used in multiple application including biomedical applications [10], environmental applications [47], cloth protectant applications, as affinity membranes [48], among others [49, 50].

Cellulose-montmorillonite electrospun nanofibers were assessed as adsorbent for wastewater treatment [47], these fibers were prepared from cellulose acetate solutions with concentrations ranging between 8% and 20 wt.% dissolved in acetone: Dimethylacetamide (DMAc) solution (2:1, v/v). Montmorillonite modified with the anion surfactant sodium dodecyl sulphate (SDS) was added to the cellulose acetate solutions in a concentration of 1–3 wt.% (relative to the weight of cellulose acetate). Cellulose-montmorillonite composite nanofibers were prepared using a voltage of 15 kV, a flow rate of 1 mL/h and a distance from needle tip to collector of 20 cm. The electrospun membranes were immersed in sodium hydroxide (NaOH) 0.1M dissolved in ethanol for 24 h to regenerate the cellulose nanofibers. The addition of montmorillonite decreased the average diameter of the nanofibers, bare cellulose nanofibers exhibited average diameter of 468 nm while composite nanofibers exhibited average diameter of 376 nm. The composite electrospun membranes exhibited good affinity toward Cr^{6+}, which evidenced the potential of this material as metal adsorbent.

Zuwei and coworkers [48] synthesized electrospun cellulose nanofibers for membrane affinity applications. Electrospun Cellulose nanofibers with average diameter ranging from 200 nm to 1 μm were prepared from 0.16 g/mL cellulose acetate solution dissolved in a mixture of acetone: dimethylformamide (DMF): trifluoroethylene in a volume ratio of 3:1:1. The electrospun fibers were prepared using a voltage of 25 kV, a flow rate of 4 mL/h and a distance from needle tip to collector of 15 cm. The electrospun membranes were heat treated at 208 °C for 1 h. Heat treatment improved the structural integrity and mechanical strength of the membranes. Cellulose nanofiber membranes were treated with NaOH to obtain regenerated cellulose. Cibacron Blue F3GA was covalently attached to the regenerated cellulose nanofibers. The membranes modified with cibacron exhibited capability to specifically bind Bovine Serum Albumin (BSA) or bilirubin with capacities of 13 mg/g and 4 mg/g, respectively.

Electrospun composite nanofibers from native cotton cellulose loaded with different hydroxyapatite concentrations were prepared by Ao and coworkers [10], for bone tissue engineering applications. Hydroxyapatite nanoparticles in concentrations of 3, 5, and 10% w/w were suspended in 8 wt.% LiCl/DMAc solutions. One gram of activated cellulose was dissolved in the hydroxyapatite suspensions to form the electrospinning solutions. The composite nanofibers were generated at a flow rate of 0.04 mL/min and a rotating velocity of the drum collector of 300 m/min.

Morphological characterization of the electrospun membranes revealed that these were constituted by aligned nanofibers. The bare cellulose nanofibers exhibited average diameter ranging between 100 and 200 nm. However, the addition of hydroxyapatite generated irregular flake-like agglomerates randomly distributed in the fibers and increased the diameter of the nanofibers, the addition of 10% hydroxyapatite generated fibers with a broad diameter distribution ranging from 100 to 500 nm.

The addition of hydroxyapatite improved the mechanical properties of the electrospun membranes. Membranes containing 5% hydroxyapatite exhibited a tensile strength and Young's modulus of 70.6 MPa and 3.12 GPa respectively.

Biocompatibility of the composite electrospun membranes was assessed by culturing human dental follicle cells (HDFCs) for 72 h. After this time, the composite scaffolds containing 10% hydroxyapatite exhibited a higher cell population with the cells evenly distributed and confluent on the scaffold surface.

The higher cell proliferation observed at higher hydroxyapatite content suggested that the cellulose hydroxyapatite electrospun composites provide an excellent platform to support cell growth for bone tissue engineering.

2.3 Alginate Nanofibers

Alginate is a linear polysaccharide, negatively charged that contains in its structure L-guluronic acid and D-mannuronic acid. Sodium alginate is water soluble, however it is able to form hydrogels upon ionic crosslinking with divalent cations such as calcium Ca^{2+}, which improves its resistance in presence of water [51].

Due to its biocompatibility and gelling properties alginate has been used in several applications such as for wound dressing [52, 53], tissue engineering [54] and drug delivery applications.

Alginate requires the use of surfactants and polymer assistants to decrease its surface tension to processing it via electrospinning. Bonino and coworkers [33] reported electrospinning of low molecular weight sodium alginate, using PEO as polymer assistant. The reported alginate: PEO wt.% ratio was higher than 80:20. Additionally, the researchers reported the use of Triton X-100 as surfactant agent. The resulting nanofibers were crosslinked with $CaCl_2$ and subsequently washed in order to remove the soluble remaining PEO.

Sodium alginate based nanofibers containing PEO with average diameter of 250 nm were also prepared from aqueous solutions by Jian and co-workers [55]. Smooth electrospun fibers were obtained from 3 wt.% solutions with varied alginate/PEO proportions ranging from 3:1 to 0:1.

Crosslinking of the fibers was conducted in two stages, the first one involves the immersion of the fibers in 2% hexamethylene diisocyanate solution containing dibutyltin dilaurate as catalyst and toluene as solvent for 24 h at room temperature. The second crosslinking stage involves the immersion of the membranes in 5% aqueous solutions of $CaCl_2$ for 24 h.

The cross-linked mats immersed in distilled water retained its morphology for 3 days, while the partially crosslinked membranes (those crosslinked only by the second stage and those crosslinked only by the first stage of crosslinking), only lasted 10 min and 1 day respectively. The authors suggested good potential of this membranes in biomedical applications.

Mohena and co-workers [52], prepared nanofiber composites with antibacterial properties for wound healing applications. These composites were prepared by impregnation of silver nanoparticles into electrospun alginate nanofibers through their complexation with chitosan.

Silver nanoparticles were synthesized by reduction of silver nitrate 10 mM with chitosan solution 0.5 wt.% dissolved in 2% acetic acid. This process was conducted under agitation for 3 h, followed by heating at 90 °C for several periods: 3, 6, 12, 24 and 48 h.

Alginate nanofibers were prepared from 3 wt.% sodium alginate-PEO (50:50) aqueous solution containing 0.5 wt.% Triton X-100 and 5 wt.% Dimethyl sulfoxide (DMSO). A flow rate of 0.8 mL/h, a voltage of 17 kV and a distance from the tip-to-collector of 17 cm were used to generate the alginate nanofibers, which were crosslinked with aqueous solutions of calcium chloride for 1 h. The crosslinked nanofibers were coated by immersion in chitosan-Ag solution containing 1 wt.% Ag NPs for 5, 10 and 15 min. The swelling of the coated membranes immersed in distilled water decreased from 276 to 150% with increasing in the coating time.

Shalumon and coworkers [53] synthesized ZnO loaded alginate-PVA electrospun nanofibers with average diameter ranging from 220 to 360 nm for wound dressing applications. A blend of aqueous solutions of sodium alginate 2 wt.% and PVA 16 wt.% in a 1:1 ratio was used to prepare the electrospinning solutions. ZnO nanoparticles in concentrations ranging from 0.5 wt.% to 5 wt.% were dispersed into the polymer solution. The electrospinning parameters used to prepare the fibers were a voltage of 17 kV, a flow rate of 0.1 mL/h and a distance between the needle tip and collector of 5 cm. The electrospun membranes were crosslinked by exposition to 2% glutaraldehyde vapor for 48 h, following the immersion of the membranes in 1% $CaCl_2$: ethanol solution for 1 h.

The bactericide effect of the electrospun fibers against *Escherichia Coli* (*E. Coli*) and *Staphylococcus Aureus* (*S. Aureus*) was assessed by mean of the disk diffusion method. All the ZnO loaded membranes exhibited an inhibition zone for both types of bacteria, however, the diameter of the observed inhibition zone was directly proportional to the ZnO concentration in the fibers.

The cell adhesion and spreading of L929 cells cultured during 96 h onto the electrospun nanofibers were used as criteria to evaluate its cytocompatibility. Good cell adhesion and spreading was observed after 48 and 96 h for the electrospun nanofibers without ZnO nanoparticles. The same scenario was observed for the membranes containing 0.5 and 1% ZnO. However, a lack of cell spreading was observed at higher ZnO concentration at 48 and 96 h, which was attributed to the slightly cytotoxic effect of ZnO nanoparticles at higher concentrations.

Sodium alginate-polylactic acid (PLA) electrospun composites were synthesized by Xu and coworkers [54] for tissue engineering scaffolds. The electrospinning solutions were prepared by the addition of aqueous solution of sodium alginate (40 mg/mL) in a varied proportion (1.2, 2.4, 3.6, 4.8 mL) to span 80 solutions (0.03 g/18 mL) dissolved in chloroform. 1.92 g of PLA was added into the solution, and the mixture was stirred at 240 rpm for 2 h to obtain uniform emulsions.

The electrospun fibers were obtained using a flow rate of 0.5 mL/h, a voltage of 15 kV, and a distance from the needle tip to collector of 15 cm as electrospinning parameters. The electrospun scaffolds were crosslinked by two methods. The first one involved the use of 0.2 g/mL $CaCl_2$ solution applied to the membrane by electrostatic spraying. The second was the immersion of the membranes in a 0.01 g/mL $CaCl_2$ solution.

It was found that variations in sodium alginate concentration does not affected the morphology and average diameter of the electrospun nanofibers which was 250 ± 90 nm. Results of Confocal Laser Scanning Microscopy (CLSM) and FTIR analysis conducted on the electrospun membranes suggested that the alginate was distributed at the surface of the fibers while the PLA in the core, which was attributed to a phase separation and migration of the alginate and the surfactant Span 80 to the surface of PLA during electrospinning.

Cell adhesion and proliferation assays demonstrate that the alginate/PLA electrospun membranes were suitable scaffolds for cell culture due to their surface roughness, which provides additional sites for adequate cells attachment and grow.

In an alternative approach Mokhena and coworkers reported the preparation of alginate-PEO electrospun nanofibers with average diameter of 114 nm [56]. The membranes were prepared from a blend of 3 wt.% PEO and 3 wt.% sodium alginate aqueous solutions. The researchers did not reported the use of surfactants, instead the aging of the electrospinning solutions for 30 days at 23 ± 2 °C and a humidity of 60 ± 5%. The researchers claim that the aging leads to a reduction in the viscosity of the solution, also to the formation of carboxylate groups in the solutions which facilitate the electrospinning process.

2.4 Collagen Nanofibers

Collagen is the most abundant protein of the mammalian tissue, it can be found in the extracellular matrix and the connective tissues [57]. Collagen type I is the most common form of this protein, which consist of long right handed triple helix conformed by three coiled subunits: two α1 chains and one α2 chains. individual collagen triple helixes can assemble in a hierarchical structure which leads to the formation of the fibrous structure observed in tissue, bone, and basement membranes [58, 59].

Collagen exhibit biodegradability, biocompatibility, high mechanical strength, weak antigenicity, anti-inflammatory properties, ability to promote tissue regeneration, which makes a suitable candidate in biomedical applications [59].

Electrospun collagen nanofibers have been prepared from multiple solvents including acetic acid solutions, trifluoroethanol, trifluoroacetic acid (TFE), and 1,1,1,3,3,3-hexafluoro-2-propanol (HFP) [59, 60].

Mathews and coworkers studied the effect of the collagen source, collagen type, and the collagen concentration of electrospinning solutions in the structural properties of collagen nanofibers [61]. Collagen type I from calfskin and type I and type III collagen from human placenta were dissolved at various concentrations in HFP. The researchers found that the collagen type influenced the optimum solution concentration for fibers formation. Type I collagen nanofibers from calfskin were only obtained at collagen concentrations higher than 0.083 g/mL. While, type I collagen fibers from human placenta with average diameter of 100 nm were obtained at concentrations below 0.083 g/mL. On the other hand, type III collagen nanofibers with average diameter of 250 nm were synthesized at concentrations of 0.04 g/mL. All the fibers exhibited the 67 nm periodic banding pattern characteristic of the native collagen. The collagen membranes crosslinked with glutaraldehyde vapor were used as scaffolds for culturing Aortic Smooth Muscle cells for 7 days. After this period, the scaffolds were densely populated and exhibited cellular infiltration into the fibrillar network.

The solvent used for electrospinning of collagen could influence the final structure of the nanofibers. Denaturation of the collagen triple helix was observed in the electrospun fibers prepared from type I collagen using trifluoroethanol and aqueous acetic acid solutions as solvent [62]. Solutions prepared with acetic acid retained 18% of the native collagen triple helix, while the solutions prepared with TFE retained 16%. Free beads fibers with average diameter of 320 nm were obtained from collagen solutions dissolved in TFE while beaded fibers with average diameter of 150 nm from collagen solutions dissolved in acetic acid. The membranes were crosslinked by immersion in 5% w/v solutions of 1-ethyl-3-(3-dimethylaminopropyl) carbodiimide (EDC) or in 1,4-butanediol diglycidyl ether (BDDGE) dissolved in ethanol, at 37 °C for 7 days. Mesenchymal stem cell from rabbit cultured for 24 h into electrospun membranes exhibited good adhesion to the scaffolds and typical morphology. Non-evident differences were observed between the cells cultured in both types of scaffold.

Collagen has been blended with others polymers such as chitosan [9, 63], PEO [64, 65], Zein [66], among others in order to enhance its spinnability and improve the mechanical properties of the electrospun membranes.

Type I collagen-PEO electrospun fibers were produced by Lei and coworkers [64] for potential applications in tissue engineering, wound healing, and as hemostatic agents. Collagen-PEO electrospinning solutions with concentrations of 1% and 2 wt.% were prepared by dissolving the polymers in HCl 10 mM, the collagen: PEO weight ratio used to prepare the solutions ranged from 10:1 to 1:2. One and two weight percent collagen solutions did not lead to the formation of fibers. However, the PEO addition in a collagen: PEO ratio of 1:2 produced free beads fibers with diameter ranging between 50 and 150 nm. Presence of beads along the fibers length were observed at collagen: PEO weight ratios of 10:1, 5:1 and 1:1.

Another approach to produce electrospun Collagen-PEO nanofibers was described by Wang and co-workers [65]. The collagen-PEO electrospun fibers were prepared from 3% w/v collagen-PEO solutions prepared by dissolving collagen type I and PEO in a 50:50 v/v solution of hexafluoro isopropanol (HFIP) and acetic acid. The weight ratios of collagen to PEO used were 90:10, 80:20 and 70:30, Free beads nanofibers were observed at all collagen: PEO ratios; however, the average diameter of the fibers increased from 263 to 353 nm with the increasing of PEO content from 20 to 30%.

FTIR characterization of the collagen-PEO nanofibers, revealed that these retained all the characteristic absorption bands of PEO and collagen, which suggest that PEO did not modify the collagen structure. However, the presence of hydrogen bonds between PEO and collagen was evidenced by the shift of the N–H stretching band from 3327 to 3298 cm^{-1}.

Electrospun collagen-chitosan nanofibers were produced by Chen and co-workers for tissue engineering applications [9]. The fibers were prepared from 8% w/v polymer solutions with collagen: chitosan weight ratios of 100:0, 80:20, 50:50, 20:80, 0:100. A mixture of HFIP/trifluoroacetic acid (TFA) (v/v, 90/10) was the solvent.

The composite fibers exhibited an average diameter ranging from 434 to 691 nm, A decreasing in the fibers diameter was observed with the increasing in chitosan content, which was attributed to the increase in the charge density on the electrospinning solutions associated to the formation of an organic salt by the reaction between TFA and the amino groups of chitosan.

Tension test conducted on dry membranes revealed that its average ultimate tensile strength decreases with the increasing in chitosan content in the fibers from approximately 2.9 MPa for pure collagen to 0.3 MPa for pure chitosan.

Electrospun nanofibers from collagen-zein blends were prepared by Jiantao and coworkers [66] to be used as wound healing dresses.

Electrospinning solutions with concentration of 60% w/v, were prepared by dissolving collagen, zein and collagen/zein in variable ratio, in acetic acid 70% (v/v) aqueous solution. Berberine was also dissolved in the collagen/zein solutions.

Collagen solution did not lead to fibers formation; while zein solution and collagen-zein solutions with weight fractions of 33, 50 and 67% produced bead-free fibers. The electrospun collagen-zein fibers exhibited smaller diameter and greater fiber flexibility than zein fibers. The increasing of zein content from 33 to 67% increased the average fiber diameter from 423 to 910 nm.

The higher content of zein also influenced the mechanical properties of the electrospun membranes. An increase in the tensile strength from 0.2 to 6.3 MPa was observed when the zein concentration was increased from 33 to 67%, while the elongation at break decreased from 39.5 to 12.8%.

Degradation studies conducted by immersion of the electrospun collagen-zein membranes in PBS (pH 7.4) at 37 °C during 7, 14, and 21 days, demonstrated that the extent of the degradation decreased with the increasing in the zein content. After 14 days of immersion the weight loss of the membranes reached 23% for those containing 67% zein, 33% for those containing 50% zein and 47% for the membranes with 33% zein. It was found also that an increase in the zein content decreased the surface wettability of the electrospun nanofibers.

2.5 Gelatin Nanofibers

Gelatin is a biopolymer derived from collagen, which has been widely used in medical [67], pharmaceutical, cosmetic applications and food industry because its biodegradability, biocompatibility, good film forming ability, non-immunogenicity, low cost and commercial availability [12].

Several compounds have been assessed as solvents for the preparation of electrospun gelatin nanofibers including THF, acetic acid solutions [68], acetic acid-ethyl acetate mixtures [69], ethanol-formic acid mixtures [70], among others.

Electrospun gelatin nanofibers and collagen type I nanofibers with average diameter of 1 μm were prepared by Balendu and coworkers using HFP as solvent. These fibers were used for endothelial and osteoblast culture [57]. The electrospinning solutions were prepared by dissolving lyophilized collagen (55 mg/mL) and gelatin (110 mg/mL) in HFP for 12 h. Additionally "recovered" electrospun membranes were prepared from dissolving previously prepared electrospun scaffolds in ice cold deionized water and adjusting the protein concentration in the solution to 1.5 mg/mL. The fibers were produced at 22 kV, a flow rate ranging from and 3 to 7 mL/h and distance from needle tip to collector of 25 cm.

The electrospun scaffolds were cross-linked in glutaraldehyde vapor for varied periods ranging from 1 to 12 h. The extent of the crosslinking influenced the fibers structure in aqueous environment. At low crosslinking time the fibers remained undissolved but exhibited coiling, which was reduced by increasing the crosslinking degree. At high crosslinking degree, the fibers exhibited a nearly linear conformation when hydrated.

The extent of the crosslinking also influenced the healing of dermal injuries inflicted to adult guinea pigs. All wounds treated with lightly crosslinked electrospun collagen and electrospun gelatin membranes underwent resolution over a similar time course. However, these wounds exhibited varying degrees of wound contraction after closing. Good fibroblast proliferation and blood vessel formation was observed in all the wounds treated with collagen membranes; while the wounds treated with gelatin membranes exhibited a lower fibroblast proliferation and the formation of foreign body giant cells at the borders of the wound, attributed to the presence of proinflammatory peptides in gelatin and its lack of intact α chains.

Electrospun gelatin nanofibers with average diameter ranging between 200 and 300 nm were prepared from 10% w/v gelatin solutions using TFE as a solvent [67]. The electrospinning parameters used to generate the fibers were: 10 kV, a flow rate of 0.8 mL/h and a distance between the needle and the collector of 13 cm. The electrospun membranes were crosslinked using glutaraldehyde vapor.

Mechanical test shows that the tensile strength of the gelatin crosslinked membranes was almost 10 times higher (12.62 ± 1.28 MPa) than the tensile strength for the non- crosslinked (1.28 ± 0.12 MPa). This behavior was also observed for the young modulus which was 424.7 ± 20.7 MPa for the crosslinked and 46.5 ± 3.82 MPa for the as-electrospun gelatin fibers. These membranes were used as scaffolds for human dermal fibroblasts culture.

Aoki and coworkers [71] evaluated a method to embed alkaline phosphatase into gelatin nanofibers using 50% aqueous solutions of *N,N*-dimethylacetamide, *N,N*-dimethylformamide and *N*-methyl-2-pyrrolidone as solvents. Fibers with variable morphology from a thick, wide and porous structure to a thin and fine structure were obtained by varying the solvent and gelatin concentration. Swiss 3T3 fibroblasts cultured onto gelatin scaffolds exhibited good adhesion.

Free beads electrospun gelatin nanofibers with average diameter of 85 nm, were obtained from 20 wt.% gelatin solutions dissolved in aqueous solutions of formic acid and ethanol [70]. The electrospun membranes were obtained at a voltage of 20 kV and a distance from needle to collector of 10 cm. The glutaraldehyde cross-linked membranes exhibited cell compatibility for mouse mesangial cells.

TFA was used as solvent for the preparation of gelatin electrospun membranes [72]. Electrospinning solutions containing 15 wt.% gelatin were dissolved in TFA, 1.5 mL of 0.05 wt.%/v glutaraldehyde diluted in PBS was added to 10 mL of the gelatin solution.

The electrospinning parameters used to generate the gelatin membranes were a voltage of 10 kV, a distance from needle tip to collector of 15 cm and flow rate of 0.5 mL/h. The electrospun membranes were heated at 100 °C during 24 h to improve the crosslinking.

The un-crosslinked gelatin electrospun fibers exhibited an average diameter ranging between 50 and 300 nm, whereas the crosslinked ones exhibited diameters between 100 and 500 nm. FTIR analysis revealed that TFA does not modify the amide groups of the gelatin and confirmed the crosslinking. The glutaraldehyde addition increased the mechanical strength of the electrospun membranes by two-fold, and allowed the formation of flexible membranes, which were insoluble after immersion in water for 1000 min. The flexibility exhibited by the crosslinked membranes and the low concentration of glutaraldehyde used for crosslinking are advantages of this method.

Gelatin has been blended with other polymers such as PEO [73], PCL [69], PVA [74], and chitosan [75] to improve thecorresponding mechanical properties and stability in aqueous environments.

Electrospun polycaprolactone (PCL)-gelatin composite nanofibers with different PCL: gelatin ratios (90:10, 80:20, 70:30, 60:40, 50:50 wt.%) were prepared from 16 wt.% PCL-gelatin solutions dissolved in acetic and ethyl acetate mixture solutions [69]. Composite nanofibers were produced at a voltage of 10 kV, a flow rate of 1 mL/h and a distance from needle tip to collector of 10 cm. Pure PCL solutions did not lead to fiber formation. However, all the gelatin containing solutions produced nanofibers. Beaded nanofibers were obtained at 10% and 20 wt.% of gelatin while free beads nanofibers with average diameter ranging between 200 to 300 nm were produced at 30–50 wt.% of gelatin. PCL nanofibers were produced using methanol-chloroform as solvent. The electrospun composite membranes containing 50% gelatin exhibited loss of its morphology after immersion in simulated body fluid (SBS) at 37 °C for 2 weeks; while the nanofibers containing 30 and 40 wt.% gelatin maintained its structural integrity. However, after immersion in SBS during 8 and 12 weeks, these fibers also exhibited loss of their structural integrity.

The PCL electrospun nanofibers retained their morphology after immersion in SBS during 12 weeks under the same conditions. These membranes were assessed as scaffolds for human mesenchymal stem cells (hMSCs) culture with promising results.

3 Nanofibers Stabilization

Despite the biocompatibility, biodegradability, and environmentally friendly features exhibited by most biopolymer-based electrospun fibers, the lack of stability in aqueous environments is a limiting factor for its use in practical applications. Several approaches have been explored in order overcome this limitation, including cross-linking by chemical methods, photochemical methods, heat treatment, plasma and combination of them. However, the induced cytotoxicity, change in the chemical characteristics of the membranes, loss of flexibility and incomplete crosslinking are limitations exhibited by some crosslinking methods.

3.1 Glutaraldehyde

Crosslinking using glutaraldehyde is a simple, easily available, cost effective and rapid method to enhance the stability of protein based electrospun fibers while preserving its biological integrity. The crosslinking is conducted by the exposition of the electrospun membranes to vapor of this compound. During the process the aldehyde groups on glutaraldehyde reacts with the amine groups in the proteins to form a Schiff's base [76, 77]. The non-bounded aldehyde groups remaining after crosslinking are responsible for the cytotoxicity exhibited by the crosslinked membranes. The risk of cytotoxicity can be reduced by using glutaraldehyde solutions of lower concentration and by mean of post crosslinking treatments such as heat treatment [67].

Glutaraldehyde has been extensively used for the crosslinking of gelatin and collagen nanofibers and films [67, 78]. Zhang and coworkers synthesized gelatin nanofibers for potential biomedical applications [67]. The membranes were crosslinked with glutaraldehyde vapor (25% v/v) at room temperature for varied time periods (6 h, 12 h, 1, 2, 3 and 4 days). The crosslinked samples were heat treated at 100 °C for 1 h and then exposed in a fume hood for 2 h in order to remove residual glutaraldehyde and enhance the crosslinking. The crosslinked membranes keep its integrity after immersion in deionized water at 37 °C for periods longer than 2 days.

The effect of the crosslinking in the Piperin release from gelatin nanofibers was studied by Laha and coworkers [79]. Piperin was added to 20% gelatin nanofibers in a concentration of 2 mg/mL. the nanofibers were crosslinked by exposition to saturated vapor of Glutaraldehyde (25% v/v) for 8 min. The fibers were exposed to glutaraldehyde in two different ways. In the first one these were exposed to glutaraldehyde only one time. In the second way, the crosslinking was conducted sequentially by exposing the first thin layer of collected fibers to glutaraldehyde for 2 min followed by further deposition of fibers and subsequent crosslinking and so on.

Degradation and swelling tests were conducted by immersing the crosslinked membranes into aqueous solutions at pH 7.4 and pH 1.2 at 37 °C for 50 h. Both types of crosslinked membranes retained their integrity during the degradation test. However, the sequentially crosslinked ones were more compact and homogenous, while the crosslinked fibers exhibited excessive shrinkage, which was attributed to a non-uniform crosslinking.

The weight loss and the swelling experienced by the membranes during the degradation test was more pronounced for the membranes immersed in PBS at pH 7.4. The Piperin release during the first hours and the cumulative drug release was lower for the membranes crosslinked sequentially especially at low pH conditions, which was attributed to less swelling of the gelatin membrane.

Glutaraldehyde has been also used for the crosslinking of chitosan based electrospun nanofibers [39, 41]. In this context Pouranvari and co-workers synthesized chitosan/PVA electrospun nanofibers for potential biomedical applications [41]. These membranes were crosslinked by exposition to 5% glutaraldehyde vapor at room temperature for 48 h. After crosslinking the membranes were washed with 2% glycine to inactivate the remaining glutaraldehyde. Chemical crosslinking of the chitosan/PVA nanofibers was confirmed by the presence of the imine band –C=N at 1586 cm^{-1} in the FTIR spectra. This research showed that glutaraldehyde even at low concentration is an effective cross-linker agent for chitosan nanofibers.

3.2 *Genipin*

Concerns regarding the cytotoxicity of glutaraldehyde have promoted the development of alternative crosslinking methods. Genipin is among the compounds evaluated for this purpose.

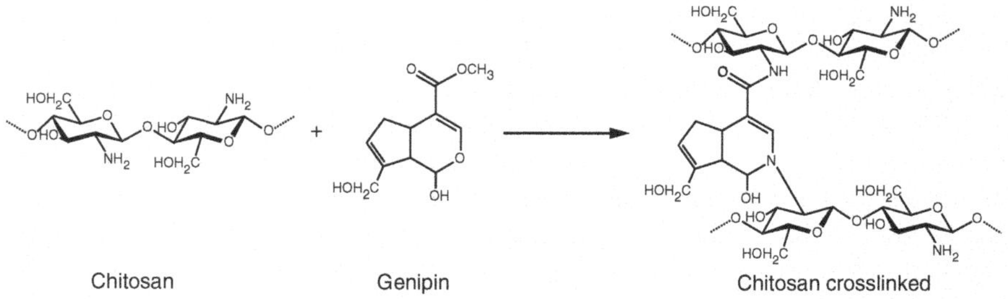

Fig. 2 Crosslinking reaction between chitosan and genipin, Reproduced from [83]

Genipin is a natural compound derived from geniposide. Because its biodegradability and low cytotoxicity it has played multiple functions such as cross-linking agent, bio-adhesive, wound-dressing material, fingerprint reagent and natural pigment for food, among the more important applications.

As a cross-linking agent genipin was used in the crosslinking of proteins such as silk [76], gelatin [80], collagen [81] and chitosan [46, 82]. The crosslinking mechanism of genipin with molecules containing primary amines is not well understood yet. However, it could proceed through the reaction between the amino group in proteins and the ester group in genipin eliminating a methanol molecule in the process. The crosslinking is reached by the interaction of two protein bound genipin molecules. The crosslinking reaction could also proceed through the nucleophilic attack of a primary amine in the protein to the C_3 carbon in genipin, binding in this way the protein to the genipin ring. The crosslinking reaction of chitosan with genipin is summarized in Fig. 2.

The crosslinking effect of genipin and glutaraldehyde in the stabilization of electrospun chitosan—PEO nanofibers was evaluated by Li and co-workers [46]. The membranes were crosslinked at 37 °C by immersion in genipin solutions with varied concentration 0.1, 0.5, and 1% w/v for 6, 12, or 24 h. For comparative purpose, the membranes were also crosslinked by immersion in 1% glutaraldehyde for 24 h.

FTIR analysis confirmed that both crosslinking methods were effective to preserve the morphology of the membranes after immersion in PBS for 72 h at 37 °C.

L929 fibroblast cells were cultured during 7 days onto the membranes in order to assess its biocompatibility after crosslinking. The genipin crosslinked scaffolds exhibited enhanced cell proliferation after 1 and 4 days of culture. The cytotoxicity assessed by MTT assay revealed that the viability of the cells cultured in the genipin crosslinked membranes was as good as the exhibited by the cells cultured in fresh medium. However, the cell viability exhibited for the glutaraldehyde crosslinked membranes was around 80%.

Panzavolta and co-workers [80] developed an optimized procedure for crosslinking of gelatin nanofibers using genipin. The optimized method involves the addition of a small amount of genipin to the gelatin electrospinning solution. The electrospun nanofibers were crosslinked in 5% w/v genipin/ethanol solution for 7 days, followed by rinsing in PBS and then air drying at 37 °C. The crosslinked nanofibers retained their morphology after immersion in Dulbecco's modified Eagle's medium (DMEM) for 1 week, and even after the culture of vascular wall mesenchymal stem cells onto the gelatin scaffolds for 7 days. The The cultured cells exhibited good cell viability and adhesion to the gelatin scaffolds.

3.3 Reactive Oxygen Species

Crosslinking by exposition to oxygen plasma is based in the formation of highly reactive oxygen species such as oxygen radicals, which promote chemical reactions such as decomposition and cross-linking on the exposed surfaces. This is considered as mild crosslinking method of low cytotoxic effect.

Gelatin scaffolds were crosslinked by mean of several methods including the exposition to vapor phase glutaraldehyde, the immersion in genipin and in D, L-glyceraldehyde solutions, as well as the exposition to reactive oxygen species [77].

For crosslinking via reactive oxygen species the scaffolds were treated for 2 min in an oxygen plasma environment.

Crosslinking with D, L-glyceraldehyde and genipin involved the immersion on solutions of these compounds for 19 h. The gelatin membranes were also crosslinked with vapor of glutaraldehyde (0.5% w/w) for 19 h.

The scaffolds crosslinked by reactive oxygen species does not exhibited change in its morphology after crosslinking. However, those crosslinked with glutaraldehyde vapor experienced changes in their morphology, mainly observed as a rubbery appearance and fusion at fiber junctions which was attributed to the presence of water vapor in the air during the cross-linking process. The fibers crosslinked with D, L-glyceraldehyde and genipin did not retain their morphology, but still preserved a porous structure. This change in morphology was most evident in the genipin crosslinked scaffolds, which was attributed to their slower reaction rate.

After crosslinking the membranes were submerged in DMEM and held at 37 °C for at least 2 weeks. The 0.5% (w/w) glutaraldehyde cross-linked fibers remained for the entire two weeks, as same as the scaffolds cross-linked with glyceraldehyde. The scaffolds cross-linked with genipin keep their integrity over the entire evaluation period, with exception of the crosslinked with 0.1% (w/w) genipin which were dissolved before the test was completed. The scaffolds cross-linked with reactive oxygen species only lasted 12 h. Summarizing, glyceraldehyde and genipin are both good options as cross-linking agents because of their low toxicity and effectivity to provide resistance to dissolution in cell culture medium at 37 °C. However, the use of reactive oxygen species is not a suitable crosslinking method to stabilize gelatin scaffolds.

3.4 Citric Acid

Citric Acid has been used for the crosslinking of hydroxyl containing monomers such as silk, starch, poly-(vinyl alcohol) (PVA), zein and polymeric blends such as alginate/PVA blends [84], chitosan/PEO blends [42], among others. Citric acid it is considered a non-toxic alternative to traditionally used crosslinking agents such as glutaraldehyde [85].

This method was used to crosslink guar gum-PVA electrospun nanofibers for targeted drug delivery applications [85]. Citric acid (5 wt.%) was added as crosslinking agent during the preparation of the PVA/guar solutions. Heat treatment at 140 °C for 2 h was conducted after electrospinning to promote esterification reactions. The crosslinking was confirmed by the systematic increase of the –C=O FTIR band (characteristic of the ester group) after heat treatment.

Jiang and coworkers reported the use of citric acid in the crosslinking of electrospun zein fibers for tissue engineering and other medical applications [86]. Zein solutions (50 wt.%) were prepared by dissolving the polymer in citric acid solutions with pH adjusted to 4.9. Citric acid solutions concentration ranged from 5.5 to 9% and ethanol at 70% was used as a solvent.

Crosslinking was conducted by aging the solutions for 48 h at room temperature, process that allows the expansion of zein molecular chains and the formation of cross-links before electrospinning. Aged solutions dissolved to 26 wt.% were used for the generation of the electrospun membranes. Heat treatment at 150 °C for 2.5 h was conducted as final step in the stabilization.

The crosslinked nanofibers not only retained their structure up to 15 days after immersion in PBS at 37 °C, these also proved to be a better template for the attachment, spreading and proliferation of mouse fibroblast cells (NIH 3T3) when compared with the uncross-linked membranes, with electrospun PLA scaffolds and with electrospun zein fibers cross-linked with a mixture of citric acid and sodium hypophosphite monohydrate (SHP). The fibroblast cells cultured for 8 h exhibited an increasing in the cell attachment to the citric acid cross-linked membranes up to sevenfold. This improvement in the cell attachment was promoted by the increasing in the citric acid concentration from 5.5 to 9% which was attributed to an increase in the stiffness of the cross-linked membranes. The lower cell attachment exhibited by the electrospun scaffolds cross-linked with SHP was attributed to a possible cytotoxic effect of this compound.

Another crosslinking method involves the use of dicarboxylic acids, which once dissociated can react with the hydroxyl groups in biopolymers to form two esters bonds. The esterification can also occur partially leaving one end of the anhydride free to become protonated. An incomplete dissociation of the dicarboxylic acid, could prevents the crosslinking reaction. Following this approach, maleic anhydride was used in the crosslinking of electrospun PVA nanofibers and PVA nanofibers containing cellulose nanocrystals (CNCs) [87]. Crosslinking was achieved by exposing the electrospun membranes to the vapor generated by the reaction of maleic anhydride (1–2 g) with few drops of HCl. The esterification reaction was carried out at 80 °C for a varied period time (15–120 min).

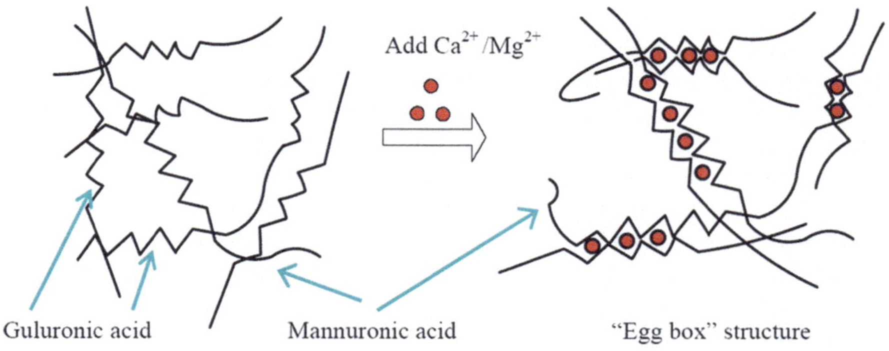

Fig. 3 Crosslinking of sodium alginate by divalent cations. Reproduced from [90]

After crosslinking, the samples were heat treated at 120 °C for 30 min. Blank samples were only heat treated. Chemical characterization of the crosslinked fibers by FTIR and Nuclear Magnetic Resonance (NMR) confirmed the formation of ester bonds in the fibers.

The stability of the crosslinked fibers in presence of different solvents were assessed by submerging them in several solvents including: water, methanol, acetonitrile, DMF, tetrahydrofuran (THF), chloroform, dioxane, DMSO, and pyridine, during varied time periods (from 2 h to 1 day). The membranes that were crosslinked for periods longer than 15 min retained their mechanical integrity when exposed to water, while the control samples exhibited some degree of resistance to dissolution but showed delamination. Both membrane types were resistant to dissolution in low polarity solvents.

3.5 *Calcium Chloride*

Calcium chloride is commonly used as crosslinking agent of sodium alginate [88], this biopolymer can exchange Na^{+} by Ca^{2+} cations from calcium chloride to form calcium alginate. The Ca^{2+} cations can selectively interact with guluronic acid monomers of the alginate chains, forming in this way junctions between guluronate blocks of adjacent polymer chains, which is termed "eggbox model" of crosslinking [88, 89]. The crosslinking of sodium alginate by divalent cations is depicted in Fig. 3.

Calcium alginate is insoluble in water, slightly soluble in ethanol, but soluble in aqueous solutions of sodium phosphate, sodium carbonate, and substances able to react with calcium ions. This property is important for its use as wound dressing and hemostat dressing [91, 92].

Alginate-PEO nanofibers and alginate-PEO nanofibers modified with the peptide glycine-arginine-glycine-aspartic acid-serine-proline (GRGDSP) were assessed as scaffolds for the culture of human dermal fibroblast [93]. These fibers were crosslinked by immersion in a 2% w/v $CaCl_2$ solution dissolved in a mixture of water: ethanol at a ratio of 1:5 for 10 s. The lyophilized crosslinked scaffolds were submerged in water at 37 °C for 120 h. FTIR analysis revealed that PEO was completely extracted from the scaffolds after 48 h of water immersion. Human dermal fibroblast cells exhibited enhanced cell attachment, spreading, and proliferation when cultured in the modified nanofibers scaffolds.

Alginate-PEO nanofibers were assessed as sorption material for the removal of heavy metals from water [56]. The crosslinking of the electrospun membranes proceeded trough their soaking in a 80 wt.% ethanol solution, followed by washing for 10 min in a 2 wt.% calcium chloride solution dissolved in ethanol and finally by the incubation in a $CaCl_2$ aqueous solution for 1 h. The crosslinked membranes were evaluated as cooper adsorbent at 25, 40, and 60 °C and at pH values of 2.0, 4.0, 6.0. The membranes effectively removed cooper ions from water under the aforementioned conditions and maintain their metal adsorption capacity over five regeneration cycles.

3.6 *Photochemical Crosslinking*

Another approach to improve the stability of water soluble electrospun nanofibers is by mean of its modification using photo-crosslinkable compounds as metacrylate [81, 94]. Following this approach Liu and coworkers stabilized electrospun collagen nanofibers using rose bengal as the photo initiator [60]. Rose bengal was added in a concentration of 0.1% (w/v) to collagen fibers prior to laser irradiation for 100 s at a wavelength of 514 nm. Acetic acid and HFP were used as solvents in electrospinning solutions. The crosslinked collagen nanofibers retained their morphology for at least 21 days immersed in PBS at 37 °C. The mass loss experimented by the membranes was 47.7 ± 7.4% at day 7 and 68.9 ± 24.7% at day 15. The researchers reported that the solvent used in electrospinning solutions affected the final structure of the crosslinked membranes. The fibers prepared with acetic acid retained the helical morphology of collagen in a greater extent than the fibers prepared with HFP. The fraction of the collagen helical structure in acetic acid electrospun fibers was 28.9 ± 5.9%, compared to 12.5 ± 2.0% for the HFP based fibers.

4 Biopolymer Based Nanofibers Applications

4.1 *Engineered Scaffolds*

Engineer designed scaffolds must be able to mimic the extracellular matrix by providing space for new tissue formation, delivering cells to the desired site until host cells can repopulate and resynthesize a new natural matrix. To achieve this function the engineered scaffolds must exhibit biocompatibility, biodegradability and non-toxicity. Collagen, chitosan, gelatin [10] and composites from these polymers [9, 66] have been used for the synthesis of gels and also nanofibers for use as cell scaffolds.

Collagen and collagen-hydroxyapatite (HAP) electrospun composites were synthesized using PBS/ethanol solutions as solvent for electrospinning solutions [95]. The electrospinning solutions were prepared by blending hydroxyapatite sol with concentration of 0.055 g/mL with a 150 mg/mL collagen solution dissolved in a mixture of PBS 20X and ethanol (PBS to ethanol ratio of 3:2 v/v) at different hydroxyapatite to collagen mass ratios. The electrospun fibers were crosslinked with the EDC/NHS at room temperature. Pure collagen fibers exhibited an average diameter of 700 nm. While the average diameter of the electrospun composite membranes decreased with the increasing of the hydroxyapatite content, from 630 to 500 nm. The crosslinking enhanced the mechanical properties of the membranes specially in those reinforced with 30% hydroxyapatite which reached a tensile strength of 147 ± 0.12 MPa and elongation of 72.3 ± 1.48%. The biocompatibility of crosslinked electrospun composites was assessed by in vitro culture of human myeloma cells (U2-OS) for 7 days. In all scaffolds, the cell population increased with the culture time. After 5 days of culture the collagen/hydroxyapatite composite fibers supported cell proliferation better than the bare chitosan fibers.

Blends of biopolymers have been also assessed as scaffolds for tissue regeneration. An example of this were the collagen-chitosan scaffolds prepared by Chen and coworkers for endothelial cells (EC) and smooth muscle cells (SMC) culture [9]. Chitosan was incorporated to the nanofibers in concentrations of 20, 50 and 100 wt.%. Good adhesion and spreading of EC and SMC cells was observed after 6 days of culturing onto electrospun collagen-chitosan nanofibers. The cells appeared to migrate and proliferate onto the scaffolds surface to form a continuous monolayer. These also migrated through the pores and interacted with the surrounding fibers. Cell proliferation studies revealed that the electrospun collagen-chitosan nanofibers containing 20, 50 and 100 wt.% chitosan exhibited better cell attachment and proliferation for both cellular lines, which would be especially useful for tissue regeneration.

Zein was also used as copolymer in the fabrication of chitosan based scaffolds for L929 cell cultures [66]. Good attachment and spreading of the cells was observed in scaffolds containing 50 wt.% zein, this behavior was superior to the observed in the membranes containing 33 wt.% zein. This improvement was attributed to the higher stiffness exhibited by the membranes with high concentration of zein. However, the membranes containing 67 wt.% zein exhibited the lowest cell adhesion and cell viability, which was attributed to the lower surface wettability of this membrane.

Gelatin has been considered as a good alternative to collagen in tissue engineered applications. Gelatin scaffolds crosslinked with glutaraldehyde were used as templates for the culture of human dermal fibroblasts [67]. Its cytotoxic effect was assessed after 1, 3, 5 and 7 days. Cells growth was almost linear in the crosslinked scaffolds during a period of 7 days. However, growth inhibition attributed to the possible presence of remaining glutaraldehyde was observed at 1 day culture. Cell proliferation in gelatin crosslinked scaffolds was inferior than that observed in scaffolds of gelatin/PCL blends but slightly superior or comparable to the cell proliferation observed on the synthetic biodegradable scaffolds of PCL.

Electrospun PCL-gelatin composite nanofibers with different PCL: gelatin ratios (90:10, 80:20, 70:30, 60:40, 50:50 wt.%) were assessed as scaffolds for hMSCs cells cultures [69]. The hMSCs cells cultured onto the composite nanofibrous scaffolds exhibited attachment and a well spread morphology after 3 h of incubation at 37 °C, while the hMSCs cultured on electrospun PCL scaffolds showed a round morphology. After 24 h of incubation the formation of well-organized actin filaments was observed in the composite scaffolds. However, less organized actin filaments were found in PCL scaffolds after the same time of incubation. Higher proliferation of hMSCs was observed on composite scaffolds containing 30 and 40 wt.% gelatin compared to plain PCL scaffolds after five days of culture.

4.2 *Wound Healing Dressings*

Healing dressings must be able to protect the wounds from pollutants of the environment, to provide a physiologically moist microenvironment while keeping the oxygen flow to the wound site, and to minimize the risk of bacterial infection. It would be also desirable for the dressings to have additional healing features such as hemostatic effect.

Some bio-polymers used for wound dressings such as chitosan have reported bactericide effect [96]. However, this effect can be enhanced by functionalizing biopolymer-based nanofibers in several ways such as the impregnation with antibiotics, addition of bactericidal agents such as honey, plant extracts and essential oils, vitamins, or by the immobilization of metal and metal oxide nanoparticles.

Arkoun and co-workers [97] reported that chitosan-PEO electrospun nanofibers containing 80 wt.% of chitosan exhibited antimicrobial effect against several strains of gram positive and gram negative bacteria. This effect was mainly bactericide rather than a bacteriostatic and was independent of the gram classification but dependent of the strain type. The antibacterial effect of the membranes was attributed to the bacterial membrane disruption and perforation as can be seen in Fig. 4. The researchers suggest that the protonation of the amino groups in the membranes was responsible for the adhesion of the bacteria to the fibers surface as the first step of the bactericide mechanism.

Chitosan has also been blended with other biopolymers such as silk fibroin, and collagen [99] to improve its biocompatibility and mechanical resistance. Electrospun composite nanofibers from Chitosan–Silk fibroin containing chitosan in weight ratios of 0, 20, 50 and 80 wt.% were synthesized by Zeng and co-workers for wound healing applications [100]. The bactericide behavior of the membranes against *S. aureus* and *E. coli* was determined conducting turbidity measurements at 570 nm. While its biocompatibility with murine fibroblast was evaluated using the MTT assay in vitro and by the staining of the cells using hematoxylin and eosin. The

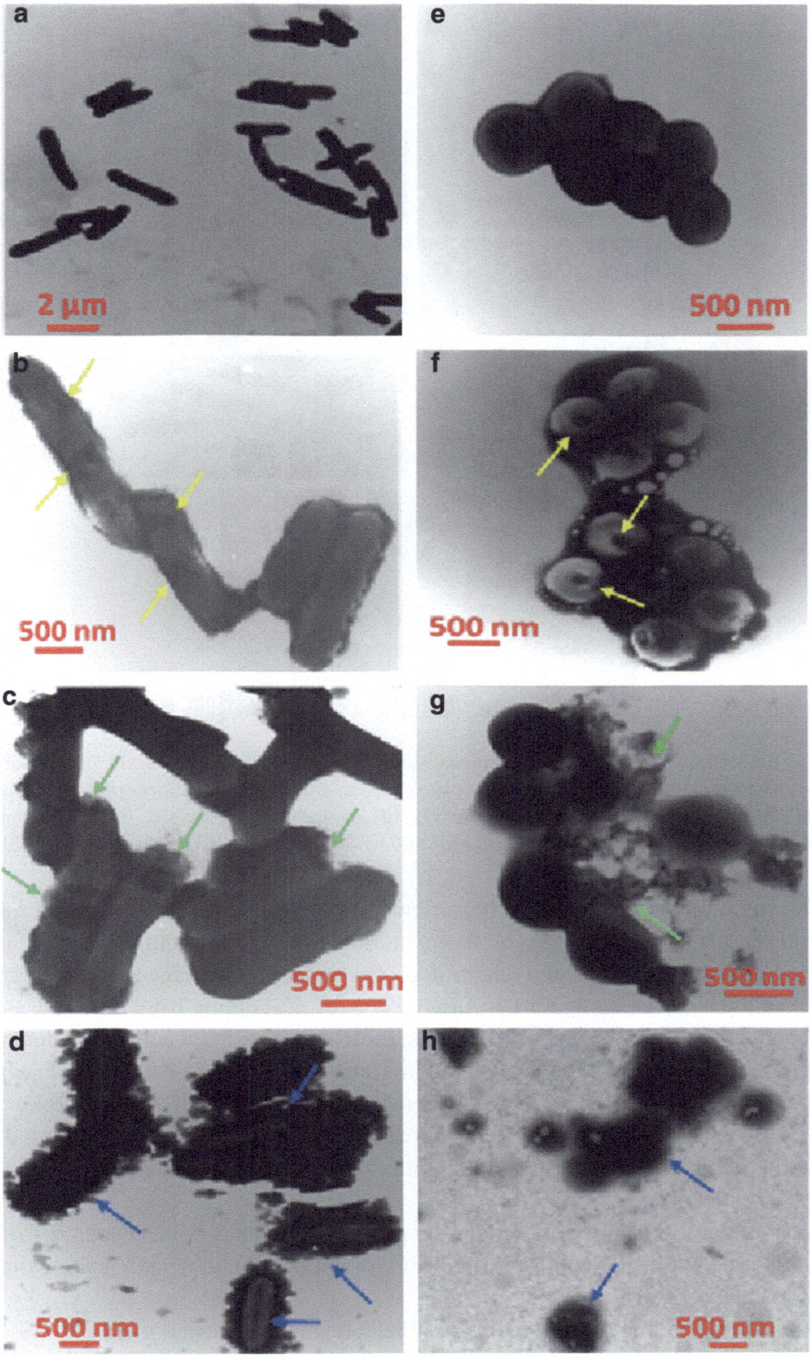

Fig. 4 Transmission electron microscopy (TEM) micrographs of *E. coli* treated with chitosan nanofibers during several period time: (**a**) 0 min, (**b**) 10 min, (**c**) 20 min, (**d**) 30 min. TEM micrographs of **Fig. 4** (continued) *S. aureus* treated with chitosan nanofibers during several period time: (**e**) 0 min, (**f**) 10 min, (**g**) 20 min and (**h**) 30 min. The yellow, green and blue arrows, point at membrane perforation, leakage of cytosol, and cell lysis respectively. Reproduced from [98]

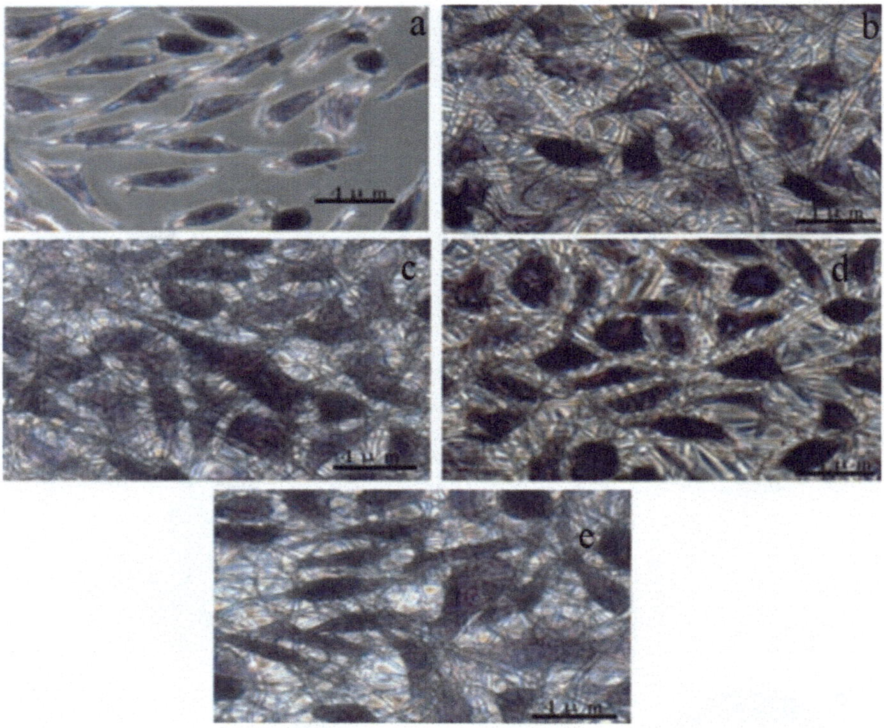

Fig. 5 Fibroblasts proliferation on composite chitosan-silk fibroin nanofibers, silk fibroin nanofibers and glass: (**a**) glass surface, (**b**) composite chitosan-silk fibroin nanofibers (80/20), (**c**) composite chitosan-silk fibroin nanofibers (50/50), (**d**) composite chitosan-silk fibroin nanofibers (20/80), (**e**) pure silk fibroin. (Eosin staining, ×400 light microscope). Reproduced from [100]

composite membranes exhibited bactericide effect against both bacteria types. For *E. coli* the bacterial growing inhibition was superior at higher chitosan concentrations. However, the chitosan content does not appear to influence the bactericide effect of the membranes against *S. Aureus*. The results coming from the MTT assay indicated that the composite membranes significantly promoted cell attachment and proliferation when compared to a plain glass surface, as can be seen in Fig. 5, where the cells growing onto the electrospun membranes exhibited a flattened morphology and polygonal shape typical of the adhered cells. The membranes containing 20% chitosan exhibited higher biocompatibility among the evaluated samples after 7 days of cells culture.

In another approach, electrospun composite membranes were prepared from a blend of type I collagen, chitosan, and PEO for application in wound healing dressing for skin regeneration [99].

The composite membranes with variable collagen to chitosan mass ratio of 1:3, 1:2, 1:1, 2:1, and 3:1 were used as templates for in vitro culturing of 3T3 fibroblasts. MTT tests were used to assess the cytotoxicity of the membranes. In the same way, in vivo tests were conducted in Sprague–Dawley rats, where dressings of electrospun membranes, gauze, and commercial collagen sponge were used to treat a wound inflicted on the back of the subject.

The cytotoxicity test conducted for 7 days shows that the composite membranes exhibited good biocompatibility and proliferation while the in vivo test results indicate that the wounds treated with the composite membranes exhibited better healing than gauze and collagen sponge. The wounds treated with the composite nanofibers were down to 95% after 21 days of treatment.

Functionalization using silver and silver ions has been extensively assessed as a way to enhance the antimicrobial effect of wound dressings [101, 102]. An example of this is the use of chitosan–Ag solution to impregnate alginate nanofibers [52]. The researchers assessed the bactericide effect of the coated membranes against *E. coli* and *S. aureus* by exposing the membrane to bacterial solutions with a concentration of 10^7 cfu/mL. During the first hours of treatment, the membranes inhibited the bacterial growth in 72% for gram negative and 98% for gram positive bacteria. While complete bactericide effect was reached for both bacteria types after 24 h of treatment. The bactericide effect was attributed to the release of silver ions into the bacterial solution trough the porous structure of the electrospun membranes. Additionally, measurements of the water vapor transmission rate (WVTR) for the alginate nanofibers and for the Ag impregnated nanofibers revealed that this parameter was within the range required for the treatment of injured skin, which ranges between 279 and 5138 g/m^2 day.

The Ag nanoparticle suspensions were also employed to impregnate electrospun silk fibroin nanofibers [103]. The functionalized fibers were evaluated as antimicrobial wound dressing against *S. Aureus* and *Pseudomona Aureginosa* (*P. Aureginosa).* The antibacterial effect was assessed by the disk diffusion method. The maximum bactericide effect against *S. Aureus* was achieved at Ag nanoparticles concentration of 1 mM. While the higher bactericide effect against *P. Aureginosa* was reached at Ag nanoparticles concentration of 2 and 4 mM. The bactericide behavior of the silk fibroin dressings was compared against commercial wound dressings, the last being more effective against both bacteria strains. However, the researchers attributed this result to the much higher Ag content in the commercial dressings, being 8000 ppm for Tegaderm Ag and 12,000 ppm for Aquacel Ag, in contrast to 432 ppm for the silk fibroin dressings.

Electrospun membranes of chitosan-PVA containing silver nanoparticles were assessed as antibacterial materials for biomedical applications by Liu and co-workers [104]. The antibacterial properties of the membranes against *S. aureus* and *E. coli* were assessed by mean of the disk diffusion test. Bare chitosan-PVA electrospun membranes exhibited an inhibition halo of 7.0 mm against *E. coli,* while the inhibition zones for the membranes containing 0.25%, 0.5% and 1 wt.% of silver nanoparticles were 7.7, 13.3 and 9.0 mm, respectively. On the other hand, the nanofibers containing 0.25 and 0.5 wt.% of Ag nanoparticles exhibited a higher bactericide effect against *S. aureus* than the observed against *E. coli.* The Silver ion release from the membranes was evaluated during 16 days by Inductively Coupled Plasma Atomic Emission Spectroscopy (ICP-AES). The results indicated that the electrospun membranes were able to release sufficient amounts of silver to exhibit a prolonged and sustained antibacterial activity, being 0.1 ppb the minimal concentration required to achieve bacteria inhibition.

Vitamins and compounds extracted from plants have been also evaluated as bactericide agents in wound dressings. Electrospun gelatin nanofibers containing vitamin A and E were assessed as antimicrobial dressings for wound healing [105]. Antibacterial tests demonstrated that the electrospun membranes loaded with vitamin E were effective bactericide agent for the growth inhibition of *E. coli* and *S. aureus*.

These exhibited a sustained release profile for 60 h.

Berberine is a compound extracted from plants that is used in the treatment of bacterial infections. This compound was loaded into electrospun collagen–zein membranes by Jiantao and co-workers to be assessed as bactericide dressings against *E. coli* and *S. aureus* [66]. The bactericide effect of the membranes was measured by the disk diffusion method. Only the electrospun nanofibers loaded with berberine exhibited an inhibition zone after 12 h of treatment for both bacterial strains. However, the inhibition zone was bigger in the electrospun membranes loaded with a higher dose of berberine. The in vivo wound healing properties of the composite membranes was evaluated in Sprague–Dawley rats. Electrospun collagen–zein membranes and cotton gauze were used to cover rectangular wounds of 1 cm^2 inflicted on the back of the subjects. The wounds covered with the electrospun membranes exhibited faster and better wound healing than normal cotton gauze and showed better fluid retaining. After 21 days treatment, the wound area retention was about 13% for the wounds covered with cotton gauze and 5% for the treated with electrospun membranes.

4.3 Water Disinfection

In addition to biomedical applications, electrospun nanofibers have potential use in the environmental field as promising materials for water remediation and disinfection. For water remediation these have been evaluated as filter membranes for micro and nano-filtration, as affinity membranes, as same as biocide agents for the treatment of water polluted with pathogenic microbes [31].

Micro and nano-filtration are based in the capability of porous membranes to retain insoluble particles, while allowing fluids to pass through their pores. The filtration capability depends on the pore size and pore density. In contrast, affinity membranes possess active sites that allow the selective immobilization and removal of contaminants from water.

Electrospun membranes for water filtration are usually fabricated from synthetic polymers. However, some of these require the use of harmful solvents for their synthesis and are not biodegradable. This has promoted the development of filtration membranes based on natural biopolymers as an alternative to synthetic polymer, despite their solubility in aqueous environments that remains as a limitation to broadening their use. Additionally, to its biodegradability, electrospun membranes from biopolymers retain in their structure functional groups that allow them to bind chemical compounds and in some cases exhibit biocide activity against pathogenic agents.

4.3.1 Pathogen Agents Treatment

Waterborne diseases are caused by water contaminated with pathogenic microorganisms, where water acts as carrier of infectious bacterial agents. Water scarcity and poor water sanitation are contributive factors to the proliferation of waterborne diseases [106].

Conventional water treatments such as activated carbon, biological treatment, oxidation and physicochemical treatments are not effective enough to treat polluted water containing pathogenic microorganisms [5].

Nanofibers from biopolymers such as chitosan exhibit intrinsic biocide activity against several microorganisms such as virus and bacteria [97]. However, the bactericide properties of biopolymers based membranes can be enhanced through the incorporation metal oxides [31], metal nanoparticles, or trough chemical modification of the polymer precursors.

Chitosan based nanofibers containing 4% of Ag and Ag/Fe nanoparticles were synthesized by photochemical reduction at 249 nm of electrospun membranes containing Ag^+ and Ag^+/Fe^{3+} ions. The fibers were irradiated in the 30–210 min range.

The bactericide effect of the composite membranes against *B. cereus, E. faecalis, E. coli, K. pneumoniae, K. oxytoca, P. aeruginosa, P. mirabilis, S. boydii, S. sonnei, and E. cloacae* strains was assessed by the disk diffusion method, while the Minimal Inhibitory Concentration (MIC) of the composite membranes was determined by the microdilution method.

Disk diffusion method was not useful in the determination of the bactericide effect; the authors attributed the lack of a noticeable inhibition disk to the incapability of the nanoparticles to diffuse from the membrane. However, some researchers have reported ion silver release from chitosan based nanofibers containing Ag nanoparticles [104]. In contrast, the microdilution method revealed that all composite membranes exhibited bactericide effect towards all the bacteria strains tested. However, the bactericide effect exhibited by the membranes containing 4% Ag, was similar to the exhibited by the membranes containing 2% Ag/2% Fe, which indicates that the Fe addition contributes to the bactericide effect. Bare chitosan nanofibers did not displayed significant bactericide behavior. Finally, the MIC was reported as 6.71 ppm for bare chitosan nanofibers, 0.27 ppm for the membranes containing Ag, and 0.27 ppm for the membranes containing Ag/Fe.

Chitosan modified electrospun nanofibers were synthesized by Mi and coworkers for adsorption and removal of *Porcine parvovirus* (*PPV*) and *Sindbis* virus from water [107]. Chitosan was functionalized with a quaternary amine to form quaternized chitosan *N*-[(2-hydroxyl-3-trimethylammonium) propyl] chitosan (HTCC), which was blended with PVA for the generation of electrospun membranes. The HTCC-PVA electrospun membranes were stabilized with glutaraldehyde exhibiting

30% of swelling after immersion in water for 6 h. The crosslinked membranes achieved a log removal value (LRV) of 3.3 for *PPV* and a 4.2 for *Sindbis*. The nanofibers exhibited higher virus removal than filter paper crosslinked with glutaraldehyde, which was used as control. These values are close to a LRV of 4, which is the stipulated by EPA in virus removal processes.

Biopolymers can be blended with synthetic polymers to improve their water stability, as in the case of Chitosan/Nylon 6 electrospun membranes which were prepared by Akram and co-workers for the disinfection and removal of metal ions from water [108]. The chitosan weight ratio in these composite fibers ranged from 0 to 30 wt.%. The researchers found that the chitosan addition improves the hydrophilicity and mechanical strength of the Nylon 6 nanofiber membranes. In the same way, the bactericide assessment conducted by the disk diffusion method showed that the composite electrospun membranes exhibited better bactericidal behavior against *E. coli* than the exhibited by the neat Nylon 6 membranes. Additionally, a rejection ratio of 87% against heavy metals ions was observed in the membranes containing 30 wt.% chitosan.

4.3.2 Pollutants Sorption

Biopolymers based electrospun nanofibers has been used as affinity membranes to adsorb and remove pollutants from water. Chitosan based nanofibers are commonly employed for the removal of heavy metal ions such as As^{5+}, Pb^{2+}, Cr^{6+} [109]. The mechanisms associated with the metal binding capability of chitosan are related with the presence of amino groups in its structure, which are able to donate an electron pair to metal ions. Wang and coworkers [110] suggested that in the formation of chitosan-metal complexes the metal ions can be connected to one or more chitosan chains trough amino groups and also by forming bridges between hydroxyl groups of neighbor chitosan molecules, as depicted in Fig. 6.

In this context electrospun chitosan nanofibers with average diameter of 235 nm were assessed as adsorbent material for the removal of Cu(II) and Pb(II) [109]. These chitosan membranes exhibited good stability in water presence and high adsorption affinity for the metal ions present in the aqueous solution. The adsorption of Cu(II) and Pb(II) followed a Langmuir model indicating that monolayer adsorption occurred on the nanofiber membranes. The equilibrium adsorption capacities (from Langmuir isotherm data) were 485.44 mg g^{-1} for Cu(II) and 263.15 mg g^{-1} for Pb (II).

Chitosan based electrospun membranes functionalized with Fe^{3+} ions were assessed as adsorbent materials for the removal of As (V) from water [13]. The membranes were constituted by nanofibers with average diameter ranging from 128 nm to 153 nm. Adsorption studies revealed that the chitosan modified electrospun membranes were highly effective for the adsorption of As (V) at neutral pH. The adsorption process followed a Freundlich model, with maximum adsorption capacity of approximately 11.2 mg/g at pH 7.2. The equilibrium was reached after 100 min of treatment. It was also found that the presence of coexisting anions such

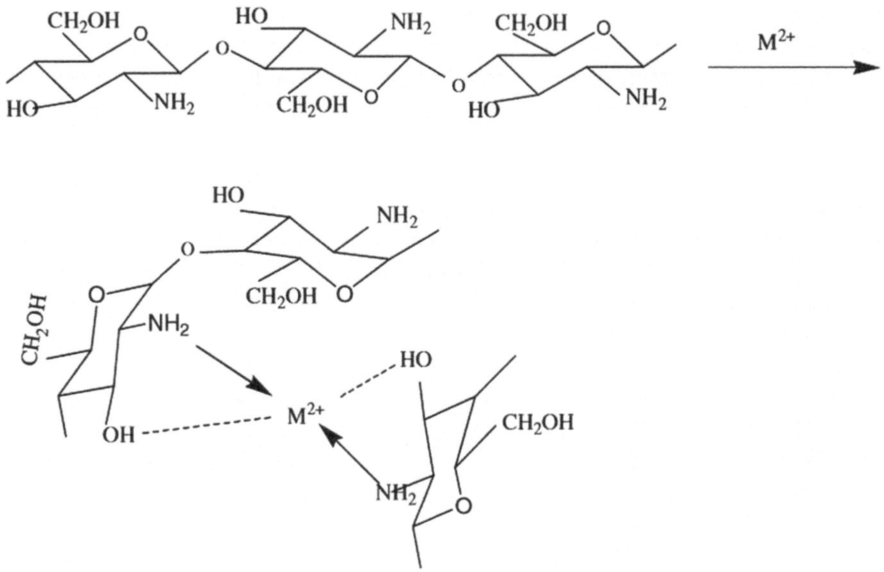

Fig. 6 Chitosan-metal complexation model proposed by Wang and coworkers. Reproduced from [110]

as Cl^{1-} and SO_4^{2-} have a negligible effect on the As (V) removal. However, the presence of PO_4^{3-} and SiO_3^{2-} reduced the As (V) adsorption because the competition for adsorption sites.

In a similar approach cerium (III) was used to functionalized chitosan based nanofibers for As (III) removal [111]. Ce was added to the nanofibers in weight concentrations ranging from 0.5 to 3.5 wt.%. Aqueous solutions containing As (III) in a concentration of 3.5 ppm were treated with the composite membranes in varied doses that ranged from 2 to 20 mg/50 mL of solution. Kinetic studies demonstrated that the adsorption of As (III) increases with the increasing of Ce content in the fibers, with the 80% of As being adsorbed between the first 10 min treatment. The adsorption process followed a Langmuir model with adsorption capacity of 18.0 mg/g. These electrospun membranes were able to remove As (III) from water below the limit concentration stablished by EPA of 15 ppm for this metal.

Alongside to chitosan other biopolymers such as cellulose have been also used as affinity membranes in water remediation. Electrospun cellulose-montmorillonite nanofibers were assessed as adsorbent materials for Cr^{6+} removal from polluted water [47]. Batch adsorption experiments were conducted by immersing electrospun cellulose - montmorillonite membranes in $K_2Cr_2O_7$ aqueous solution under constant agitation. The effects of the pH solution (pH 3–11), temperature (25–65 °C), and contact time (10–240 min) on the removal of Cr^{6+} were evaluated.

The adsorption capacity of electrospun composite nanofibers was higher than the exhibited by bare cellulose nanofibers at the same experimental conditions. The Cr^{6+} removal capability of the composite nanofibers decreased with the increasing of the initial pH, with optimal removal capability achieved at pH 3.0. The Cr^{6+} removal capability of the composite nanofibers increased with increasing in the contact time until reaching equilibrium approximately at 120 min. Overall the composite electrospun cellulose-montmorillonite nanofibers exhibited high selectivity toward Cr^{6+}.

Electrospun membranes of cellulose and chitosan were used as packed material for the removal of Cd, Pb, Cu, Cr and Ni from natural water and treated water [112]. The retention columns were able to pre-concentrate the metal ions present in the water samples in a 20×fold. The treated waste water exhibited a content of lead and copper of 217 μg/mL and 782 μg/mL respectively, while the river samples contained lead and copper in a concentration of 58 μg/mL and 81 μg/mL respectively.

Electrospun alginate-PEO nanofibers were evaluated as adsorbent material for the removal of Cu^{2+} from water [56]. The metal removal capability of the membranes was assessed at different temperatures (25, 40, and 60 °C and different pH values (2.0, 4.0, 6.0). The concentration of the membranes used in the experiment was 0.005 g/mL, while the concentration of the Cu^{2+} in the tested solutions ranged from 0 to 1000 mg/L. The adsorption tests revealed that the alginate-PEO membranes effectively removed Cu^{2+} ions from water with a maximum adsorption capability of 15.6 mg/g at 25 °C and a pH 4.0. The adsorption process followed a Langmuir model.

The use of biopolymer based nanofibers is not only limited to the removal of heavy metals from water, these had been also successfully used in the removal of organic compounds. As an example, cellulose nanofibers functionalized with TiO_2, TiO_2/Ag and TiO_2/Au were used in the photodegradation of metilene blue [113]. Under irradiation at 664 nm the membranes containing TiO_2/Ag and TiO_2/Au were able to degrade in 75% the organic compound. Additionally, the incorporation of Ag and Au contributes to minimize the mechanical deterioration in the composites membrane extending in this way its lifetime.

4.3.3 Filtration

The use of biopolymers in water filtration is mainly in the chemical modification of synthetic polymers to improve properties such as resistance to biofouling formation or impart hydrophilicity [31]. However, some researchers have reported the use of cellulose acetate based electrospun nanofibers for this application [114].

This ester derived from natural cellulose was used in the fabrication of electrospun membranes coated with 5% chitin nanocrystals for filtration applications [115]. Mechanical characterization revealed that the surface modification of the cellulose acetate nanofibers increased its strength by 131% and its stiffness by 340%. Additionally, the coated membranes also exhibited higher hydrophilicity. These presented a contact angle of 0°, compared to 132° for the uncoated membranes. The resistance to fouling formation of the coated membranes was determined after the filtration of bovine serum albumin and humic acid solutions at a flux of 14,217 L m^{-2} h^{-1} and 0.5 bar of pressure. The chitin coated membranes exhibited higher resistance to fouling formation compared to the bare cellulose acetate membranes. The researchers also reported reduction in biofouling and biofilm formation.

The efficiency of the filtration process not only relies on the pore size and pore distribution of the filters, but in its resistance to fouling formation. Chitosan has been incorporated into membrane filters to helps to reduce biofouling formation [116].

Chitosan - iron oxide nanoparticles were coated onto polyacrylonitrile hollow fibers to modify its antibacterial and antifouling properties [116]. *P. aeruginosa* and *S. aureus* were used in this research for their capacity to develop biofilms. The coated membranes demonstrated to be effective for the bacterial removal after 5 h of filtration. The bacterial removal efficiency was attributed in part to the positively charged surface of the ferrite nanoparticles. It was not found evidence of fouling formation after 10 h of treatment.

In an alternative approach chitosan-PCL electrospun membranes were used to filtrate particles sized in 300 nm as same as to remove *S. Aureus* from aqueous media [38]. The electrospun membranes containing 25% chitosan supported a water flux of 7000 L m^{-2} h^{-1} while preserving their integrity. The addition of chitosan in a 25 wt.% reduced by 50% the adhesion of *S. Aureus* to the membranes when compared to the neat PCL membranes.

5 Conclusion

Electrospinning is a manufacturing technique for the synthesis of tridimensional structures constituted by fibers. The morphological features and size of electrospun fibers are mainly influenced by processing, environmental and solution factors. However, its chemical behavior relies not only in the chemical properties of the fibers precursors, or in its functionalization treatments but also in the size of the fibers which could affect their reactivity.

The careful control of these factors during the synthesis allows to produce electrospun nanofibers for a wide number of applications including: tissue engineering, filtration, affinity membranes, disinfection, catalysis, sensors, protective clothing, drug delivery, wound healing, food packaging, among others.

The current environmental preservation necessity, along with the intrinsic properties of biopolymer based nanofibers has motivated its use in most of the aforementioned applications. However, their low stability in aqueous environments is a challenge to overcome in order to broadening its applications.

Several stabilization approaches have been used to overcome this problem. However, most of them exhibit toxicity at some extent or does not provide the adequate stabilization to the fibers. Crosslinking with glutaraldehyde remains as the more effective method regarding its reported cytotoxicity. Incorporating glutaraldehyde into the electrospinning solutions at low concentrations and assisting the crosslinking process with heat treatment are feasible strategies to minimize its cytotoxic effect.

Cytotoxicity could be also minimized by mean of greener synthesis process. For example, the use of mild toxic solvents such as acetic acid, ethanol, ethyl acetate or blends of those instead of the traditionally used TFA, THF, HFIP mainly used in the electrospinning of collagen and chitosan.

Functionalization by addition of appropriated co-polymers, active nanoparticles, antibiotics, vitamins, plant extracts, etc., or by chemical modification, are a widely used alternatives to improve or impart desired characteristics to the electrospun membranes. Some examples are the improvement of the bactericidal properties by immobilization of Ag nanoparticles, metal oxide nanoparticles, and antibiotics into the electrospun membranes, the generation of virus adsorbent chitosan membranes after its chemical modification with a quaternary amine, the improvement of cells proliferation in collagen and in gelatin scaffolds by addition of hydroxyapatite, or the improvement in the antifouling properties of filters after modification with chitosan, among others.

Despite the high extent of the literature focused in electrospinning of biopolymers and their applications, there is still room to research in critical subjects including electrospinning of biopolymers for water treatment, the development of formulations that allows to minimize the integrity loss of the native structure of biopolymers during processing, also in the use of modern characterization techniques that allows to elucidate the interactions between biopolymer based nanofibers, living organisms and environment.

References

1. Boakye M, Rijal N, Adhikari U, Bhattarai N (2015) Fabrication and characterization of electrospun PCL-MgO-keratin-based composite nanofibers for biomedical applications. Materials 8(7):4080–4095
2. Ramakrishna S, Fujihara K, Teo WE, Lim TC, Ma Z (2007) An introduction to electrospinning and nanofibers. World Scientific Publishing, Singapore
3. Sha M, El-chami I, Rintoul L, Bahreyni B (2017) Morphology of electrospun poly(ethylene oxide) ultra-fine fibres with incorporated MoO_3 nanoparticles. Mater Des J 113:76–83
4. Al Aani S, Wright CJ, Atieh MA, Hilal N (2016) Engineering nanocomposite membranes: addressing current challenges and future opportunities. Desalination 401:1–15
5. Schiffman JD, Schauer CL (2008) A review: electrospinning of biopolymer nanofibers and their applications. Polym Rev 48(2):317–352
6. Ahmed FE, Lalia BS, Hashaikeh R (2015) A review on electrospinning for membrane fabrication: challenges and applications. Desalination 356:15–30
7. Wang Z, Crandall C, Sahadevan R, Menkhaus TJ, Fong H (2017) Microfiltration performance of electrospun nanofiber membranes with varied fiber diameters and different membrane porosities and thicknesses. Polymer 114:64–72
8. Mohammad AW, Teow YH, Ang WL, Chung YT, Oatley-radcliffe DL, Hilal N (2015) Nano filtration membranes review: recent advances and future prospects. Desalination 356:226–254
9. Chen ZG, Wang PW, Wei B, Mo XM, Cui FZ (2010) Electrospun collagen-chitosan nanofiber: a biomimetic extracellular matrix for endothelial cell and smooth muscle cell. Acta Biomater 6(2):372–382

10. Ao C, Niu Y, Zhang X, He X, Zhang W, Lu C (2017) Fabrication and characterization of electrospun cellulose/nano-hydroxyapatite nanofibers for bone tissue engineering. Int J Biol Macromol 97:568–573
11. Yu D-G (2009) Electrospun nanofiber-based drug delivery systems. Health 1(2):67–75
12. Zhao Y, Qiu Y, Wang H, Chen Y, Jin S, Chen S (2016) Preparation of nanofibers with renewable polymers and their application in wound dressing. Int J Polym Sci 2016:4672839
13. Min L-L, Zhong L-B, Zheng Y-M, Liu Q, Yuan Z-H, Yang L-M (2016) Functionalized chitosan electrospun nanofiber for effective removal of trace arsenate from water. Sci Rep 6:32480. https://doi.org/10.1038/srep32480
14. Kanmani P, Jeyaseelan A, Kamaraj M, Sureshbabu P, Sivashanmugam K (2017) Environmental applications of chitosan and cellulosic biopolymers: a comprehensive outlook. Bioresour Technol 242:295–303
15. Kim C, Yang K-S, Lee W-J (2004) The use of carbon nanofiber electrodes prepared by electrospinning for electrochemical supercapacitors. Electrochem Solid-State Lett 7(11):A397–A399
16. Faccini M, Vaquero C, Amantia D (2012) Development of protective clothing against nanoparticle based on electrospun nanofibers. J Nanomater 2012:1–9. https://doi.org/10.1155/2012/892894
17. Lovell A (2015) Electrospun materials for hydrogen storage - from electrospinning for advanced energy and environmental applications. CRC Press, Boca Raton, pp 77–106
18. Haider A, Haider S, Kang I (2015) Review: a comprehensive review summarizing the effect of electrospinning parameters and potential applications of nanofibers in biomedical and biotechnology. Arab J Chem. https://doi.org/10.1016/j.arabjc.2015.11.015
19. Dadvar S, Tavanai H, Morshed M (2014) Fabrication of nanocomposite PAN nanofibers containing MgO and Al_2O_3 nanoparticles. Polym Sci Ser A 56(3):358–365
20. Beachleya V, Wena X (2009) Effect of electrospinning parameters on the nanofiber diameter and length. Mater Sci Eng C Mater Biol Appl 29(3):663–668
21. Zdraveva E, Pejnovic N, Mijovic B (2011) Electrospinning of polyurethane nonwoven fibrous mats. Scientists from the Faculty of Textile. Technology 1:55–60
22. Eleyas A et al (2017) The effect of flow rate, concentration, and voltage on diameter of pan precursor fiber by electrospinning technique. J Built Environ Technol Eng 2:71–78
23. A. G. Şener, A. S. Altay, and F. Altay (2011) Effect of voltage on morphology of electrospun nanofibers. 7th International Conference on Electrical and Electronics Engineering, vol January, p I324–I328
24. Rodoplu D, Mutlu M (2012) Effects of electrospinning setup and process parameters on nanofiber morphology intended for the modification of quartz crystal microbalance surfaces. J Eng Fiber Fabr 7(2):118–123
25. Heikkilä P, Harlin A (2009) Electrospinning of polyacrylonitrile (PAN) solution: effect of conductive additive and filler on the process. Express Polym Lett 3(7):437–445
26. Okutan N, Terzi P, Altay F (2014) Affecting parameters on electrospinning process and characterization of electrospun gelatin nanofibers. Food Hydrocoll 39:19–26
27. Tang XP, Si N, Xu L, Liu HY (2014) Effect of flow rate on diameter of electrospun nanoporous fibers. Therm Sci 18(5):1447–1449
28. Hasan M, Nayem KA, Hossain MB, Nahar S (2014) Production of tissue engineering scaffolds from poly caprolactone (PCL) and its microscopic analysis. Int J Text Sci 3(3):39–43
29. Nezarati RM, Eifert MB, Cosgriff-Hernandez E (2013) Effects of humidity and solution viscosity on electrospun fiber morphology. Tissue Eng Part C Methods 19(10):810–819
30. Nasefa MM, Abbasia A, Faridi-majidib R, Takeshic M (2012) The Effect of solution viscosity and concentration on morphological properties of electrospun nylon-6 nanofibers. CONCEPT 2012, vol 115, 2009–2010
31. Mokhena TC, Jacobs V, Luyt AS (2015) A review on electrospun bio-based polymers for water treatment. Express Polym Lett 9(10):839–880
32. Kusumastuti Y, Putri NRE, Dary AR (2016) Electrospinning optimization and characterization of chitosan/alginate/polyvinyl alcohol nanofibers. AIP Conf Proc 1755:150007

33. Bonino CA et al (2011) Electrospinning alginate-based nanofibers: from blends to cross-linked low molecular weight alginate-only systems. Carbohydr Polym 85(1):111–119
34. Rinaudo M (2006) Chitin and chitosan. Properties and applications. Prog Polym Sci 31(7):603–632
35. Ifuku S (2014) Chitin and chitosan nanofibers: preparation and chemical modifications. Molecules 19(11):8367–18380
36. Ahmed S, Ikram S (2016) Chitosan based scaffolds and their applications in wound healing. Achiev Life Sci 10(1):27–37
37. Salehi E, Daraei P, Arabi Shamsabadi A (2016) A review on chitosan-based adsorptive membranes. Carbohydr Polym 152:419–432
38. Cooper A, Oldinski R, Ma H, Bryers JD, Zhang M (2013) Chitosan-based nanofibrous membranes for antibacterial filter applications. Carbohydr Polym 92(1):254–259
39. Ali FA, Haider S, Al-masry W, Al-zeghaye Y Fabrication of chitosan nanofibers membrane and its treatment. College of Engineering, King Saud University
40. Geng X, Kwon OH, Jang J (2005) Electrospinning of chitosan dissolved in concentrated acetic acid solution. Biomaterials 26(27):5427–5432
41. Pouranvari S, Ebrahimi F, Javadi G, Maddah B (2016) Chemical cross-linking of chitosan/polyvinyl alcohol electrospun nanofibers. Mater Technol 50(5):663–666
42. Martinová L, Lubasová D (2008) Electrospun chitosan based nanofibers. Technical University of Liberec, Liberec
43. Lemma SM, Bossard F, Rinaudo M (2016) Preparation of pure and stable chitosan nanofibers by electrospinning in the presence of poly(ethylene oxide). Int J Mol Sci 17(11):E1790
44. Sun K, Li ZH (2011) Preparations, properties and applications of chitosan based nanofibers fabricated by electrospinning. Express Polym Lett 5(4):342–361
45. Su P et al (2011) Electrospinning of chitosan nanofibers: the favorable effect of metal ions. Carbohydr Polym 84(1):239–246
46. Li Q et al (2015) Genipin-crosslinked electrospun chitosan nanofibers: determination of crosslinking conditions and evaluation of cytocompatibility. Carbohydr Polym 130:166–174
47. Cai J et al (2017) Electrospun composite nanofiber mats of cellulose-organically modified montmorillonite for heavy metal ion removal: design, characterization, evaluation of absorption performance. Compos Part A 92:10–16
48. Ma Z, Kotaki M, Ramakrishna S (2005) Electrospun cellulose nanofiber as affinity membrane. J Membr Sci 265:115–123
49. de Oliveira Barud HG et al (2016) A multipurpose natural and renewable polymer in medical applications: bacterial cellulose. Carbohydr Polym 153:406–420
50. Ohkawa K (2015) Nanofibers of cellulose and its derivatives fabricated using direct electrospinning. Molecules 20(5):9139–9154
51. Jeong SI, Krebs MD, Bonino CA, Samorezov JE, Khan SA, Alsberg E (2011) Electrospun chitosan-alginate nanofibers with in situ polyelectrolyte complexation for use as tissue engineering scaffolds. Tissue Eng Part A 17:59–70
52. Mokhena TC, Luyt AS (2017) Electrospun alginate nanofibres impregnated with silver nanoparticles: preparation, morphology and antibacterial properties. Carbohydr Polym 165:304–312
53. Shalumon KT, Anulekha KH, Nair SV, Nair SV, Chennazhi KP, Jayakumar R (2011) Sodium alginate/poly(vinyl alcohol)/nano ZnO composite nanofibers for antibacterial wound dressings. Int J Biol Macromol 49(3):247–254
54. Xu W, Shen R, Yan Y, Gao J (2017) Preparation and characterization of electrospun alginate/PLA nanofibers as tissue engineering material by emulsion eletrospinning. J Mech Behav Biomed Mater 65:428–438
55. Lu JW, Zhu YL, Guo ZX, Hu P, Yu J (2006) Electrospinning of sodium alginate with poly(ethylene oxide). Polymer 4(23):8026–8031
56. Mokhena TC, Jacobs NV, Luyt AS (2017) Electrospun alginate nanofibres as potential biosorption agent of heavy metals in water treatment. Express Polym Lett 11(8):652–663

57. Simpson DG et al (2011) Electrospun collagen: a tissue engineering scaffold with unique functional properties in a wide variety of applications. J Nanomater 2011:1–15. https://doi.org/10.1155/2011/348268
58. Shoulders MD, Raines RT (2010) Collagen structure and stability. Annu Rev Biochem 78:929–958
59. Babitha S et al (2017) Electrospun protein nano fibers in healthcare: a review. Int J Pharm 523:52–90
60. Liu T, Teng WK, Chan BP, Chew SY (2010) Photochemical crosslinked electrospun collagen nanofibers: synthesis, characterization and neural stem cell interactions. J Biomed Mater Res Part A 95A(1):276–282
61. Li Y (2013) Electrospinning of core-shell collagen nanofibers. Electron Thesis Diss Repository 1:232–238
62. Fiorani A et al (2014) Comparative performance of collagen nanofibers electrospun from different solvents and stabilized by different crosslinkers. J Mater Sci Mater Med 2(10):2313–2321
63. Chen Z, Mo X, He C, Wang H (2008) Intermolecular interactions in electrospun collagen-chitosan complex nanofibers. Carbohydr Polym 72(3):410–418
64. Huang L, Nagapudi K, Apkarian RP, Chaikof EL (2001) Engineered collagen-PEO nano bers and fabrics. J Biomater Sci Polym 12(9):979–993
65. Xinzhe Z, Jing C, Lu W, Fujun W, Wenbin W (2015) Collagen-PEO composite nanofibers by electrospinning. Chem Fibers Int 65(1):50–51
66. Lin J, Li C, Zhao Y, Hu J, Zhang LM (2012) Co-electrospun nanofibrous membranes of collagen and zein for wound healing. ACS Appl Mater Interfaces 4(2):1050–1057
67. Zhang YZ, Venugopal J, Huang ZM, Lim CT, Ramakrishna S (2006) Crosslinking of the electrospun gelatin nanofibers. Polymer 47(8):2911–2917
68. Horuz Tİ, Belibağlı KB (2017) Production of electrospun gelatin nanofibers: an optimization study by using Taguchi's methodology. Mater Res Express 4(1):15023
69. Binulal NS, Natarajan A, Menon D, Bhaskaran VK, Mony U, Nair SV (2014) PCL-gelatin composite nanofibers electrospun using diluted acetic acid-ethyl acetate solvent system for stem cell-based bone tissue engineering. J Biomater Sci Polym Ed 25:325–340
70. Chen H, Jao W, Yang M (2009) Characterization of gelatin nanofibers electrospun using ethanol/formic acid/water as a solvent. Polym Adv Technol 20(2):98–103
71. Aoki H, Miyoshi H, Yamagata Y (2015) Electrospinning of gelatin nanofiber scaffolds with mild neutral cosolvents for use in tissue engineering. Polym J 47(3):267–277
72. Nguyen T, Lee B (2010) Fabrication and characterization of cross-linked gelatin. J Biomed Sci Eng 3:1117–1124
73. Righi TM, Almeida RS, d'Ávila MA (2012) Electrospinning of gelatin/PEO blends: influence of process parameters in the nanofiber properties. Macromol Symp 319(1):230–234
74. Linh NTB, Lee B-T (2012) Electrospinning of polyvinyl alcohol/gelatin nanofiber composites and cross-linking for bone tissue engineering application. J Biomater Appl 27(3):255–266
75. Zhuang X, Cheng B, Kang W, Xu X (2010) Electrospun chitosan/gelatin nanofibers containing silver nanoparticles. Carbohydr Polym 82(2):524–527
76. Kiernan J (2000) Formaldehyde, formalin, paraformaldehyde and glutaraldehyde: what they are and what they do. Micros Today 12:8–12
77. Sisson K, Zhang C, Farach-Carson MC, Chase DB, Rabolt JF (2009) Evaluation of cross-linking methods for electrospun gelatin on cell growth and viability. Biomacromolecules 10(7):1675–1680
78. Zhan J, Morsi Y, Ei-Hamshary H, Al-Deyab SS, Mo X (2016) In vitro evaluation of electrospun gelatin-glutaraldehyde nanofibers. Front Mater Sci 10(1):90–100
79. Laha A, Sharma CS, Majumdar S (2016) Electrospun gelatin nanofibers as drug carrier: effect of crosslinking on sustained release. Mater Today Proc 3(10):3484–3491
80. Panzavolta S, Gioffrè M, Focarete ML, Gualandi C, Foroni L, Bigi A (2011) Electrospun gelatin nanofibers: optimization of genipin cross-linking to preserve fiber morphology after exposure to water. Acta Biomater 7(4):1702–1709

81. Rabiatul AR, Lokanathan Y, Rohaina CM, Chowdury SR, Aminuddin BS, Ruszymah BHI (2014) Collagen coated PMMA nanofiber for the development of respiratory. Regener Res 3(2):86–87
82. Butler MF, Ng YF, Pudney PDA (2003) Mechanism and kinetics of the crosslinking reaction between biopolymers containing primary amine groups and genipin. J Polym Sci Part A Polym Chem 41(24):3941–3953
83. Fernandes M, Gonçalves IC, Nardecchia S, Amaral IF, Barbosa MA, Martins MCL (2013) Modulation of stability and mucoadhesive properties of chitosan microspheres for therapeutic gastric application. Int J Pharm 454(1):116–124
84. Stone SA, Gosavi P, Athauda TJ, Ozer RR (2013) In situ citric acid crosslinking of alginate/polyvinyl alcohol electrospun nanofibers. Mater Lett 112:32–35
85. Shi J, Yang E (2016) Post-electrospinning crosslinking of guar/polyvinyl alcohol membrane. Therm Sci 20(1):1–5
86. Jiang Q, Reddy N, Yang Y (2010) Cytocompatible cross-linking of electrospun zein fibers for the development of water-stable tissue engineering scaffolds. Acta Biomater 6(10):4042–4051
87. Peresin MS et al (2014) Crosslinked PVA nanofibers reinforced with cellulose nanocrystals: water interactions and thermomechanical properties. J Appl Polym Sci 131(11):1–12
88. Sood A, Granick MS, Tomaselli NL (2014) Wound dressings and comparative effectiveness data. Adv Wound Care 3(8):511–529
89. Aadil KR, Prajapati D, Jha H (2016) Improvement of physcio-chemical and functional properties of alginate film by Acacia lignin. Food Packag Shelf Life 10:25–33
90. Wan LQ, Jiang J, Arnold DE, Guo XE, Lu HH, Mow VC (2009) Calcium concentration effects on the mechanical and biochemical properties of chondrocyte-alginate constructs. Cell Mol Bioeng 1(1):93–102
91. Barnett SE, Varley SJ (1987) The effects of calcium alginate on wound healing. Ann R Coll Surg Engl 69(4):153–155
92. Groves AR, Clawrence J (1986) Alginate dressing site haemostat. Ann R Coll Surg Engl 68:1–2
93. Jeong SI, Krebs MD, Bonino CA, Khan SA, Alsberg E (2010) Electrospun alginate nanofibers with controlled cell adhesion for tissue engineering. Macromol Biosci 10(8):934–943
94. Oktay B, Kayaman-Apohan N, Erdem-Kuruca S (2014) Fabrication of nanofiber mats from electrospinning of functionalized polymers. IOP Conf Ser Mater Sci Eng 64:12011
95. Zhou Y, Yao H, Wang J, Wang D, Liu Q, Li Z (2015) Greener synthesis of electrospun collagen/hydroxyapatite composite fibers with an excellent microstructure for bone tissue engineering. Int J Nanomedicine 10:3203–3215
96. Goy RC, de Britto D, Assis OBG (2009) A review of the antimicrobial activity of chitosan. Polímeros 19(3):241–247
97. Arkoun M, Daigle F, Heuzey MC, Ajji A (2017) Mechanism of action of electrospun chitosan-based nanofibers against meat spoilage and pathogenic bacteria. Molecules 22(4):E585
98. Arkoun M, Daigle F, Heuzey MC, Ajji A (2017) Antibacterial electrospun chitosan-based nanofibers: a bacterial membrane perforator. Food Sci Nutr 5:865–874. https://doi.org/10.1002/fsn3.468
99. Chen JP, Chang GY, Chen JK (2008) Electrospun collagen/chitosan nanofibrous membrane as wound dressing. Colloids Surfaces A Physicochem Eng Asp 314:183–188
100. Cai ZX et al (2010) Fabrication of chitosan/silk fibroin composite nanofibers for wound-dressing applications. Int J Mol Sci 11(9):3529–3539
101. Kumar-Krishnan S et al (2015) Chitosan/silver nanocomposites: Synergistic antibacterial action of silver nanoparticles and silver ions. Eur Polym J 67:242–251
102. Li C et al (2013) Silver nanoparticle/chitosan oligosaccharide/poly (vinyl alcohol) nanofibers as wound dressings: a preclinical study. Int J Nanomedicine 8:4131–4145. https://doi.org/10.3390/molecules22040585
103. Uttayarat P, Jetawattana S, Suwanmala P, Eamsiri J, Tangthong T, Pongpat S (2012) Antimicrobial electrospun silk fibroin mats with silver nanoparticles for wound dressing application. Fibers Polym 13(8):999–1006

104. Liu Y, Liu Y, Liao N, Cui F, Park M, Kim HY (2015) Fabrication and durable antibacterial properties of electrospun chitosan nanofibers with silver nanoparticles. Int J Biol Macromol 79:638–643
105. Li H et al (2016) Electrospun gelatin nanofibers loaded with vitamins A and E as antibacterial wound dressing materials. RSC Adv 6(55):50267–50277
106. Nel LH, Markotter W (2004) In: Cloete T, Joan B, Rose L, Nel H, Ford T (eds) Microbial Waterborne Pathogens. IWA Publishing, London, p 55
107. Mi X, Vijayaragavan KS, Heldt CL (2014) Virus adsorption of water-stable quaternized chitosan nanofibers. Carbohydr Res 387(1):24–29
108. Jabur AR, Abbas LK, Moosa SA (2016) Fabrication of electrospun chitosan/nylon 6 nanofibrous membrane toward metal ions removal and antibacterial effect. Adv Mater Sci Eng 2016:1–10
109. Haider S, Park SY (2009) Preparation of the electrospun chitosan nanofibers and their applications to the adsorption of Cu(II) and Pb(II) ions from an aqueous solution. J Membr Sci 328(2):90–96
110. Wang X, Du Y, Fan L, Liu H, Hu Y (2005) Chitosan- metal complexes as antimicrobial agent: synthesis, characterization and structure-activity study. Polym Bull 55:105–113
111. Sharma R, Singh N, Gupta A, Tiwari S, Tiwari SK, Dhakate SR (2014) Electrospun chitosan–polyvinyl alcohol composite nanofibers loaded with cerium for efficient removal of arsenic from contaminated water. J Mater Chem A 2(39):16669–16677
112. Musyoka SM, Ngila JC, Mamba BB (2013) Remediation studies of trace metals in natural and treated water using surface modified biopolymer nanofibers. Phys Chem Earth 66:45–50
113. Snyder A, Bo Z, Moon R, Rochet JC, Stanciu L (2013) Reusable photocatalytic titanium dioxide-cellulose nanofiber films. J Colloid Interface Sci 399:92–98
114. Omollo E, Zhang C, Mwasiagi JI, Ncube S (2016) Electrospinning cellulose acetate nanofibers and a study of their possible use in high-efficiency filtration. J Ind Text 45(5):716–729
115. Goetz LA, Jalvo B, Rosal R, Mathew AP (2016) Superhydrophilic anti-fouling electrospun cellulose acetate membranes coated with chitin nanocrystals for water filtration. J Membr Sci 510:238–248
116. Mukherjee M, De S (2017) Investigation of antifouling and disinfection potential of chitosan coated iron oxide-PAN hollow fiber membrane using gram-positive and gram-negative bacteria. Mater Sci Eng C 75:133–148

Biomaterials Produced via Green Electrospinning

Liliana Liverani, Lars Vester, and Aldo R. Boccaccini

Abstract Green electrospinning, starting from its definition, achieved great interest in the scientific community for the fabrication of electrospun products for biomedical, filtration, pharmaceutical and cosmetics applications. The use of green electrospinning results particularly relevant for the fabrication of biomaterials, in particular for tissue engineering and regenerative medicine applications. The focus of the present book chapter is on the definition of green electrospinning and the description of the pivotal parameters affecting the electrospinning process with particular focus on the solvent selection. The process parameters and the potential final applications of obtained electrospun mats starting from natural and synthetic polymers, their blends and composites will be also reported and highlighted.

Keywords Electrospinning • Green electrospinning • Benign solvents • Biomaterials • Electrospun scaffolds

1 Introduction

The electrospinning process is a technique used for the fabrication of nano and micro-sized fibrous mats, widely used in the biomedical field, as scaffolds for tissue engineering (TE), regenerative medicine applications [1], and other applications in pharmaceutical, cosmetic, food packaging, filtration, sensor and agriculture fields [2].

For applications in the biomaterials field, there are electrospun products available on the market, such as coronary balloon expandable stent system (Papyrus, Biotronik AG and BiowebTM, Zeus Inc.), vascular access graft (AVflo™, Nicast Ltd), dural patch (ReDuraTM, Medprin Regenerative Medical Technologies Co. Ltd) and synthetic bone (Rebossis, Ortho ReBirth Co. Ltd).

In the electrospinning process, fiber formation is driven by electrostatic force, starting from a polymeric solution or a melt. The process is regulated by the application of high voltage between two electrodes with electric charges of opposite

L. Liverani (✉) • L. Vester • A.R. Boccaccini
Institute of Biomaterials, Department of Materials Science and Engineering, Friedrich-Alexander-University Erlangen-Nuremberg, Erlangen, Germany
e-mail: liliana.liverani@fau.de

J. Almodovar (ed.), *Electrospun Biomaterials and Related Technologies*,
DOI 10.1007/978-3-319-70049-6_5

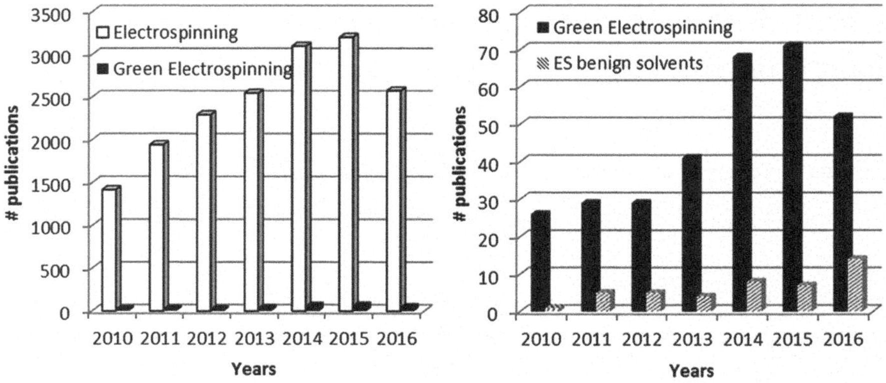

Fig. 1 Bar chart illustrating annual research number of published papers in the field of electrospinning, green electrospinning and benign solvents for electrospinning in article title, abstract or keywords between 2010 and 2016 (Search data from http://www.scopus.com using the keywords electrospinning, green electrospinning and benign solvents for electrospinning. Updated October 2016)

polarity. In fact, the surface tension of the charged polymeric solution is overcome by the electrostatic forces. Fibers are collected on the grounded (or negative charged) target after the polymer jet is ejected from the Taylor cone (conical shape instability region due to drop elongation, related to the electrical potential). During the time of flight between the needle and the fiber collector, the solvent evaporates and polymeric fibers are collected on the target. It could happen that solvents residuals are still detected in the obtained fibrous mats, which should be treated in order to remove or neutralize these residuals. Considering that the most commonly used solvents for the electrospinning are organic toxic solvents, the presence of these residuals will affect the field of application of the obtained electrospun mats, which are not suitable for any biomedical applications.

The type of solvents used for the electrospinning and their toxicity could represent a pivotal issue not just for what concerns final application of the obtained electrospun mats (i.e. biomedical or pharmaceutical applications), but also for the safety and environmental impact related to the use of these solvents. For this reason, the concept of "Green Electrospinning" has been introduced. In 2010, Green Electrospinning was defined as "an approach which would alleviate concerns regarding safety, toxicology and environmental problems and it could be accomplished by electrospinning from aqueous suspensions" [1, 2].

In the last years, an increasing number of research papers have been focused on electrospinning, also an increasing number of publications deal with green electrospinning. However, the number of these publications represents a low percentage of the overall number of publications on electrospinning, as reported in Fig. 1. Even lower is the amount of publications dealing with the benign solvents for the electrospinning (see Sect. 3.1), but the trend is increasing in the last year.

This book chapter is focused on the rationale of the use of green electrospinning and its applications in the biomedical field. In details, devoted paragraphs will report the processing of natural and synthetic polymers, as well as composites. Post-processing of the obtained electrospun samples will be the focus of a separated paragraph. Conclusions and future perspectives of the applications of green electrospinning will be also reported.

2 Electrospinnability

Many different parameters affect the electrospinning process. They could be grouped in three categories: solution property parameters, electrospinning process parameters and environmental parameters, briefly summarized in Table 1.

It is possible to define "electrospinnability" as the regime, related to the solution properties, in which continuous uniform fibers are obtained [3]. It was reported that electrospinnability correlates with the polymeric solution viscoelastic properties [4, 5]. In particular, this evidence has been widely reported for natural polymers, which have the tendency to form viscous gel-like solutions at low concentrations, inhibiting the possibility for these solutions to be processed by electrospinning [4]. Moreover, it has been also proven that there is a correlation between chain entanglements in the polymer solution and electrospinnability [6]. The degree of chain entanglements is affected by the length of polymer molecular chain and the conformation of polymer chain in solution. These characteristics are related to the solvents quality in polymer solution rheology. In particular, it is possible to define good solvents, those in which polymer–solvent interactions are favored, while in poor solvents mainly polymer–polymer self-interactions are favored. It is relevant to underline that the correlation reported by Shenoy et al. is valid only for good solvents and under the hypothesis that the polymer–polymer interactions are negligible.

Table 1 Summary of the classification of the main parameters affecting the electrospinning process

Parameter group	Parameter list
Solution property parameters	Selection of suitable solvent (or solvent system), polymeric solution concentration, addition of organic particles and/or drugs, solution conductivity, viscosity, density, surface tension, solvent volatility and dielectric effect
Electrospinning process parameters	Applied voltage, feed rate of polymeric solution, solution temperature, capillary-to-collector distance, effect and type of collector, needle diameter and the configuration of the nozzle Additional tools added to the standard setup and their settings (i.e. accessory for the stabilization of Taylor Cone by solvent vaporization close to the needle; parameter to regulate: solvents system and inert gas flow rate)
Environmental parameters	Temperature, humidity, pressure and type of atmosphere

Recently, LeCorre-Bordes et al. reported that, by using a new spinneret design, it is possible to improve the electrospinnability of high-viscosity polymers, by applying shear forces to the polymer solution at the spinneret. In this way, they demonstrated the electrospinnability of solutions that are not spinnable with a standard setup [4].

Recently, several works have been focused on the use of novel techniques in order to make electrospinnable solutions that are usually not spinnable with the standard electrospinning setup, like using atmospheric pressure non-equilibrium plasma treatment [7], introduction of salt additives [8] or by the addition of bifunctional hydrogen-bonding molecules, allowing the formation of crosslinks in the electrospinning solution, improving the electrospinnability of low molecular weight polymers [9].

3 Effect of Solvent Selection

Among the other parameters affecting the obtained electrospun mats, the selection of a proper solvent plays a pivotal role [10], in particular the two solvents properties which affect mostly the electrospinning process are boiling point and the dielectric constant. In fact, these characteristics could promote solvent evaporation during the time of flight and could be triggered by solvent mixtures in order to obtain porosity on the fiber surface [11]. Some research works in the literature have focused on the investigation of the effects using different solvents on the electrospinning of a target polymer, the most investigated one being polystyrene, poly (ε-caprolactone) (PCL) and poly lactic acid [12–15]. In particular for PCL, considering the relevance of polymer solubility in correlation with the polymeric solution electrospinnability, Luo et al. developed a map, reported in Fig. 2, which include the investigations of 32 solvents for PCL and the electrospinnability of the relative solutions, providing an effective tools for the development of optimized solution for the electrospinning [14].

Beside the extensive evaluation of different solvents, statistical Taguchi design of experiments (DoE) was used to analyze the effects of polymeric solution concentration and electrospinning parameters on polymethyl methacrylate (PMMA) fibers morphology to obtain optimized process conditions for getting the smaller fiber diameter possible. These conditions were evaluated, by using the following electrospinning variables: voltage, polymeric solution feed rate, tip-to-collector distance, needle diameter and polymer solution concentration. The smallest fiber diameter (about 228 nm) was obtained with the following values of the electrospinning variables: 20 kV, 1 mL/h, 15 cm distance, 19 needle gauge and 15 wt% polymer concentration. Among these variables, the most significant factor on fiber diameter was the polymeric solution concentration. The experimental results were closed to the predicted values, demonstrating the validity of the Taguchi DoE [16].

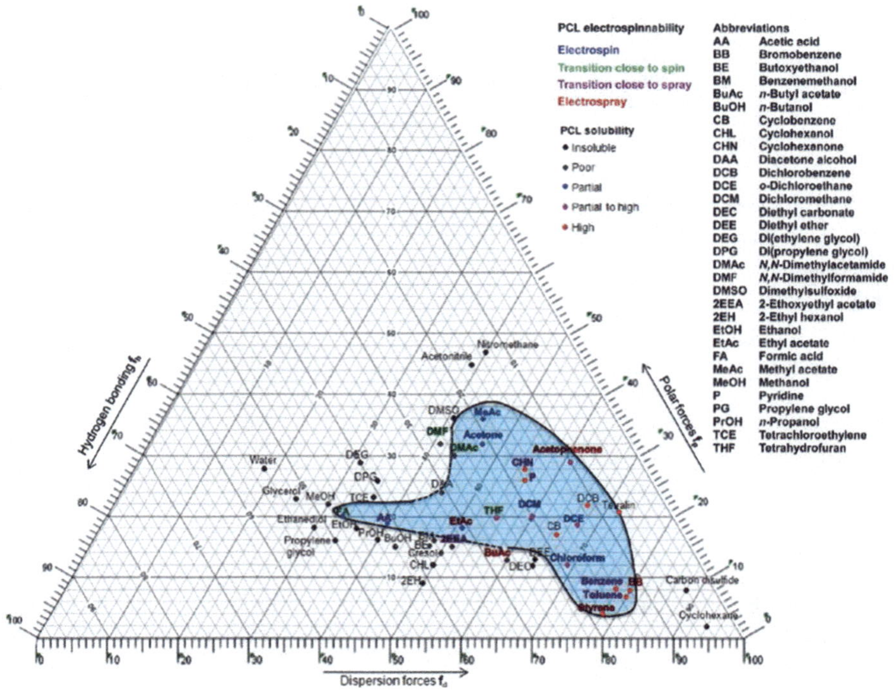

Fig. 2 Map based on the Teas graph of electrospinnability–solubility for PCL in 32 different solvents. Reprinted with permission from Luo et al. [14]. Copyright 2012 American Chemical Society

3.1 Solvent Classification and Evaluation of Solvents Residuals in Electrospun Mats

Besides, the relevance of the solvent selection and its influence on the fibers morphology, the concerns about solvent toxicity, in particular for biomedical applications of the electrospun mats, have been a pivotal topic in the last years [17]. In fact, Xie et al. stated that the residual toxic solvents in the electrospun fibers should be removed, by post-processing the samples (i.e. freeze drying or drying under vacuum) and the content of these residuals should be less than the exposure limits reported in the regulations [17].

According to the classification reported in ICH (International Conference on Harmonisation of Technical Requirements for Registration of Pharmaceuticals For Human Use) harmonised tripartite guideline – Q3C: Impurities: guideline for residual solvents, there are 3 classes of solvents [18].

These three classes are grouped as follows: Class 1 (Solvents to Be Avoided, related to known human carcinogens, strongly suspected human carcinogens, and environmental hazards), Class 2 (Solvents to Be Limited, related to non-genotoxic animal carcinogens or possible causative agents of other irreversible toxicity such as neurotoxicity or teratogenicity. Solvents suspected of other significant but

Table 2 Most used electrospinning solvents properties [19] and class, according to ICH guidelines [18]

	Boiling point [°C]	Dielectric constant [ε, 25 °C]	Solvent class according to ICH [18]
Dichloromethane	40	9.1	Class 2
Chloroform	62	4.8	Class 2
Dimethylformamide	153	37	Class 2
Tetrahydrofuran	66	7.6	Class 2
Methanol	64.7	33	Class 2
Ethyl acetate	77.1	6	Class 3
Ethanol	78.3	25	Class 3
Acetic acid	117.9	6	Class 3
Formic acid	100.6	59	Class 3
Acetone	56.3	21	Class 3
Dimethylsulfoxide	189	47	Class 3
Water	100	79	–

reversible toxicities) and Class 3 (Solvents with Low Toxic Potential, including solvents with low toxic potential to man and no health-based exposure limit is needed. Class 3 solvents have a permitted daily exposure (PDE) of 50 mg or more per day) [18].

In Table 2, the solvents properties and class are briefly summarized, according to ICH, for the most used electrospinning solvents. Most of the solvents used for the electrospinning belong to class 1 or 2, like chloroform and dichloromethane, while acetic acid and formic acid belong to class 3. For this reason, the solvents belonging to class 3 could also be defined "benign solvents" for electrospinning and their use is increasing in the last years, as already reported in Fig. 1.

In particular, when toxic solvents are used, but also in case of benign solvents, the detection of possible solvents residuals in the electrospun mats is crucial. Several techniques could be used for the evaluation of solvents residual, such as Fourier transform infrared (FTIR) analysis [20, 21], scanning electron microscopy (SEM) analysis to evaluate the effects on fiber surface due to the different rate of solvent evaporation [22] and electro-spray ionization mass spectroscopy, able to provide a quantitative measurement of the solvent content in the electrospun mats [23]. In particular, in this research work developed by Nam et al., a correlation between the solvent retention and electrospun polymer was highlighted. In fact, PCL resulted more stable with less retained solvent, having less affinity with the solvent 1,1,1,3,3,3-hexafluoro-2-propanol (HFIP), respect to gelatin; while in the blend PCL-gelatin samples, the amount of retained solvent increased with the increase of the amount of gelatin in the blend [23].

4 Emulsion and Suspension Electrospinning for Biomedical Applications

Emulsion and suspension electrospinning has been already correlated to green electrospinning in literature by Agarwal and Greiner [24]. Emulsion electrospinning allows encapsulation and sustained release of water soluble drugs in hydrophobic/amphiphilic polymers [25]. In fact, by using this technique it has been reported the successful encapsulation of nerve growth factor (NGF) [26], proteinase K [27] and lysozyme [28]. The process is constituted by emulsification to form water/oil emulsion, dissolution of polymer (suitable for the electrospinning process) and the electrospinning of the obtained emulsion. This technique still required the use of toxic solvents, which even if diluted, could represent a critical issue for biomedical applications.

A promising area closer to the green electrospinning concept is represented by suspension electrospinning [24]. It is based on the electrospinning of a suspension obtained by emulsion polymerization (primary latex suspension) mixed with a small amount of fiber forming water-soluble polymer (like PEO), used as template and removed after the obtainment of the electrospun mats [29]. Some disadvantages are also reported by the use of this technique, like the use of toxic tensides and the limited number of polymers from which it is possible to synthesize primary lattices suitable for electrospinning. At the same time, it is very interesting the electrospinning of water-based secondary suspensions, namely dispersion in water of water-insoluble polymers, of biodegradable polyesters [2]. The requirements for this process are the use of water-insoluble polyester with low melting point and the chemical attachment of water-soluble polymer to the polyester block, to ensure the suspendability of the polyester in water, suitable for the electrospinning process. In Table 3, a summary of main applications of emulsion and suspension electrospinning, including details about polymers, solvents, fiber diameter, in vitro tests and applications is reported.

5 Electrospun Fiber Mats from Benign Solvents

In the last years, the use of benign solvents for electrospinning was focused mainly on the electrospinning of collagen, because benign solvents could preserve the collagen triple helical structure [33, 34]. In fact, Zeugolis et al. [35] demonstrated that the use of fluoroalcohols (i.e. 1,1,1,3,3,3-hexafluoro-2-propanol (HFIP) or 2,2,2-trifluoroethanol (TFE)) as electrospinning solvents, lead to the denaturation of collagen, and formation of gelatin, characterized by destroyed α-chain and disrupted triple-helical structure. Similar results were obtained also by the co-electrospinning of collagen with PCL with the same solvents [35]. Also other natural polymers, like gelatin, silk, chitosan, soy and zein have been successfully electrospun using benign solvents as, ethanol, acetic acid and formic acid [36–38]. In

Table 3 Summary of polymers or blends, solvent systems, fiber diameter, crosslinking process, in vitro tests, indication of application in tissue engineering (TE) field and reference, fabricated by emulsion or suspension electrospinning

Polymer or blend	Solvent system	Average fiber diameter [nm]	Emulsion or suspension electrospinning	In vitro tests (cell type)	Indication of application	Reference
PCL/Polyvinyl alcohol (PVA)	Water/ toluene	600–1000	Emulsion	Human breast cancer cell line (MCF7)	TE applications	Pal et al. [30]
PLACL/NGF	Chloroform/ PBS	Without protein: 910 ± 20 With protein: 610/630 ± 20	Emulsion	Rat pheochromocytoma cells (PC12)	Bio-functional TE scaffold	Li et al. [26]
PS/BSA	L-Limonene/ water	PS_{75k} 596 ± 58, PS_{192k} 868 ± 96, PS_{280k}1175 ± 106, PS_{350k} 1385 ± 162	Emulsion	–	–	Wang et al. [31]
Polypropylene carbonate/ polyethylene oxide (PEO)/ sericin	Ethyl acetate/ water	~1000	Emulsion	MDA-MB-231 breast cancer cells and KELLY neuroblastoma cells	TE applications	Spano et al. [32]
Poly(ethylene glycol)-poly(L-lactide) Proteinase K	Chloroform/ water	30–300	Emulsion	–	–	Li et al. [27]
Poly(hexamethylene adipate)-PEO (PHA-b-PEO)	Water	350–550	Suspension	–	TE applications	Sun et al. [2]

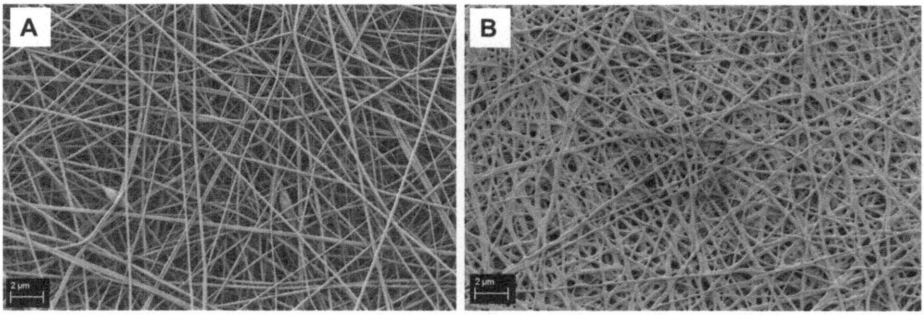

Fig. 3 SEM micrographs of electrospun silk fibroin in formic acid (12 wt%) as-spun (**a**) and after immersion in methanol (**b**), (electrospun mats fabricated and characterized by the authors, unpublished data, with an optimized protocol from Min et al. [37])

particular, several works, even before the introduction of the term "Green Electrospinning", were focused on the electrospinning of silk fibroin and the avoidance of solvents like HFIP and hexafluoroacetone (HFA) [39]. In fact, it is possible to obtain electrospun neat silk fibroin fibers using formic acid, as obtained by the authors and reported in Fig. 3, by optimizing the protocol reported by Min et al. [37], and even the use of water was reported as solvents for the blend of silk fibroin and PEO [40].

A summary of polymers or blends, solvents used, crosslinking process, cell line for the in vitro tests and indication of the TE field of application is reported in the following Table 4.

Even if only few studies have been published investigating the electrospinning of polyesters using benign solvents, like PCL and PLLA [53, 54, 60, 61], recently there is an increased interest in the fabrication of electrospun scaffolds of their blends with natural polymers. Positive results were reported from in vitro and in vivo studies on the electrospun mats obtained using benign samples. One example was reported by Da Silva et al. [55]. In fact they reported promising results from the implant of PCL electrospun nanofibers into the vitreous cavity of female Lewis rat for 10 days, as showed in Fig. 4.

6 Composites Electrospun Scaffolds from Benign Solvents

In order to well describe the topic of this paragraph, it is mandatory to define a composite as a material composed by a "matrix" and a "reinforcement" phase, usually in the form of particles, incorporated in the matrix. There are not molecular bonding between the two phases, but the interface between them plays a pivotal role in the definition of the composite properties [62].

Currently, it has not been widely investigated the possible interaction of inorganic particles and the solvents for electrospinning, in the same papers dealing with the incorporation of inorganic particles in polymeric matrix before the electrospinning

Table 4 Summary of polymers or blends, solvent systems, fiber diameter, crosslinking process, in vitro tests, indication of application in tissue engineering (TE) field and reference, fabricated by using benign solvents

Polymer or blend	Solvent system	Average fiber diameter [nm]	Crosslinking	In vitro tests (cell type)	Indication of application	Reference
Collagen type I	Acetic acid/dimethyl sulfoxide	200–1100	1-Ethyl-3 (3-dimethylaminopropyl) carbodiimide hydrochloride (EDC)	L6 rat skeletal myoblasts	Cardiac TE	Elamparithi et al. [34]
Collagen type I	Phosphate-buffered saline and ethanol	10× PBS: 540 ± 210 20×: 210 ± 60	EDC and *N*-hydroxysuccinimide (NHS)	–	–	Dong et al. [33]
Collagen type I/ PEO and NaCl	10 mM hydrochloric acid solution	400 ± 50	EDC/NHS	Smooth muscle cells	Vascular graft	Buttafoco et al. [41]
Human-like collagen (HLC)/ PEO and HLC/ chitosan/PEO and NaCl	Acetic acid	HLC/PEO: 260–415 HLC/chitosan/ PEO: 110–210	Glutaraldehyde solution	Rabbit bone marrow stromal cells	Blood vessel, skin or cornea tissue engineering	Chen et al. [42]
Collagen type I/ PCL	Acetic acid	100–200	–	–	Skin tissue engineering	Chakrapani et al. [43]
Gelatin	Acetic acid	271.0 ± 0.1	–	–	–	Choktaweesap et al. [44]
Gelatin	Ethyl acetate and acetic acid in water	47–145				Song et al. [45]
Silk fibroin	Formic acid	Less than 100	–	–	–	Sukigara et al. [46]
Silk fibroin	Formic acid	30–120	Methanol	Human keratinocytes and fibroblasts	Wound dressing	Min et al. [37]
Silk fibroin/PEO	Water	800 ± 50	Methanol	–	–	Wang et al. [40]

Polymer or blend	Solvent system	Average fiber diameter [nm]	Crosslinking	In vitro tests (cell type)	Indication of application	Reference
Silk fibroin	Formic acid	140–590	Genipin added to the solution before ES	–	Wound dressings	Silva et al. [47]
Silk fibroin/PEO	Water	530 ± 100	Methanol and decoration with fibronectin	Primary human mesenchymal stem cells	–	Meinel et al. [48]
Silk fibroin	Citric acid–sodium hydroxide (NaOH)–hydrochloric acid (HCl) buffer	1500–2500	–	–	–	Zhu et al. [49]
Zein	Ethanol	600–1000	Glyoxal	Murine fibroblasts (L929) and human foreskin fibroblasts (HFF)	–	Suwantong et al. [38]
Zein	Acetic acid	35 wt%: 153 ± 18 45 wt%: 344 ± 27	–	Primary human dermal fibroblasts	Skin regeneration	Lin et al. [50]
Soy protein isolate/PEO	NaOH aqueous solution	50 and 270	EDC/NHS	hMSCs	TE applications	Ramji and Shah [51]
PCL	Acetic acid and sodium acetate	150–2000	–	–	–	Lawson et al. [52]
PCL	Acetic acid	1360 ± 330	–	Vero cells	–	Ferreira et al. [53]
PCL	Acetic acid/formic acid	266 ± 39	–	–	–	Van der Schueren et al. [54]

(continued)

Table 4 (continued)

Polymer or blend	Solvent system	Average fiber diameter [nm]	Crosslinking	In vitro tests (cell type)	Indication of application	Reference
PCL	Acetic acid/formic acid	130 ± 62	–	ARPE-19 and Müller glial cell (MIO-M1 cell) and in vivo on rat eyes	Drug carrier for ophthalmic use	Da Silva et al. [55]
PLA	Acetone	757 ± 275	–	–	–	Casasola et al. [12]
Chitosan/PCL	Formic acid/acetone	102 ± 24 nm	–	–	–	Shalumon et al. [56]
Chitosan/PEO	Acetic acid	About 500	Genipin	Human fibroblast cells	Wound dressing	Mirzaei et al. [57]
Chitosan/PVA	Acetic acid	160 ± 20	Glyoxal	–	–	Liu et al. [58]
Chitosan/PCL	Acetic acid (AA)/ formic acid (FA)	3/7 AA/FA: 203 ± 44 5/5 AA/FA: 367 ± 118	–	–	–	Van Der Schueren et al. [59]

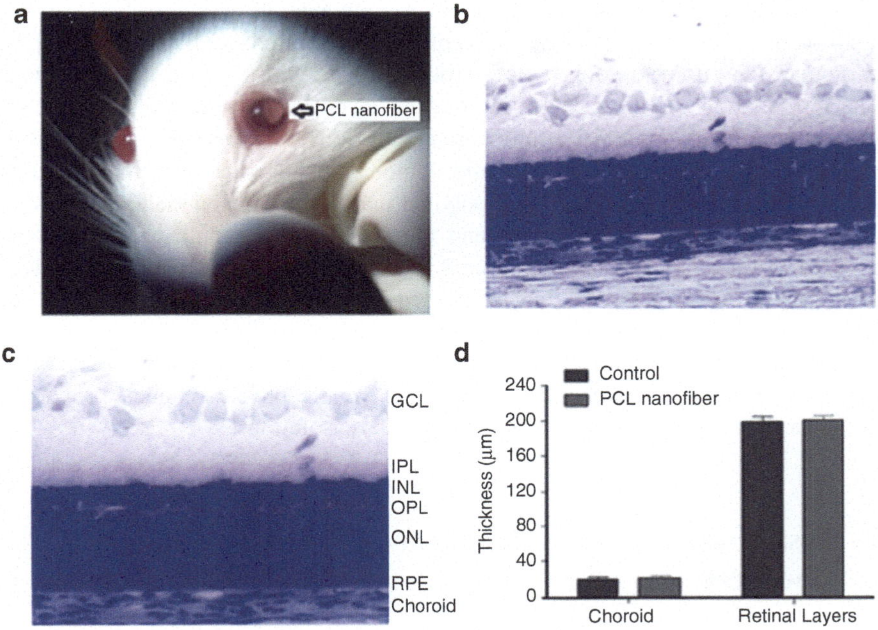

Fig. 4 Digital image of the rat with PCL implant, 10 days after the implantation (**a**); histological examination of the posterior and anterior segments of the eye with the PCL nanofibers after 10 days (**b**); histological examination of the same segment, without PCL nanofibers (**c**); Choroid and retinal layers thickness of the animals of both groups (**d**). Reprinted with permission from Da Silva et al. [55]

process. In this paragraph, an overview about the fabrication of composites by using benign solvents, will be reported.

Ghosal et al. reported the addition of TiO_2 nanoparticles (20 nm size) to PCL, in order to obtain scaffolds with antibacterial properties. The obtained PCL fibers were coated with collagen to modulate and enhance their mechanical properties, suitable for skin substitutes applications [63].

As already reported in literature for toxic solvents, also for benign solvents, the addition of particles in the electrospinning solution affects the solution properties and influences the obtained fibers morphology. In particular, a reduction in the obtained average fibers diameter was reported by Liverani and Boccaccini [21] after the addition of bioactive glass (BG) particles (45S5 composition), with sizes around 2 µm, into the PCL solution before electrospinning. The incorporation of the particles in the electrospun polymeric mats was investigated by SEM and EDX analyses, in vitro mineralization was observed on the samples after immersion in SBF solution, demonstrating that the incorporation of the BG particles in the PCL, does not affect their bioactivity, highlighting the suitability of these scaffolds for applications related to bone tissue engineering or interface tissue engineering.

A similar method for the obtainment of composite electrospun fibers, by the addition of BGs particles in PCL/gelatin blend was reported by Gönen et al. [64]. In particular, they used ion-doped BG particles, doped with Copper and Strontium. The introduction of BG particles induced an increase in average fiber diameter, while after immersion in SBF solution, not all the ions-doped samples showed mineralization on their surface. These results demonstrated that the bioactivity depends on the amount of BG particles in the polymer matrix, but also on the ions release.

A summary of polymers matrix and reinforcement inorganic phase, solvents used, fiber diameter, crosslinking process, cell line for the in vitro tests and indication of the TE field of applications is reported in the following Table 5.

The promising results obtained by using benign solvents for the fabrication of composites laid down the basis for further studies oriented on the investigation of the possible influence that the solvent could have on the physic-chemical properties of the particles after their addition in the polymeric solution.

7 Post Processing: Crosslinking and the Use of Toxic Chemicals

As already reported before, the electrospinning of natural polymers often leads to the fabrication of mats that are not stable in aqueous solution. For this reason, crosslinking is mandatory and also useful for enhancing electrospun mats mechanical properties, but often toxic crosslinkers, like glutaraldehyde, glyoxal, etc. [38, 65] are used to increase the fiber mats stability in aqueous environment, as reported in Tables 4 and 5. The future trends and research focus are actually oriented on the decrease of the use of these toxic crosslinkers and towards the use of less or non-toxic one, like citric acid and genipin [47, 57, 70]. Beyond the solvents and the crosslinkers, the same trend could involve also all the other chemicals relevant for the fabrication of composite samples, like the selection of natural extracts as reducing agents [71].

8 Conclusions and Future Perspectives

Green electrospinning and related improvements for the reduction in the use of toxic harsh solvents is a relevant topic and focus of many research publications in the last years. Even if the use of less suitable solvents implies longer and more accurate optimization of the process parameters, promising results have been obtained for what concerns the electrospinning of natural, synthetic polymers and their blends. The use of these solvents resulted also suitable for the fabrication of composites for tissue engineering applications.

Table 5 Summary of composite, solvent systems, fiber diameter, crosslinking process, in vitro tests, indication of application in tissue engineering (TE) field and reference

Polymer matrix and inorganic phase	Solvent system	Average fiber diameter [nm]	Post-treatments	In vitro tests (cell type)	Application	Reference
Collagen/ hydroxyapatite (HA)	Acetic acid	342 ± 67 nm	Crosslinking with glutaraldehyde vapor	MC3T3-E1	Bone TE	Castilla-Casadiego et al. [65]
Silk fibroin/nHA	Formic acid	242 ± 34 nm	Mineralization obtained after series of calcium and phosphate treatments	MC3T3-E1	Bone graft applications	Wei et al. [66]
PCL/TiO_2 nanoparticles	Acetic acid	200–800	Collagen coated fibers	–	Skin substitutes	Ghosal et al. [63]
PCL/bioactive glass (BG) particles (2 μm)	Acetic acid	500 ± 200	–	–	Bone TE	Liverani and Boccaccini [21]
PCL/bioactive glass (BG) particles (<25 μm)	Acetic acid	800–3500	–	–	Bone TE	Lepry et al. [67]
PCL/gelatin/BG particles ion doped (Cu and Sr)	Acetic acid/formic acid	Gt/PCL/Sr-BG: 2.5 wt%BG: 448 ± 111 7.5 wt%BG: 532 ± 190 Gt/PCL/Cu-BG: 2.5 wt%BG: 400 ± 71 7.5 wt%BG: 463 ± 107	Crosslinking with glutaraldehyde vapor	–	Bone TE	Gönen et al. [64]
Gelatin/silver nanoparticles (AgNPs)	Acetic acid	Neat: 230 ± 30 and with Ag: 280 ± 40	Crosslinking with glutaraldehyde vapor	–	Wound-dressing materials with antibacterial activity	Rujitanaroj et al. [68]
Chitosan/gelatin AgNPs	Acetic acid	220–400	–	–	General indication	Zhuang et al. [69]

Future perspectives will be oriented in the avoidance or reduction in the use not only of the toxic solvents for the electrospinning, but also of toxic crosslinkers or surfactants, in order to completely fulfill the initial definition of Green Electrospinning.

Acknowledgements Liliana Liverani acknowledges funding from the European Union's Horizon 2020 research and innovation programme under the Marie Skłodowska-Curie grant agreement No. 657264.

References

1. Krishnan R, Sundarrajan S, Ramakrishna S (2013) Green processing of nanofibers for regenerative medicine. Macromol Mater Eng 298:1034–1058. https://doi.org/10.1002/mame.201200323
2. Sun J, Bubel K, Chen F, Kissel T, Agarwal S, Greiner A (2010) Nanofibers by green electrospinning of aqueous suspensions of biodegradable block copolyesters for applications in medicine, pharmacy and agriculture. Macromol Rapid Commun 31:2077–2083. https://doi.org/10.1002/marc.201000379
3. Shenoy SL, Bates WD, Wnek G (2005b) Correlations between electrospinnability and physical gelation. Polymer (Guildf) 46:8990–9004. https://doi.org/10.1016/j.polymer.2005.06.053
4. LeCorre-Bordes D, Tucker N, Huber T, Buunk N, Staiger MP (2016) Shear-electrospinning: extending the electrospinnability range of polymer solutions. J Mater Sci 51:6686–6696. https://doi.org/10.1007/s10853-016-9955-y
5. McClements DJ (2000) Comments on viscosity enhancement and depletion flocculation by polysaccharides. Food Hydrocoll 14:173–177. https://doi.org/10.1016/S0268-005X(99)00065-X
6. Shenoy SL, Bates WD, Frisch HL, Wnek GE (2005a) Role of chain entanglements on fiber formation during electrospinning of polymer solutions: good solvent, non-specific polymer-polymer interaction limit. Polymer (Guildf) 46:3372–3384. https://doi.org/10.1016/j.polymer.2005.03.011
7. Colombo V, Fabiani D, Focarete ML, Gherardi M, Gualandi C, Laurita R, Zaccaria M (2014) Atmospheric pressure non-equilibrium plasma treatment to improve the electrospinnability of poly(L-lactic acid) polymeric solution. Plasma Process Polym 11:247–255. https://doi.org/10.1002/ppap.201300141
8. Arayanarakul K, Choktaweesap N, Aht-ong D, Meechaisue C, Supaphol P (2006) Effects of poly(ethylene glycol), inorganic salt, sodium dodecyl sulfate, and solvent system on electrospinning of poly(ethylene oxide). Macromol Mater Eng 291:581–591. https://doi.org/10.1002/mame.200500419
9. Wang X, Pellerin C, Bazuin CG (2016) Enhancing the electrospinnability of low molecular weight polymers using small effective cross-linkers. Macromolecules 49:891–899. https://doi.org/10.1021/acs.macromol.5b02670
10. Pillay V, Dott C, Choonara YE, Tyagi C, Tomar L, Kumar P, Toit LC, Ndesendo VMK (2013) A review of the effect of processing variables on the fabrication of electrospun nanofibers for drug delivery applications. J Nanomater 2013:22. https://doi.org/10.1155/2013/789289
11. Megelski S, Stephens JS, Chase DB, Rabolt JF (2002) Micro-and nanostructured surface morphology on electrospum polymer fibers. Macromolecules 35:8456–8466. https://doi.org/10.1021/ma020444a
12. Casasola R, Thomas NL, Trybala A, Georgiadou S (2014) Electrospun poly lactic acid (PLA) fibres: effect of different solvent systems on fibre morphology and diameter. Polymer (Guildf) 55:4728–4737. https://doi.org/10.1016/j.polymer.2014.06.032
13. Jarusuwannapoom T, Hongrojjanawiwat W, Jitjaicham S, Wannatong L, Nithitanakul M, Pattamaprom C, Koombhongse P, Rangkupan R, Supaphol P (2005) Effect of solvents on

electro-spinnability of polystyrene solutions and morphological appearance of resulting electrospun polystyrene fibers. Eur Polym J 41:409–421. https://doi.org/10.1016/j.eurpolymj.2004.10.010
14. Luo CJ, Stride E, Edirisinghe M (2012) Mapping the influence of solubility and dielectric constant on electrospinning polycaprolactone solutions. Macromolecules 45:4669–4680. https://doi.org/10.1021/ma300656u
15. Qin X, Wu D (2012) Effect of different solvents on poly(caprolactone) (PCL) electrospun nonwoven membranes. J Therm Anal Calorim 107:1007–1013. https://doi.org/10.1007/s10973-011-1640-4
16. Mohammad Khanlou H, Chin Ang B, Talebian S, Muhammad Afifi A, Andriyana A (2014) Electrospinning of polymethyl methacrylate nanofibers: optimization of processing parameters using the Taguchi design of experiments. Text Res J 85:356–368. https://doi.org/10.1177/0040517514547208
17. Xie J, Li X, Xia Y (2008) Putting electrospun nanofibers to work for biomedical research. Macromol Rapid Commun 29:1775–1792. https://doi.org/10.1002/marc.200800381
18. ICH (2011) Q3C guideline for residual solvents (R5). Int Conf Harmon Tech Requir Regist Pharm Hum Use 29
19. Bruice PY (2011) Organic chemistry. Prentice Hall
20. Ahn Y, DH H, Hong JH, Lee SH, Kim HJ, Kim H (2012) Effect of co-solvent on the spinnability and properties of electrospun cellulose nanofiber. Carbohydr Polym 89:340–345. https://doi.org/10.1016/j.carbpol.2012.03.006
21. Liverani L, Boccaccini A (2016) Versatile production of poly(epsilon-caprolactone) fibers by electrospinning using benign solvents. Nanomaterials 6:75. https://doi.org/10.3390/nano6040075
22. Putti M, Simonet M, Solberg R, Peters GWM (2015) Electrospinning poly(ε-caprolactone) under controlled environmental conditions: influence on fiber morphology and orientation. Polymer (Guildf) 63:189–195. https://doi.org/10.1016/j.polymer.2015.03.006
23. Nam J, Huang Y, Agarwal S, Lannutti J (2008) Materials selection and residual solvent retention in biodegradable electrospun fibers. J Appl Polym Sci 107:1547–1554. https://doi.org/10.1002/app.27063
24. Agarwal S, Greiner A (2011) On the way to clean and safe electrospinning-green electrospinning: emulsion and suspension electrospinning. Polym Adv Technol 22:372–378. https://doi.org/10.1002/pat.1883
25. McClellan P, Landis WJ (2016) Recent applications of coaxial and emulsion electrospinning methods in the field of tissue engineering. Biores Open Access 5:212–227. https://doi.org/10.1089/biores.2016.0022
26. Li X, Su Y, Liu S, Tan L, Mo X, Ramakrishna S (2010a) Encapsulation of proteins in poly(l-lactide-co-caprolactone) fibers by emulsion electrospinning. Colloids Surf B Biointerfaces 75:418–424. https://doi.org/10.1016/j.colsurfb.2009.09.014
27. Li X, Zhang H, Li H, Yuan X (2010b) Encapsulation of proteinase K in PELA ultrafine fibers by emulsion electrospinning: preparation and in vitro evaluation. Colloid Polym Sci 288:1113–1119. https://doi.org/10.1007/s00396-010-2235-5
28. Yang Y, Li X, Qi M, Zhou S, Weng J (2008) Release pattern and structural integrity of lysozyme encapsulated in core-sheath structured poly(DL-lactide) ultrafine fibers prepared by emulsion electrospinning. Eur J Pharm Biopharm 69:106–116. https://doi.org/10.1016/j.ejpb.2007.10.016
29. Buruaga L, Sardon H, Irusta L, González A, Fernández-Berridi MJ, Iruin JJ (2010) Electrospinning of waterborne polyurethanes. J Appl Polym Sci 115:1176–1179. https://doi.org/10.1002/app.31219
30. Pal J, Sharma S, Sanwaria S, Kulshreshtha R, Nandan B, Srivastava RK (2014) Conducive 3D porous mesh of poly(ε-caprolactone) made via emulsion electrospinning. Polymer (United Kingdom) 55:3970–3979. https://doi.org/10.1016/j.polymer.2014.06.067
31. Wang X, Yuan Y, Huang X, Yue T (2015) Controlled release of protein from core-shell nanofibers prepared by emulsion electrospinning based on green chemical. J Appl Polym Sci 132:n/a-n/a. https://doi.org/10.1002/app.41811

32. Spano F, Quarta A, Martelli C, Ottobrini L, Rossi RM, Gigli G, Blasi L (2016) Fibrous scaffolds fabricated by emulsion electrospinning: from hosting capacity to in vivo biocompatibility. Nanoscale 8:9293–9303. https://doi.org/10.1039/C6NR00782A
33. Dong B, Arnoult O, Smith ME, Wnek GE (2009) Electrospinning of collagen nanofiber scaffolds from benign solvents. Macromol Rapid Commun 30:539–542. https://doi.org/10.1002/marc.200800634
34. Elamparithi A, Punnoose AM, Kuruvilla S (2016) Electrospun type 1 collagen matrices preserving native ultrastructure using benign binary solvent for cardiac tissue engineering. Artif Cells Nanomed Biotechnol 44:1318–1325. https://doi.org/10.3109/21691401.2015.1029629
35. Zeugolis DI, Khew ST, Yew ESY, Ekaputra AK, Tong YW, Yung LYL, Hutmacher DW, Sheppard C, Raghunath M (2008) Electro-spinning of pure collagen nano-fibres - just an expensive way to make gelatin? Biomaterials 29:2293–2305. https://doi.org/10.1016/j.biomaterials.2008.02.009
36. Liu Q, Zhou Y, Zhao Z, Li Z (2013b) Environmentally benign synthesis of electrospun collagen-I in acetic acid solution. Optoelectron Adv Mater Rapid Commun 7:272–275
37. Min BM, Lee G, Kim SH, Nam YS, Lee TS, Park WH (2004) Electrospinning of silk fibroin nanofibers and its effect on the adhesion and spreading of normal human keratinocytes and fibroblasts in vitro. Biomaterials 25:1289–1297. https://doi.org/10.1016/j.biomaterials.2003.08.045
38. Suwantong O, Pavasant P, Supaphol P (2011) Electrospun zein fibrous membranes using glyoxal as cross-linking agent: preparation, characterization and potential for use in biomedical applications. Chiang Mai J Sci 38:56–70
39. Zhang X, Reagan MR, Kaplan DL (2009) Electrospun silk biomaterial scaffolds for regenerative medicine. Adv Drug Deliv Rev 61:988–1006. https://doi.org/10.1016/j.addr.2009.07.005
40. Wang M, Jin H, Kaplan D, Rutledge G (2004) Mechanical properties of electrospun silk fibers. Macromolecules:6856–6864
41. Buttafoco L, Kolkman NG, Engbers-Buijtenhuijs P, Poot AA, Dijkstra PJ, Vermes I, Feijen J (2006) Electrospinning of collagen and elastin for tissue engineering applications. Biomaterials 27:724–734. https://doi.org/10.1016/j.biomaterials.2005.06.024
42. Chen L, Zhu C, Fan D, Liu B, Ma X, Duan Z, Zhou Y (2011) A human-like collagen/chitosan electrospun nanofibrous scaffold from aqueous solution: electrospun mechanism and biocompatibility. J Biomed Mater Res A 99A:395–409. https://doi.org/10.1002/jbm.a.33202
43. Chakrapani VY, Gnanamani A, Giridev VR, Madhusoothanan M, Sekaran G (2012) Electrospinning of type I collagen and PCL nanofibers using acetic acid. J Appl Polym Sci 125:3221–3227. https://doi.org/10.1002/app.36504
44. Choktaweesap N, Arayanarakul K, Aht-ong D, Meechaisue C, Supaphol P (2007) Electrospun gelatin fibers: effect of solvent system on morphology and fiber diameters. Polym J 39:622–631. https://doi.org/10.1295/polymj.PJ2006190
45. Song J-H, Kim H-E, Kim H-W (2008) Production of electrospun gelatin nanofiber by water-based co-solvent approach. J Mater Sci Mater Med 19:95–102. https://doi.org/10.1007/s10856-007-3169-4
46. Sukigara S, Gandhi M, Ayutsede J, Micklus M, Ko F (2003) Regeneration of Bombyx mori silk by electrospinning - Part 1: processing parameters and geometric properties. Polymer (Guildf) 44:5721–5727. https://doi.org/10.1016/S0032-3861(03)00532-9
47. Silva SS, Maniglio D, Motta A, Mano JF, Reis RL, Migliaresi C (2008) Genipin-modified silk-fibroin nanometric nets. Macromol Biosci 8:766–774. https://doi.org/10.1002/mabi.200700300
48. Meinel AJ, Kubow KE, Klotzsch E, Garcia-Fuentes M, Smith ML, Vogel V, Merkle HP, Meinel L (2009) Optimization strategies for electrospun silk fibroin tissue engineering scaffolds. Biomaterials 30:3058–3067. https://doi.org/10.1016/j.biomaterials.2009.01.054
49. Zhu J, Shao H, Hu X (2007) Morphology and structure of electrospun mats from regenerated silk fibroin aqueous solutions with adjusting pH. Int J Biol Macromol 41:469–474. https://doi.org/10.1016/j.ijbiomac.2007.06.006
50. Lin L, Perets A, el H-e Y, Varma D, Li M, Lazarovici P, Woerdeman DL, Lelkes PI (2013) Alimentary "green" proteins as electrospun scaffolds for skin regenerative engineering. J Tissue Eng Regen Med 7:994–1008. https://doi.org/10.1002/term.1493

51. Ramji K, Shah RN (2014) Electrospun soy protein nanofiber scaffolds for tissue regeneration. J Biomater Appl 29:411–422. https://doi.org/10.1177/0885328214530765
52. Lawson C, Stanishevsky A, Sivan M, Pokorny P, Lukáš D (2016) Rapid fabrication of poly(ε-caprolactone) nanofibers using needleless alternating current electrospinning. J Appl Polym Sci 133:n/a-n/a. https://doi.org/10.1002/app.43232
53. Ferreira JL, Gomes S, Henriques C, Borges JP, Silva JC (2014) Electrospinning polycaprolactone dissolved in glacial acetic acid: fiber production, nonwoven characterization, and in vitro evaluation. J Appl Polym Sci 131:41086. https://doi.org/10.1002/app.41068
54. Van der Schueren L, De Schoenmaker B, Kalaoglu ÖI, De Clerck K (2011) An alternative solvent system for the steady state electrospinning of polycaprolactone. Eur Polym J 47:1256–1263. https://doi.org/10.1016/j.eurpolymj.2011.02.025
55. Da Silva GR, Lima TH, Oréfice RL, Fernandes-Cunha GM, Silva-Cunha A, Zhao M, Behar-Cohen F (2015) In vitro and in vivo ocular biocompatibility of electrospun poly(ε-caprolactone) nanofibers. Eur J Pharm Sci 73:9–19. https://doi.org/10.1016/j.ejps.2015.03.003
56. Shalumon KT, Anulekha KH, Girish CM, Prasanth R, Nair SV, Jayakumar R (2010) Single step electrospinning of chitosan/poly(caprolactone) nanofibers using formic acid/acetone solvent mixture. Carbohydr Polym 80:414–420. https://doi.org/10.1016/j.carbpol.2009.11.039
57. Mirzaei E, Faridi-Majidi R, Shokrgozar MA, Asghari F (2014) Genipin cross-linked electrospun chitosan-based nanofibrous mat as tissue engineering scaffold. Nanomed J 1:137–146. https://doi.org/10.7508/nmj.2014.03.003
58. Liu Y, Park M, Shin HK, Pant B, Park SJ, Kim HY (2014) Preparation and characterization of chitosan-based nanofibers by ecofriendly electrospinning. Mater Lett 132:23–26. https://doi.org/10.1016/j.matlet.2014.06.041
59. Van Der Schueren L, Steyaert I, De Schoenmaker B, De Clerck K (2012) Polycaprolactone/chitosan blend nanofibres electrospun from an acetic acid/formic acid solvent system. Carbohydr Polym 88:1221–1226. https://doi.org/10.1016/j.carbpol.2012.01.085
60. Gholipour Kanani A, Bahrami SH (2011) Effect of changing solvents on poly(ε-caprolactone) nanofibrous webs morphology. J Nanomater 2011:1–10. https://doi.org/10.1155/2011/724153
61. Katsogiannis KAG, Vladisavljević GT, Georgiadou S (2015) Porous electrospun polycaprolactone (PCL) fibres by phase separation. Eur Polym J 69:284–295. https://doi.org/10.1016/j.eurpolymj.2015.01.028
62. Liu H, Ding X, Zhou G, Li P, Wei X, Fan Y, Liu H, Ding X, Zhou G, Li P, Wei X, Fan Y (2013a) Electrospinning of nanofibers for tissue engineering applications. J Nanomater 2013:1–11. https://doi.org/10.1155/2013/495708
63. Ghosal K, Thomas S, Kalarikkal N, Gnanamanis A (2014) Collagen coated electrospun polycaprolactone (PCL) with titanium dioxide (TiO2) from an environmentally benign solvent: preliminary physico-chemical studies for skin substitute. J Polym Res 21:2–6. https://doi.org/10.1007/s10965-014-0410-y
64. Gönen SÖ, Taygun ME, Küçükbayrak S (2016) Fabrication of bioactive glass containing nanocomposite fiber mats for bone tissue engineering applications. Compos Struct 138:96–106. https://doi.org/10.1016/j.compstruct.2015.11.033
65. Castilla-Casadiego DA, Maldonado M, Sundaram P, Almodovar J (2016) "Green" electrospinning of a collagen/hydroxyapatite composite nanofibrous scaffold. MRS Commun:1–6. https://doi.org/10.1557/mrc.2016.43
66. Wei K, Kim BS, Kim IS (2011) Fabrication and biocompatibility of electrospun silk biocomposites. Membranes (Basel) 1:275–298. https://doi.org/10.3390/membranes1040275
67. Lepry WC, Smith S, Liverani L, Boccaccini AR, Nazhat SN (2016) Acellular bioactivity of sol-gel derived borate glass-polycaprolactone electrospun scaffolds. Biomed Glass 2:88–98
68. Rujitanaroj P, Pimpha N, Supaphol P (2008) Wound-dressing materials with antibacterial activity from electrospun gelatin fiber mats containing silver nanoparticles. Polymer (Guildf) 49:4723–4732. https://doi.org/10.1016/j.polymer.2008.08.021
69. Zhuang X, Cheng B, Kang W, Xu X (2010) Electrospun chitosan/gelatin nanofibers containing silver nanoparticles. Carbohydr Polym 82:524–527. https://doi.org/10.1016/j.carbpol.2010.04.085

70. Jiang Q, Reddy N, Yang Y (2010) Cytocompatible cross-linking of electrospun zein fibers for the development of water-stable tissue engineering scaffolds. Acta Biomater 6:4042–4051. https://doi.org/10.1016/j.actbio.2010.04.024
71. Arvand M, Mirzaei E, Derakhshan MA, Kharrazi S, Sadroddiny E, Babapour M, Faridi-Majidi R (2015) Fabrication of antibacterial silver nanoparticle-modified chitosan fibers using Eucalyptus extract as a reducing agent. J Appl Polym Sci 132:42133. https://doi.org/10.1002/app.42133

Electrospun Biomaterials for Cancer Research

Akshay A. Narkhede and Shreyas S. Rao

Abstract Electrospun biomaterials have been extensively utilized in tissue engineering and drug delivery; however, more recently, these biomaterials have been employed to address various challenges in cancer research. Tunable topography, morphology and surface chemistry coupled with bio-compatibility/degradability, render electrospun biomaterials suitable for the development of anti-cancer drug delivery systems. The ability of electrospun biomaterials to preserve the properties of genetic materials has led to their applications in targeted gene therapies for cancer. In addition, electrospun biomaterials can be engineered to recapitulate the tumor microenvironment and have been found to regulate cellular processes; making them suitable for the development of disease models and cell based therapies in cancer. Electrospun biomaterials are also being investigated as biosensors for early detection and diagnosis of cancer. This chapter explores the utility of electrospun biomaterials in the field of cancer research and highlights its future prospects.

Keywords Electrospun biomaterials • Cancer research • Drug delivery • Gene delivery • Disease models • Tumor microenvironment • Cell based therapy • Biosensors

List of Abbreviations

Fiber Components

CNT	Carbon nanotube
coPLA	Poly(L-lactide-co-D,L-lactide)
HA	Hyaluronic acid
HCPD	2-Hydroxypropyl-β-cyclodextrin

A.A. Narkhede • S.S. Rao (✉)
Department of Chemical and Biological Engineering, The University of Alabama, Tuscaloosa, AL, USA
e-mail: srao3@eng.ua.edu

J. Almodovar (ed.), *Electrospun Biomaterials and Related Technologies*,
DOI 10.1007/978-3-319-70049-6_6

P(LL-CL)	Poly(L-lactic acid-co-ε-caprolactone)
P(NIPAAm-MAA-VP)	Poly(N-isopropylacrylamide-acrylamide-vinyl pyrrolidone)
PCEC	Poly(3-caprolactone)-poly(ethylene glycol)-poly-(3-caprolactone)
PCL	Poly(ε-caprolactone)
PDMS	Poly(dimethyl siloxane)
PEG	Poly(ethylene glycol)
PELA	Poly(D,L-lactic acid)-poly(ethylene glycol)
PEO	Poly(ethylene oxide)
PES	Poly(ether sulphone)
PGC-C18	Poly(glycerol monostearate-co-ε-caprolactone)
PHBV	Poly(3-hydroxybutyrate-co-3-hydroxyvalerate)
PLA	Poly(L-lactide)
PLGA	Poly(D,L-lactic-co-glycolic acid)
QCh	Quaternized chitosan
UNCPS	$NaGdF_4$:Yb/Er@$NaGdF_4$:Yb@mSiO_2-polyethylene glycol

Fiber Alignment

A	Aligned
R	Random

Drugs

5-FU	5-Fluorouracil
BCNU	1,3-Bis(2-chloroethyl)-1-nitrosourea
HCPT	Hydroxycamptothecin

Miscellaneous

BMP-7	Bone morphogenetic protein-7
EMT	Epithelial to mesenchymal transition
GBM	Glioblastoma multiforme
OB	Osteoblasts

1 Introduction

Cancer is one of the leading causes of death worldwide [1]. WHO, a leading health organization, projects a substantial rise in cancer cases from 14 million in 2012 to 22 million in next two decades [1]. Currently, cancer therapy regimen typically includes surgery, radiation, and administration of chemo or targeted therapies based on the type and staging of the disease. However, delayed cancer diagnosis, inefficient delivery of chemotherapeutics, drug resistance and poor prognosis of the disease often leading to disease recurrence, continue to be key challenges in cancer treatment [2, 3]. For example, failure in detecting cancer in its early stages makes it difficult to manage the disease as it has already entered the advanced stage; asymptotically decreasing the survival chances of the patient. Similarly, inefficient delivery of chemotherapeutics has been associated with many side-effects involving damage to normal cells inducing toxicity [3]. Moreover, recent studies have showed that certain sub-population of cancer cells can enter dormancy and become resistant to radio-/chemotherapies resulting in relapse years after surgical resection of primary tumor [4]. These therapeutic challenges have motivated research in the development of targeted anti-cancer drug delivery systems, physiologically relevant in vitro cancer models for robust anti-cancer drug screening, and technologies for early cancer diagnosis. With advances in material science and nanotechnology, biomaterials have become a key platform for addressing challenges in cancer research [5]. In particular, electrospun biomaterials (i.e., electrospun fibers) have been widely investigated for applications in targeted cancer therapy, in vitro cancer modeling and disease sensing because of their biocompatibility and biodegradability (Fig. 1).

The ability to encapsulate a variety of drugs with high loading efficiencies combined with high surface areas make electrospun fibers an attractive platform for targeted drug delivery as well as delivery mediated by external stimuli [3]. These

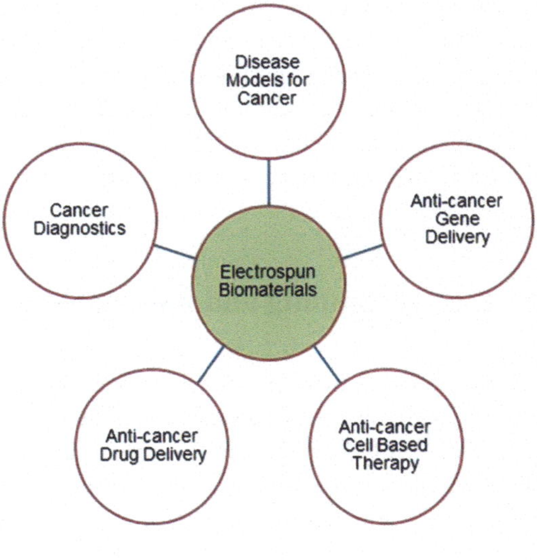

Fig. 1 Applications of electrospun biomaterials in cancer research

materials could also be designed to provide controlled and sustained drug release for extended periods of time and to deliver multiple therapeutics. A wide variety of polymers (natural and synthetic) provides flexibility in choosing a suitable polymer for delivery of a particular drug. In addition, electrospun fibers have been utilized as carriers for gene delivery owing to their ability in retaining the properties of genetic material such as plasmid DNA and viral vectors [6]. Moreover, electrospinning technique is economically efficient enabling development of low cost targeted drug delivery systems.

In addition to anti-cancer drug and gene delivery, electrospun fibers have been employed as disease models to study cancer biology. Rapid development in the fields of tissue engineering, molecular biology and genomics have enabled studies of cancer cell behaviors in physiologically relevant environments in vitro. It is well appreciated that the tumor microenvironment plays a vital role in cancer progression in addition to genetic mutations [7]. The tumor microenvironment mediates the behavior of cancer cells by providing them with biochemical, cellular, and biomechanical cues. Biochemical cues consist of growth factors, extracellular matrix (ECM) proteins and soluble factors that regulate tumor-stroma interactions and can activate various signaling pathways in the cancer cells due to binding of cell surface receptors [8]. Biomechanical cues include matrix stiffness which trigger mechanotransduction pathways in cancer cells ultimately regulating cell fate [9]. Owing to their large surface area, cyto-compatibility, tunability of mechanical stiffness and versatile chemistry; electrospun fibers have been utilized as a scaffold for studying cancer cell behaviors [10]. In particular, electrospun fibers enable recapitulation of specific topographical aspects in the tumor microenvironment like the fibrous collagen network evidenced in vivo [11]. Electrospun fibrous scaffold facilitates cell adhesion, cell-cell/matrix interactions and provides mechanical cues to the cancer cells recapitulating the microenvironment in vitro to study cancer cell-microenvironment interactions.

More recently, researchers have been investigating electrospun fibrous scaffolds for their application in anti-cancer stem cell based therapies [12]. Apart from their application in disease modeling and delivering anti-cancer therapeutics, electrospun fibers have also been employed in fabrication of biosensors for cancer diagnostics [3]. The following discussion focuses on the key process parameters for fabrication of electrospun biomaterials and the applications of electrospun biomaterials in diagnosing, understanding, and treating cancer.

2 Electrospinning and Associated Process Parameters

Electrospinning utilizes electro-hydrodynamic forces to draw fibers from a polymer solution with diameters ranging from hundreds of micrometers to a few nanometers [13]. Fibers of desired morphologies, surface or mechanical properties can be obtained via optimization of electrospinning process parameters (discussed below) based on the polymer material chosen.

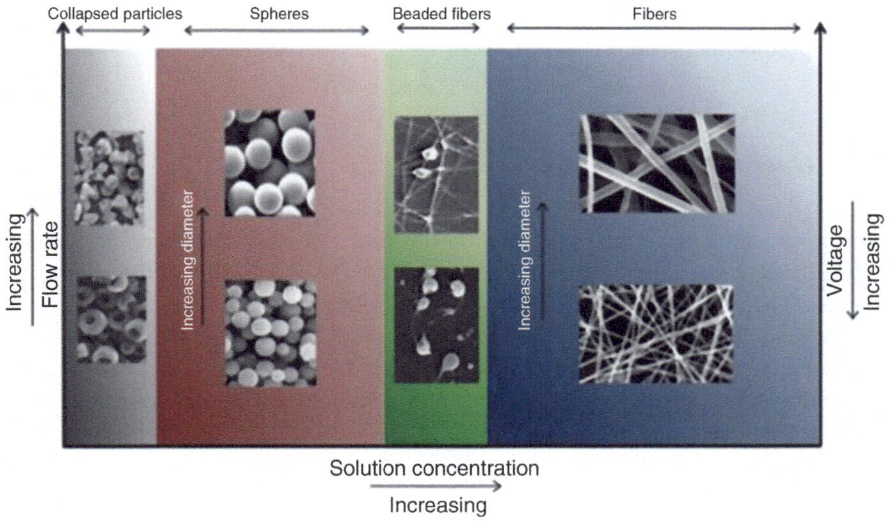

Fig. 2 Effects of electrospinning solution flowrate, concentration and the applied voltage on the fiber morphology (reprinted with permission from Zamani et al. [13])

2.1 Solution Properties

Solution properties mainly consist of solution concentration, viscosity, conductivity, polydispersity index of the polymer and solvent volatility. Solution properties dictate the morphology of the fibers. For example, if the polymer solution concentration is low, then the fibers may collapse into fragments forming rings, discs and beads before reaching the collector [14] (Fig. 2). However, if the solution concentration is increased beyond this critical value, it leads to blockage of capillary tip yielding undesired beaded polymer fibers [15]. Solution concentration also dictates the viscosity of the polymer solution and therefore must be carefully optimized to get smooth fibers [16]. Similarly, polymer solution conductivity is crucial as the electrospinning process is dependent on the charges building upon the polymer solution. Higher the conductivity of the polymer solution, lower is the diameter of the fibers [17, 18]. In some cases, salts are added to the polymer solution to increase their conductivity making them feasible for electrospinning [19]. Polydispersity index has also been found to dictate the diameter distribution of the fibers [3]. Lower the polydispersity index, narrower is the distribution of fiber diameter yielding fibers with more or less uniform diameters [3]. All the above solution properties are unique to the polymer being used; however, it has been observed that solvent volatility also impacts the fiber morphology. Generally, a solvent with a moderate boiling point is used for electrospinning applications as the solvents with low or very high boiling points hamper the drying characteristics resulting in undesired fiber morphology [14]. Very recently, researchers have been using a solvent system instead of a single solvent, as the differences in the boiling point of solvents in the system can yield highly porous fibers [20, 21].

2.2 Operating Parameters

Operating parameters consist of solution flowrate as well as distance and applied voltage between capillary and collector. Solution flowrate affects the drying of fibers and hence the fiber morphology. For example, an increase in the solution flow rate beyond certain point can cause the formation of beads in the fiber due to inefficient drying [14, 22] (Fig. 2). Similar to solution flowrate, the distance between the collector and the capillary tip is also crucial for efficient drying of the fiber. A larger distance is suitable for better drying. Typical distance between the capillary and the tip is in the range of 10–20 cm. The applied voltage between capillary and collector significantly impacts the fiber diameter. Generally, increasing the applied voltage is known to decrease the fiber diameter [20] (Fig. 2). Typically, a voltage of 10–25 kV is applied between the capillary tip and the collector.

2.3 Ambient Factors

Ambient factors like ambient temperature and humidity are known to affect the fiber diameter and morphology [3]. Generally, the diameter of fiber decreases with the increase in the humidity [23, 24]. Humidity also plays a crucial role in forming porous fibers [25]. Similar to humidity, an increase in the ambient temperature leads to a decrease in the fiber diameter [3, 26].

The parameters described above generally apply for most electrospinning processes; however, one must conduct rigorous trials for observing the trends between fiber morphology and stated parameters and optimize the system accordingly to yield fibers suitable for a particular application in cancer research.

3 Electrospun Biomaterials in Anticancer Drug and Gene Delivery

3.1 Drug Delivery

The concept of drug delivery itself evolved around half a century ago [2]. Modern era of drug delivery focuses on delivery of a wide spectrum of drugs possessing differences in physicochemical parameters and having different sources viz. synthetic as well as naturally occurring therapeutic molecules [27]. Another aspect critical for drug delivery is the ability to provide controlled release and retain the drug without losing its efficacy over a desired time within the body. Traditional cancer therapies were not targeted which led to toxicity in normal cells, affecting other vital functions like cardiovascular functioning. With advances in drug delivery, efficient systems for targeted delivery of anti-cancer therapeutics have been developed. Electrospun fibers have been traditionally and extensively utilized for delivery of a variety of drugs targeting various diseases. Further discussion emphasizes on the application of electrospun fibers in developing targeted anticancer therapies.

As electrospun fibers possess high surface area and high drug encapsulation/retention efficiency, they serve as an ideal platform for engineering anti-cancer drug delivery systems [3]. With better understanding of electro-hydrodynamics followed by advancement in electrospinning techniques, it is now possible to engineer the fiber topology, morphology, and surface chemistry for designing efficient anti-cancer drug delivery systems. In addition, the low costs associated with electrospinning techniques have potential to yield drug delivery systems affordable to patients across wide socio-economic strata. For developing anti-cancer drug delivery systems based on electrospun fibers, choosing an appropriate polymer and technique for drug loading is vital as they impact the drug release kinetics and mechanisms.

3.1.1 Polymer Selection

The biodegradability of polymer decides the release mechanism of the anti-cancer drug [20]. The drug release is guided by two mechanisms: (a) Diffusion and (b) Polymer degradation. Generally, diffusion controlled drug release is desired as it gives precise control over the release kinetics [20]. However, if the selected polymer undergoes faster biodegradation, then it adds complexity to the drug release kinetics and also results in sudden initial burst release. Therefore, both the aspects viz. degradation kinetics of polymer and desired drug release kinetics as well as the biocompatibility of the polymer should be taken into consideration [20].

3.1.2 Techniques for Anti-cancer Drug Loading on Electrospun Fibers

Current electrospinning techniques offer variety of strategies to load anti-cancer drugs onto the fibers. The drug can be loaded (or encapsulated) onto the electrospun fibers by using techniques such as co-electrospinning, co-axial electrospinning, emulsion electrospinning, side by side electrospinning, multi-jet electrospinning and surface immobilization [2]. Of these, co-electrospinning is the most widely used technique to load anti-cancer drugs onto the electrospun fibers followed by emulsion and co-axial electrospinning (Fig. 3). Every technique has its own merits and based on the requirements such as extent of drug loading and drug release characteristics, the most suitable technique can be adopted.

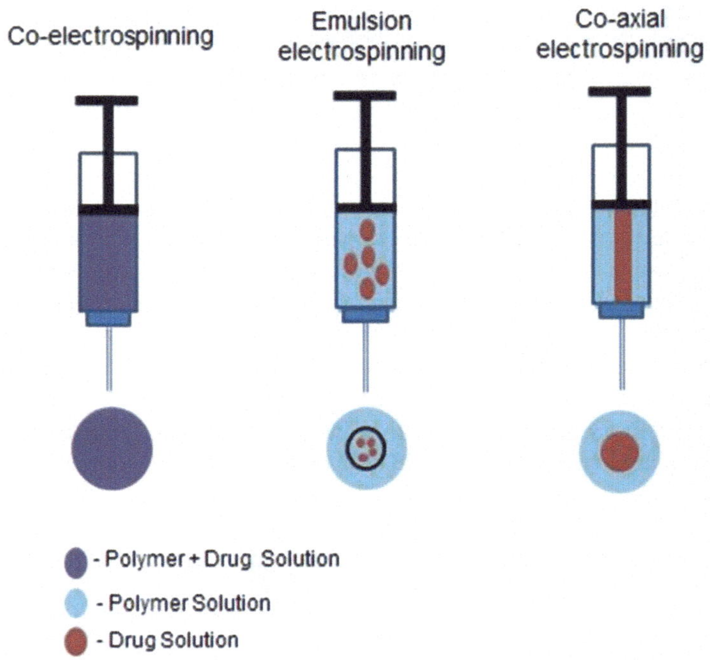

Fig. 3 Schematic representation of most commonly used techniques for encapsulating anti-cancer drugs into electrospun fibers

Co-electrospinning Technique

Co-electrospinning/blending technique is the most common technique used to load an anti-cancer drug onto the electrospun fiber mainly due to its simplicity and low costs associated with the equipment setup [13]. In this technique, the polymer solution and the drug solution (may or may not be in same solvent) are mixed together to form a homogenous electrospinning solution from which the fibers are drawn. This technique yields fibers having uniform drug distribution throughout with a high extent of drug loading. This loading strategy can result in an initial burst release of drugs which may or may not be desirable. Moreover, co-electrospinning does not entirely shield the drug from organic solvents and high voltage during electrospinning process which may have detrimental effects on sensitive drugs [28].

Co-electrospinning has been used with various drug/polymer systems to develop electrospun fiber based delivery systems (Table 1). In one such study, co-electrospinning technique was used to produce electrospun fiber mats from PEG-PLA copolymer which were loaded with BCNU [29]. The average diameter of the fibers was reported to be less than 1500 nm. It was observed that with an increase in BCNU loading, release rate and the initial burst also increased. However, the anti-tumor activity of BCNU loaded on fibers against rat glioma C6 cells lasted for more than 3 days whereas the free BCNU lost its activity in around 2 days [29]. To address

Table 1 Electrospun fibers utilized for anti-cancer drug delivery

Polymer used for electrospinning	Drugs encapsulated	Targeted cancer	Reference
Co-electrospinning			
PCL doped with hydrophobic PGC-C18	CPT-11, SN-38	Colorectal	[56]
PLA	Doxorubicin hydrochloride, Paclitaxel, Doxorubicin base	N/A	[30]
PLGA	Paclitaxel	Glioma	[31]
PLA	Titanocene dichloride	Lung	[57]
PEG-PLA	BCNU	Glioma	[29]
QCh-coPLA	Doxorubicin	Graffi myeloid tumor	[36]
PLA & PLA-PLGA	Cisplatin	Glioma	[58]
PLGA	Paclitaxel	Glioma	[59]
PLA	Dichloroacetate	Cervical	[32]
PLA	Sodium dichloroacetate, diisopropylamine dichloroacetate	Colorectal	[33]
PELA with HCPD	HCPT	Breast	[60]
PEG-polybutylene succinate	Curcumin, Triclosan	N/A	[61]
QCh-coPLA	Doxorubicin hydrochloride	HeLa cells (Cervical)	[35]
PVA-Chitosan	Curcumin	Breast, Hepatoma	[34]
PCEC	Curcumin	Glioma	[62]
PCL containing UNCPS nanoparticles	Doxorubicin, Indomethacin	Hepatoma	[63]
PLGA	Doxorubicin hydrochloride immobilized on CNT	HeLa cells (Cervical)	[64]
PCL-Gelatin	Piperine	HeLa cells (Cervical), Breast	[65]
PLA	5-FU, Oxaliplatin	Colorectal	[66]
Emulsion electrospinning			
PEG-PLA	Paclitaxel, Doxorubicin hydrochloride	Glioma	[39–41]
PELA with HCPD	HCPT	Hepatoma	[42]
Co-axial electrospinning			
P(LLA-CL)	Paclitaxel	HeLa cells (Cervical)	[43]
Other technique (surface adsorption)			
Chitosan with HA coating	Paclitaxel	Prostate	[67]
Smart anti-cancer drug delivery systems			
P (NIPAAm-MAA-VP)	Doxorubicin	Lung	[45, 46]
P (NIPAAm-MAA-VP)	Doxorubicin	Melanoma	[47]
Tetra-peptide substrates of cathepsin	Doxorubicin	N/A	[48]

(continued)

Table 1 (continued)

PCL poly(ε-caprolactone), *PGC-C18* poly(glycerol monostearate-co-ε-caprolactone), *PLA* poly(L-lactide), *PLGA* poly(D,L-lactic-co-glycolic acid), *PEG* poly(ethylene glycol), *QCh* quaternized chitosan, *coPLA* poly(L-lactide-co-D,L-lactide), *PELA* poly(D,L-lactic acid)-poly(ethylene glycol), *HCPD* 2-hydroxypropyl-β-cyclodextrin, *PCEC* poly(3-caprolactone)-poly(ethylene glycol)-poly(3-caprolactone), *P(LL-CL)* poly(L-lactic acid-co-ε-caprolactone), *PEO* poly(ethylene oxide), *HA* hyaluronic acid, *P(NIPAAm-MAA-VP)* poly(N-isopropylacrylamide-acrylamide-vinyl pyrrolidone), *UNCPS* $NaGdF_4$:Yb/Er@$NaGdF_4$:Yb@$mSiO_2$-polyethylene glycol, *CNT* carbon nanotube

BCNU 1,3-bis(2-chloroethyl)-1-nitrosourea, *HCPT* hydroxycamptothecin, *5-FU* 5-fluorouracil

the issue of initial burst release associated with co-electrospinning approach used for drug loading, Zeng et al. showed that by using a compatible drug/polymer/solvent system, the initial burst can be avoided [30]. Paclitaxel, doxorubicin hydrochloride and doxorubicin base were used as model drugs and were loaded onto PLA electrospun fibers using co-electrospinning. An initial burst release was observed in case of doxorubicin hydrochloride as it localizes near the surface of fibers, whereas no initial burst was observed in case of paclitaxel and doxorubicin base suggesting the role of drug/polymer/solvent compatibility in overcoming the initial burst [30]. Using the same drug loading technique, the impact of fiber diameter on the drug release kinetics was demonstrated. For example, paclitaxel loaded PLGA nanofibers had a faster drug release rate releasing 80% of the drug in 60 days compared to 60% release in case of microfibers [31]. In a recent study, Liu et al., reported a drug delivery system wherein dichloroacetate loaded onto PLA fiber mats using co-electrospinning technique proved to attack the aerobic glycolysis process in cervical cancer cells in vivo inducing necrosis [32]. Almost 50% mice were found to recover from the disease in less than 3 weeks following implantation of dichloroacetate loaded PLA fiber mats [32] (Fig. 4). In a follow up study, the same group validated this drug delivery system against colon cancer [33].

Chitosan and its derivative (quaternized chitosan) have been found to possess intrinsic anti-tumor, anti-bacterial, anti-fungal and wound healing properties [34–36]. Therefore, recent studies have reported drug delivery systems utilizing chitosan based electrospun fibers loaded with anti-cancer drugs using co-electrospinning approach for targeted anticancer therapy against breast cancer, hepatoma and cervical cancer [34–36]. However, it is relatively difficult to electrospin chitosan alone due to the repulsive forces generated by positive charges in the polymer; therefore it is generally electrospun in presence of another polymer such as coPLA or PEO [37, 38]. Such a drug delivery system has potential in targeting both the cancer cells as well as microbial infections following tumor resection and implantation of drug loaded mats [34]. Studies utilizing co-electrospinning technique to encapsulate the anti-cancer drugs have been summarized in Table 1.

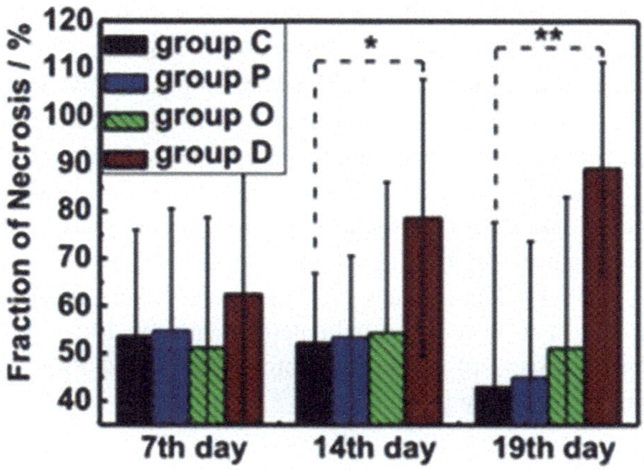

Fig. 4 In vivo anti-tumor activity of dichloroacetate loaded polylactide (PLA) electrospun fibrous mats against cervical cancer. The fraction of necrosis clearly indicates the anti-tumor activity of dichloroacetate loaded PLA electrospun fibrous mats against cervical cancer (Group D) over a period of 19 days compared to the control group (Group C). Group O represents the tumors in the mice treated by oral administration of dichloroacetate and Group P represents the tumors in the mice implanted with blank PLA mats (reprinted with permission from Liu et al. [32])

Emulsion Electrospinning

To protect anti-cancer drug against direct interactions with the body fluids and avoid undesired release and interactions, emulsion electrospinning technique for encapsulating drugs into the fibers is widely utilized [2]. In this technique, the drug and polymer are dissolved in two different solvents. The two solutions are later mixed together to form an emulsion which is used for electrospinning. When this emulsion is electrospun, it generally yields fibers with a core-shell structure, with drug largely in the core part surrounded by the polymer. In doing so, the initial burst release of drugs is also reduced. This technique can be adopted easily with no or minimal modifications required in the conventional electrospinning setup and parameters rendering it economically efficient. The main disadvantage of the process is that it does not yield a perfect core-shell structure, with the diffused core containing drug as well as some polymer and vice versa in case of the shell [2].

Emulsion electrospinning has been used to produce electrospun fibers from PEG-PLA copolymer loaded with doxorubicin hydrochloride [39]. The polymer dissolved in chloroform formed oil or organic phase and aqueous solution of doxorubicin hydrochloride formed the water phase. The two solutions were mixed to form an emulsion, which were then electrospun yielding fibers having doxorubicin hydrochloride entirely encapsulated within them. The in vitro drug release studies suggested that in the initial timeframe, drug release was mediated via diffusion, which later shifted towards the enzymatic degradation mechanism of the polymer. In vitro cytotoxicity studies revealed that the drug loaded PEG-PLLA fibers showed similar cytotoxicity against C6 glioma cells as that of free drug but with controlled release. A follow up study revealed the core-shell structure of the doxorubicin hydrochloride loaded PEG-PLA fibers and showed that the release kinetics followed Fick's second law of diffusion [40]. Interestingly, it was observed that with an increase in the drug loading, the release rate decreased indicating a reservoir type delivery system. Emulsion electrospinning can be used to load multiple drugs with different physio-chemical properties on to the fibers. For example, paclitaxel (hydrophobic) and doxorubicin hydrochloride (hydrophilic) were loaded onto PEG-PLA polymer simultaneously forming a multidrug delivery system for combination therapies [41] (Table 1). The efficacy of drug loaded fibers formed by emulsion electrospinning has been tested and validated in vivo by Luo et al., where they observed inhibitory effects of hydroxycamptothecin loaded PELA fibers against H22 hepatoma tumors in mice [42].

Co-axial Electrospinning

In order to overcome the inability of emulsion electrospinning in forming anti-cancer drug loaded fibers with a distinct core-shell structure, co-axial electrospinning can be employed. In the co-axial spinning method, the polymer and drug containing solutions are spun separately using a specialized spinneret having two separate co-axial reservoirs. The inner reservoir consists of the drug solution surrounded by the polymer solution in the co-axial outer reservoir. The electrospinning of these solutions yields fibers with distinct core-shell structure with drug trapped in the core [28]. Another advantage of this method is that the drugs are shielded from the high electric field by the polymer solution; therefore this technique can be used for encapsulating sensitive drugs [2]. This technique has been used to develop paclitaxel loaded P(LLA-CL) fibers [43] (Table 1). An initial short burst release of paclitaxel was observed in vitro followed by a steady release of paclitaxel over time period of 60 days and the efficacy of this system against HeLa cells was demonstrated in vitro.

3.1.3 Smart Anti-cancer Drug Delivery Systems Based on Electrospun Fibers

The onset of cancer is often followed by physicochemical changes in the ECM such as pH and temperature changes [44]. Efforts have been directed to take advantage of these changes in the tumor microenvironment to develop smart drug delivery systems which release drug in response to external stimuli [45, 46]. For example, doxorubicin loaded poly (N-isopropylacrylamide-acrylamide-vinylpyrrolidone) electrospun fibers have been developed for pH sensitive drug delivery [45, 46]. These fibers retain their morphology in slightly alkaline pH (7.4) but collapse in acidic pH delivering the drug [45, 46]. Similarly, a temperature sensitive drug delivery system based on doxorubicin loaded poly (N-isopropylacrylamide-acrylamide-vinylpyrrolidone) electrospun fibers bearing magnetic nanoparticles has been

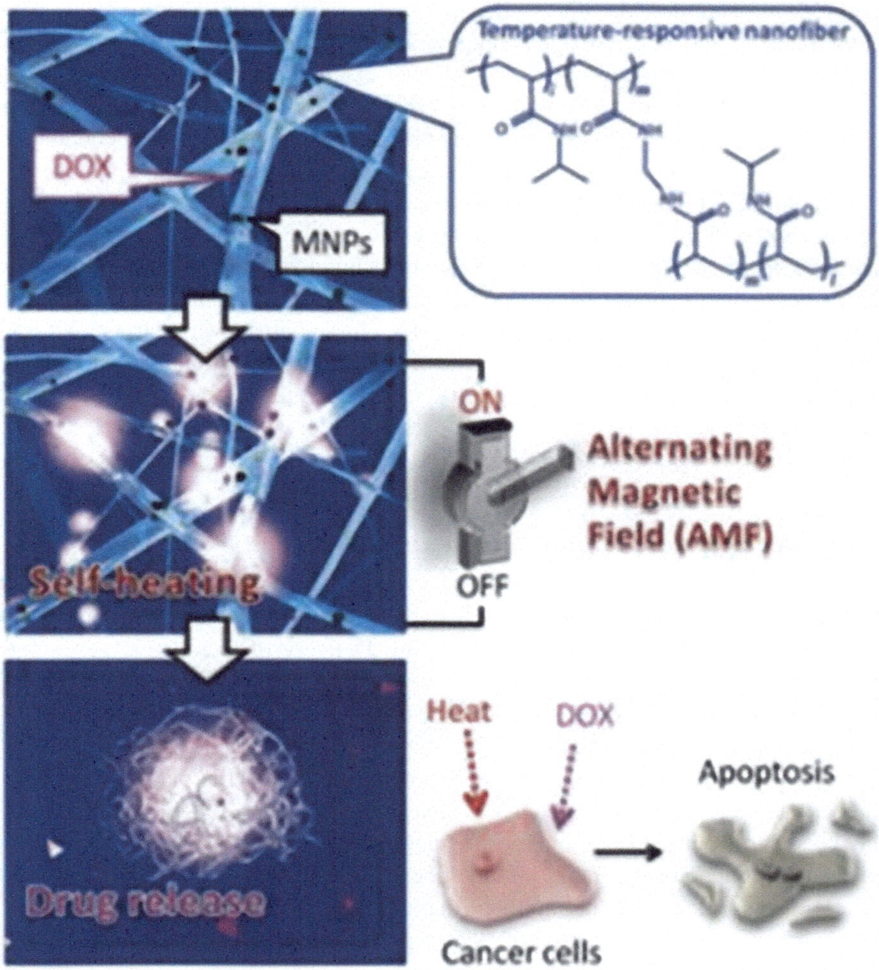

Fig. 5 Schematic of a smart drug delivery system based on electrospun fibers. Doxorubicin loaded poly (N-isopropylacrylamide-acrylamide-vinylpyrrolidone) electrospun fibers bearing magnetic nanoparticles were developed which release doxorubicin in response to the external alternating magnetic field (AMF) due to self-heating caused by magnetic nanoparticles (reprinted with permission from Kim et al. [47])

developed [47] (Fig. 5, Table 1). When the fibers were exposed to an alternating magnetic field (AMF), interaction between magnetic nano-particles and AMF led to self-heating of the fibers triggering drug release acting as 'on' and 'off' switch [47].

In addition to physicochemical changes in cancerous tissue, secretion of various ECM proteins and enzymes also occurs. Taking advantage of these secreted enzymes, electrospun fibers which respond to the secretion of cathepsin enzyme and release anti-cancer drugs have been reported [48]. The drug loaded fibers are made up of peptides which serve as substrates for the cathepsin enzyme. As the cathepsin degrades the peptides, the fibers release the anti-cancer drug. More recently, the biomimetic aspects of electrospun nanofibers have been coupled with drug delivery to design a smart therapeutic system [49]. For example, aligned electrospun PCL fibers mimicking white matter tracts were used to trick the highly migratory glioma cells to migrate along the fibers away from primary tumor towards a drug containing sink [49].

Overall, electrospun fibers serve as a versatile and cost effective platform to develop anti-cancer drug delivery systems. Owing to the ability to tune the electrospinning process parameters and drug loading techniques coupled with a wide variety of polymers to choose from; electrospun fibers can be engineered to deliver a variety of anti-cancer drugs. They also provide close control over pharmacokinetic properties such as the drug release kinetics. The use of electrospun biomaterials as smart anti-cancer drug delivery systems is an emerging area in modern drug delivery.

3.2 Electrospun Biomaterials in Targeted Gene Delivery

Recent advances in genetic engineering have resulted in employing gene delivery as a powerful platform for anti-cancer therapies. Delivery of therapeutic genes can potentially activate apoptosis pathways in cancerous cells [6]. However, direct administration of these therapeutic genes causes toxicity in normal cells and can lead to a hyper immune response [6]. This necessitated the development of targeted gene delivery carriers. Electrospun fibers prove to be a versatile platform for gene delivery as they present a high surface area to volume ratio, tunable morphology, and chemistry. With advances in electrospinning techniques, different strategies of encapsulating therapeutic genes into the fibers were devised; protecting them from external environment, retaining their therapeutic properties and maintaining a steady release rate. The potential of electrospun fibers in encapsulating and immobilizing bio-macromolecules like enzymes have been investigated almost a decade ago [50]. However, it was not until recently that they were investigated for potential applications in anti-cancer gene delivery [2, 6].

Co-electrospinning is a very easy and cost effective way of encapsulating genetic material on to the fibers. In this technique, the genetic material and polymers are dissolved together in the solvent forming a homogeneous electrospinning solution [6]. The release rate of genetic material from the polymer is modulated by bulk degradation and surface erosion of polymer by enzymes [6]. Achille et al., recently demonstrated the anti-proliferative effects of electrospun PCL fibers encapsulating plasmid DNA encoding Cdk2 silencing shRNA against MCF-7 breast cancer cells [51] (Fig. 6). The plasmid DNA was encapsulated in fiber by co-electrospinning and ~20–60 ng/mL of plasmid DNA was released over 21 days causing 40% decrease in the proliferation of MCF-7 breast cancer cells [51]. Nanoparticles containing genetic material can also be encapsulated within fibers which provide additional control over release of the genetic material and additional protection from external

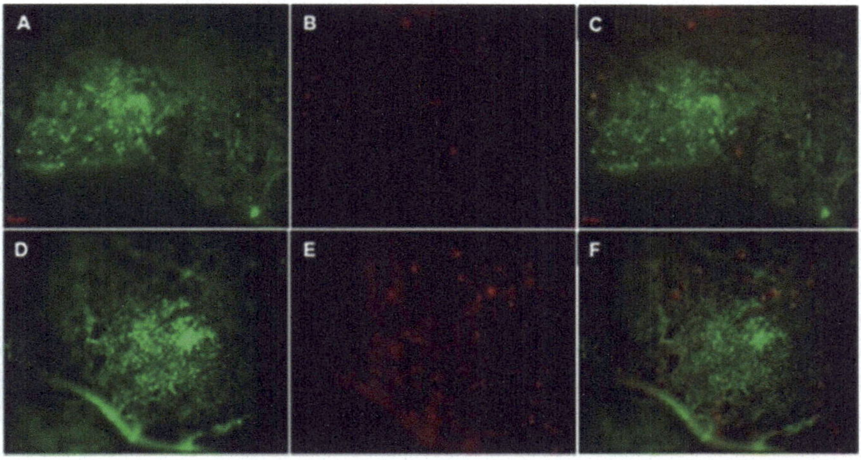

Fig. 6 Electrospun PCL fibers encapsulating plasmid DNA encoding Cdk2 silencing shRNA exhibit anti-proliferative effects against MCF-7 breast cancer cells (bottom panel). **a**, **d**—live cells; **b**, **e**—dead cells; **c**, **f**—merged images of live and dead cells (reprinted with permission from Achille et al. [51])

environment [6]. To this end, PLGA electrospun fibers with encapsulated chitosan/siRNA nanoparticles enabled silencing of EGFP activity by 50% in H1299 lung cancer cells within 48 h of transfection [52]. Similarly, miRNA-145 containing polyethyleneimine (PEI) nanoparticle coated electrospun PCL fibers loaded with paclitaxel using co-electrospinning have been reported for the treatment of liver cancer with simultaneous drug and gene delivery [53].

Co-axial electrospinning can be utilized to create a core-shell fiber which can be used for simultaneous gene and drug delivery. By using co-axial electrospinning, core-shell PEO fibers with the core containing cytotoxic drug 5-fluorocytosine and the shell containing RhoB-bPEI-pDNA, a plasmid DNA polyplex, which functions as efficient suicide gene have been developed and shown to exert cytotoxic effects against A549 lung cancer cells in vitro [54].

In addition to the above techniques, Layer-by-layer deposition of multiple DNA layers on the fibers have also been employed for enhanced gene delivery [6]. Using this approach, multiple DNA films were deposited on the electrospun PCL fibrous matrices due to poly(β-amino esters) in the PCL fibers which interact with anionic DNA molecule leading to DNA adsorption [55]. Iterative adsorption steps yielded a multi layered pEGFP-N1 DNA coating onto the PCL fibers. The sustained release of pEGFP-N1 DNA from the fibers was accompanied by 60–80% transfection efficiency in primary human glioblastoma cells (GB-319) [55].

In sum, electrospun fibers provide a versatile template for targeted gene delivery which can help overcome the challenges associated with direct administration of genes. Innovation in targeted gene therapy strategies can play a pivotal role in widening the applications of gene therapy using electrospun fibers.

4 Electrospun Biomaterials as Disease Models for Cancer

In addition to drug and gene delivery, electrospun fibers are extensively used in tissue engineering applications [10]. Because of their versatile surface chemistry, tunability of mechanical properties and surface topology, high surface area to volume ratio, and low cost of production with flexibility of choosing appropriate polymer, electrospun biomaterials are attractive for applications in tissue engineering as scaffolds [10]. Scaffolds serve as three dimensional (3D) templates recapitulating the key aspects of the tissue environment in vitro. Cells are known to behave differently in a 3D context compared to a 2D context (i.e., in traditional 2D culture flasks) which closely mimics the cellular behavior in vivo [68]. In particular, electrospun fibers have been shown to modulate several aspects of cell behaviors including morphology, adhesion, and expression of cytoskeletal proteins such as vinculin, actin and focal adhesion kinases [10]. These aspects are critical for tissue engineering applications in creating a relevant microenvironment and directing them to form tissue. Inspired by their tissue engineering applications, more recently, electrospun fibers are being investigated for their potential in recapitulating the tumor microenvironment in vitro. This has led to the development of engineered 3D cancer models for investigating cancer cell behavior in vitro, resulting in better understanding of cancer cell biology as well as robust drug screening platforms (Table 2).

4.1 Electrospun Fibers for Recapitulating Tumor Microenvironments In Vitro

Electrospun fibers provide a relevant 3D context in vitro which recapitulates the tumor microenvironment evidenced in vivo in terms of topography, mechanics and chemistry [69, 70]. In order to serve as 3D substrates for cancer cells, it is essential for the electrospun fibers to promote cell adhesion, proliferation and viability. To this end, Szot et al., demonstrated that PCL/collagen electrospun fibers promoted cell adhesion and proliferation across a variety of cancer cell types (viz. prostate cancer cell line PC-3, murine renal cancer cell line RENCA and human breast cancer cell line MDA-MD-231) over conventional bacterial cellulose fibers [71]. The polymer concentration in the electrospinning solution and fiber diameter affected the cancer cell proliferation and viability [71]. In addition to cancer cells, the PCL/collagen electrospun fibers supported the adhesion and proliferation of HMEC-1 endothelial cells, one of the main components in the tumor stroma [71].

The alignment, stiffness and surface chemistry of electrospun fibers strongly influence cancer cell morphology, migration and biomechanics. Generally, the cancer cells (viz. MDA-MB-231, U-87, U-251) elongate and stretch in the direction of the fiber orientation specifically in the case of aligned fibers, whereas they tend to take up a flat stellar or round shape on random fibers [72–75] (Fig. 7). However,

Table 2 Electrospun fibers as disease models to study cancer biology

Polymer	Fiber orientation	Cell type	Type of cancer	Key aspects of the study	Reference
PCL (Shell) & gelatin, PES, PDMS (Core) (Core-shell)	A	Patient derived GBM cells	Glioma	Employed core-shell nanofibers to recapitulate white matter tracts	[79]
PCL	A	Patient derived GBM cells and rat astrocytes	Glioma	Modeled interactions between GBM cells and astrocytes (stromal cells)	[81]
PCL	A	Patient derived GBM stem like cells	Glioma	Modeled the shift in the invasion strategy of GBM cells	[80]
PCL	R, A	MCF-7, MCF-7 10A, MDA-MB-231	Breast	Fiber orientation has significant impact on migration	[76]
Chitosan-PCL	R, A	U-87 MG	Glioma	Aligned fibers lead to elongated cell morphology, increased migration and a mesenchymal shift	[73]
PCL	R, A	U-87 MG, A172	Glioma	Fiber orientation dictates cytoskeletal stiffness, aligned fiber increase the expression of migratory genes	[78]
PCL	R, A	U-87, U-251	Glioma	Upregulation of JAK/STAT pathway in cells on aligned fibers	[75]
PCL	R, A	U-251, Patient derived GBM stem like cells	Glioma	Fiber orientation has significant impact on cell migration	[77]
PCL	R	Co-culture of CT-26 cells and bone marrow dendritic cells	Colorectal	Co-culture recapitulates cancer cell-immune cell interaction	[93]
PCL	A	H605	Breast	Fibers induce EMT in H605	[72]
PCL	R, A	MDA-MB-231, T47D	Breast	Aggressive breast cancer cells enter dormancy	[74]
PCL	R	MCF-7, T47-D, SK-Br-3, MDA-MB-231	Breast	Upregulation in stem cell markers OCT3/4, Sox2 following EMT	[11]

(continued)

Table 2 (continued)

Polymer	Fiber orientation	Cell type	Type of cancer	Key aspects of the study	Reference
PLGA & PLA-PEG copolymer	R	LLC1, PC-3, B16, MCF-10A, MDA-MB-231, BG1	Lung, Prostate, Melanoma, Breast & Ovarian	Cells underwent EMT followed by an increase in drug resistance	[87]
PCL-Chitosan	A	MCF-7, T47-D	Breast	Fibers enriched the population of stem like breast cancer cells, an increase in drug resistance towards docetaxel and doxorubicin	[89]
PCL	R	TC-71	Ewing's Sarcoma	Increased drug resistance to doxorubicin	[85]
Collagen	R	C4-2B	Prostate	Increased drug resistance towards docetaxel and camptothecin	[84]
Silk	R	HN-12	Head and neck squamous carcinoma	Increased drug resistance to paclitaxel in cells seeded on the fiber compared to 2D cultures	[88]
PHBV-Collagen peptide	R	MKN-28	Gastric cancer	Increased drug resistance to 5-FU, oxaliplatin, cisplatin, paclitaxel, irinotecan in cells seeded on fibers	[86]
PCL & PCL-gelatin	R	C4-2B	Prostate	Electrospun fibers were functionalized with perlecan heparin sulphate to recapitulate bone marrow architecture for studying metastasis	[90]
PCL	A	Human OBs, MDA-MB-231 bone seeking, MCF-7 10A	Breast	The PCL fibers functionalized with BMP-7 were used to create tissue engineered bone serving as a humanized metastatic site in mice	[91, 92]

(continued)

Table 2 (continued)

PCL poly(ε-caprolactone), *PLA* poly(L-lactide), *PLGA* poly(D,L-lactic-co-glycolic acid), *PEG* poly(ethylene glycol), *PES* poly(ether sulphone), *PDMS* poly(dimethylsiloxane), *PHBV* poly(3-hydroxybutyrate-co-3-hydroxyvalerate)
R random, *A* aligned
GBM glioblastoma multiforme, *OB* osteoblasts
EMT epithelial to mesenchymal transition, *BMP-7* bone morphogenetic protein-7

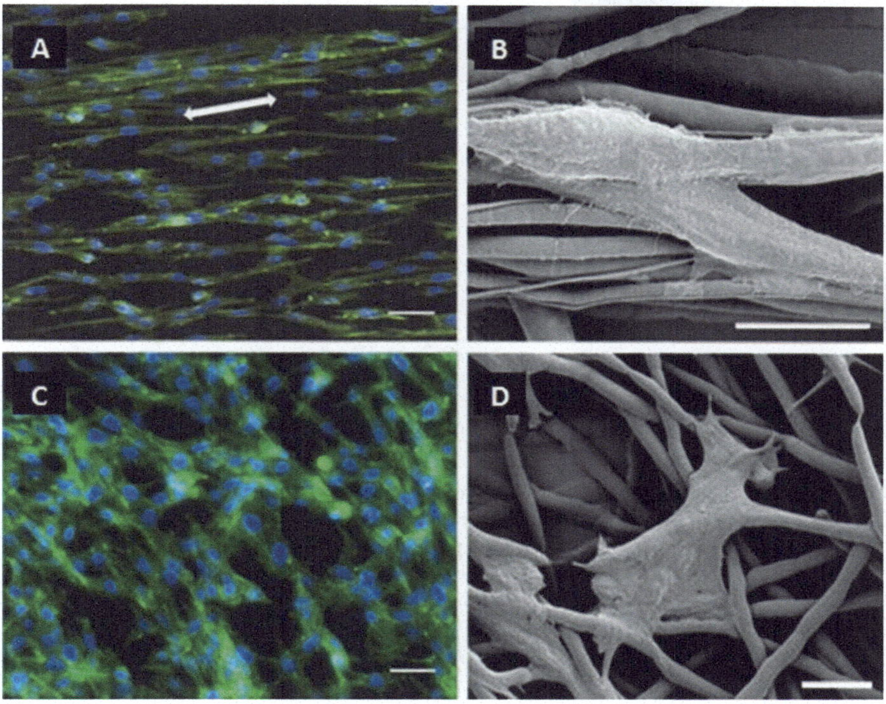

Fig. 7 Effect of fiber alignment on cancer cell morphology. H605 rat mammary tumor cells exhibit a spindle like elongated morphology on the aligned electrospun PCL fibers as examined via fluorescence microscopy (**a**) and SEM (**b**). In contrast, H605 cells exhibit a flat stellar morphology on the random electrospun PCL fibers as examined via fluorescence microscopy (**c**) and SEM (**d**) (fluorescence microscopy images: scale bar = 40 μm, SEM images: scale bar = 20 μm) (reprinted with permission from Saha et al. [72])

random fibers can recapitulate the initial random orientation of collagen fibers in the ECM before the cancer cells reorganize them into an aligned orientation [11].

Apart from influencing morphology, fiber alignment has been shown to strongly influence cancer cell migration and cytoskeletal stiffness. For example, Nelson et al. found that the random PCL electrospun fibers did not affect the MDA-MB-231 breast cancer cell migration in presence of CXCL12 gradient whereas the same cells on aligned PCL electrospun fibers displayed 82% higher migration in presence of CXCL12 gradient [76]. Similar trends between fiber alignment and cell migration

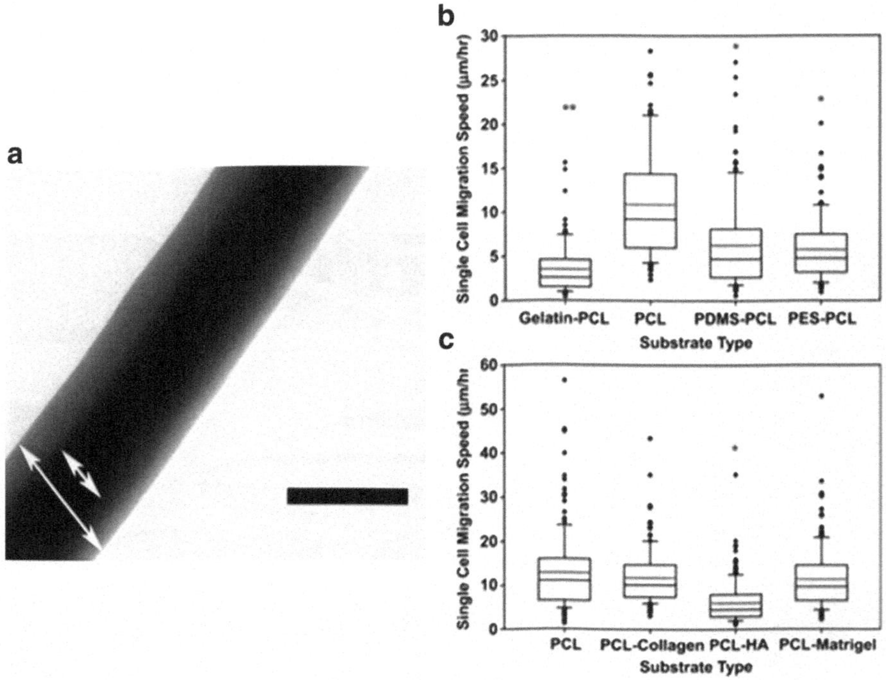

Fig. 8 Effect of fiber stiffness and surface chemistry on the migration of GBM cells. (**a**) TEM image of core-shell electrospun fiber for studying effect of surface chemistry and fiber stiffness on the patient derived GBM cells. Scale bar = 0.2 μm. (**b**) GBM cells exhibit enhanced migration on the electrospun PCL fibers of moderate stiffness of 8 MPa compared to gelatin-PCL (2 MPa), PDMS-PCL (33.3 MPA) and PES-PCL (28.6 MPa). (**c**) Hyaluronic acid as a shell decreases the GBM migration speed indicating the role of surface chemistry in mediating GBM migration (reprinted with permission from Rao et al. [79])

has been observed with glioblastoma (GBM) cells [73, 77]. A recent study provided mechanistic insight into the effect of fiber alignment on migration by demonstrating activation of JAK/STAT pathway in U87 and U251 GBM cells when they were seeded on aligned PCL electrospun fibers [75]. Inhibition of STAT3 in these cells hindered migration on aligned fibers [75]. Cytoskeletal stiffness of cancer cells could also be modulated by fiber alignment. For example, cytoskeletal stiffness of U-87MG and A172 GBM cells was lower when they were seeded on the aligned PCL electrospun fibers compared to the random fibers [78]. Pro-migratory genes viz. SNAI1 and NOTCH1 were found to be upregulated in GBM cells seeded on aligned PCL fibers [78]. Overall, the aligned electrospun fibers support cancer cell migration and can be used to recapitulate the in vivo migratory pathways.

The surface chemistry and electrospun fiber stiffness have been shown to profoundly impact cancer cell behaviors. The effect of surface chemistry and electrospun fiber stiffness on the behavior of patient derived GBM cells was demonstrated in a study by Rao et al., which employed aligned electrospun core-shell nano-fibers to recapitulate the migration of GBM cells over white matter tracts [79] (Fig. 8). To study the effect of fiber stiffness, core-shell fibers with common shell material (PCL) and varying core materials (viz. Gelatin, PES and PDMS) were used. GBM cells were found to migrate faster on fibers with intermediate stiffness (8 MPa) compared to those with higher (30 MPa) or lower (2 MPa) stiffness respectively [79]. To study the effect of surface chemistry, core-shell fibers with common core material (PCL) and varying shell material (viz. hyaluronic acid, collagen and Matrigel®) were used. Hyaluronic acid reduced cell attachment and migration of GBM cells on the fibers whereas the collagen and Matrigel® did not have any significant effect [79]. Electrospun fibers closely recapitulate the topography of white matter tracts evidenced in vivo; making them suitable candidates for recapitulating the migration of GBM cells over the white matter tracts in the brain [69]. Effect of stromal cells (viz. astrocytes) and ECM components (viz. hyaluronic acid) on the migration of patient derived GBM cells over the electrospun fibers has also been studied; indicating the potential of electrospun fibers in providing a versatile platform for modeling the tumor stroma by incorporation of relevant stromal components [80, 81].

Overall, electrospun fibers prove to be excellent substrates for recapitulating the tumor microenvironment in vitro. A variety of tunable parameters associated with electrospun fibers enable researchers to modify the cancer cell behavior in vitro by providing them with a relevant 3D context and unravel various aspects of cancer biology.

4.2 Electrospun Fibers for Studying Phenotypic Changes in Cancer Cells

Most of the cancers are characterized by a phenotypic shift conferring them with an invasive potential or dormancy. For instance, the Epithelial to Mesenchymal Transition (EMT) process has implications in cancer progression and metastasis as it confers cancer cells with an invasive mesenchymal phenotype [82]. Recently, it was found that the electrospun fibers can induce EMT in breast cancer and GBM cells [72, 73]. In a study by Saha et al., H605 mouse mammary tumor cells were found to undergo EMT when seeded on aligned PCL fibers [72]. Induction of EMT was confirmed by increase in the expression of genes such as Ck14, Sma, TGF-β, Snail, STAT3, Fsp-1 and SMAD3, which are implicated in EMT [72] (Fig. 9). An increase in the expression of the similar genes was observed when U87MG GBM cells were seeded on the aligned Chitosan-PCL nanofibers indicating a mesenchymal shift [73]. Furthermore, Feng et al. noted an upregulation in stem cell markers in MCF-7, T47D and SK-BR-3 breast cancer cells seeded onto electrospun random PCL fibers following EMT linking EMT to stemness in cancer cells [11]. The stemness in breast cancer cells was characterized by an increased expression of SOX4 and CD49f [11]. Similarly, electrospun fibers have been used to create a 3D model of dormancy wherein the aggressive MDA-MD-231 cells were found to become

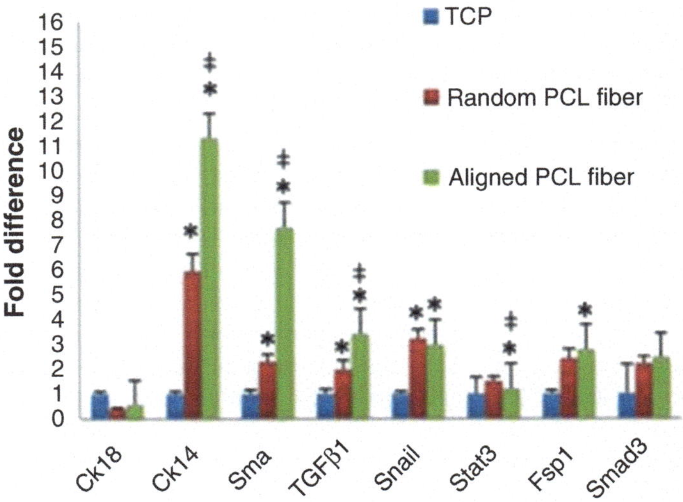

Fig. 9 Aligned electrospun PCL fibers induce EMT in H605 rat mammary tumor cells indicated by an increase in the expression of genes implicated in the EMT process such as Ck14, Sma, TGFβ1, Snail, Stat3 and Fsp1 compared to Random PCL fibers or tissue culture polystyrene (TCP) (reprinted with permission from Saha et al. [72])

dormant when seeded on PCL fibers [74]. Cell cycle analysis revealed that the cells were arrested in S phase of cell cycle [74]. Dormancy in cancer cells is also associated with drug resistance and metastasis [4]. Overall, 3D in vitro models based on electrospun fibers provide a platform to study and target the phenotypic changes in cancer cells which are typically difficult to recapitulate in traditional 2D cultures.

4.3 Electrospun Fibers for Studying Drug Resistance in Cancer Cells

Drug resistance continues to be a key challenge in treating many types of cancer. Traditionally, anti-cancer drug screening is performed on conventional 2D cultures. These 2D cultures do not provide cancer cells with a relevant context they experience in vivo ultimately resulting in an inefficient and unreliable drug screen leading to poor drug performance in vivo and eventually drug resistance [83]. To overcome the drawbacks of 2D in vitro drug screening, novel 3D in vitro drug screening models are being developed. Electrospun scaffolds have proved to be promising in developing 3D models for anti-cancer drug screening. For example, cancer cells seeded onto the electrospun scaffolds exhibited higher drug resistance compared to cells cultured on 2D surfaces. In particular, Hartman et al. showed that C4-2B prostate cancer cells seeded onto electrospun collagen fibers exhibited higher drug resistance towards docetaxel and camptothecin compared to the cells grown on tissue

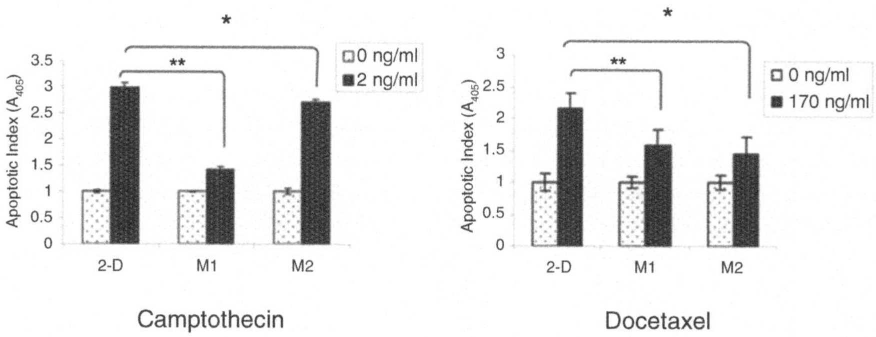

Fig. 10 Cancer cells exhibit enhanced drug resistance when seeded onto electrospun fibers. C4-2B prostate cancer cells exhibit enhanced resistance to camptothecin and docetaxel when cultured on electrospun collagen fibers (M1 & M2) compared to 2-D TCPS controls as indicated by apoptotic index (M1—collagen type I dissolved in a 1,1,1,3,3,3-hexafluoro-2-propanol (HFIP); M2—collagen type I in 2,2,2-trifluoroethanol (TFE)) (reprinted with permission from Hartman et al. [84])

culture polystyrene (TCPS) [84] (Fig. 10). Similarly, subsequent studies showed increased drug resistance in TC-71 human Ewing's sarcoma cells to doxorubicin and in MKN28 gastric cancer cell line to 5-FU, Oxaliplatin and Cisplatin when seeded onto electrospun PCL and PHBV fibers, respectively [85, 86].

Drug resistance is also associated with EMT shift in cancer cells [82]. Cancer cells having mesenchymal phenotype are found to be resistant to drugs. Girard et al. showed that the LLC1 lung cancer cells undergo EMT on PLGA/PLA-PEG copolymer electrospun fibers exhibiting higher drug resistance [87]. Similarly, Bulysheva et al. showed increased chemo-resistance in HN12 head and neck squamous carcinoma cells towards paclitaxel when seeded onto electrospun silk fibers following EMT [88]. Certain subpopulation of cancer cells exhibit properties similar to stem cells. These are referred to as cancer stem like cells (CSCs). They are implicated in cancer metastasis and are known to exhibit radio/chemo-resistance [82]. Electrospun fibers have been used to enrich the CSCs population in cancer cells [11]. For example, aligned PCL electrospun scaffold were found to enrich the breast cancer CSCs population when MCF-7 and T-47D breast cancer cells were seeded onto the fibrous scaffold [89]. These breast cancer CSCs exhibited higher drug resistance to docetaxel and doxorubicin [89]. Overall, electrospun fibrous scaffolds can be used to create robust 3D models for anti-cancer drug screening as they provide a relevant 3D context to the cancer cells and recapitulate cellular behavior typically observed in vivo.

4.4 Electrospun Fibers for Studying Cancer Metastasis

The spread of cancer from a primary tumor site to a distant organ resulting in metastasis is a critical event which defines the most advanced stages of cancer. The mechanisms underlying cancer metastasis are not well understood, primarily because of lack of 3D in vitro models to recapitulate the complex metastatic cascade. Because of their ability to support growth of a wide range of cell types and versatile chemistry allowing functionalization with various ECM proteins, electrospun fibers provide versatile platforms for recapitulating the 3D microenvironment a cancer cell experiences at a metastatic site. To this end, electrospun PCL-gelatin fibers functionalized with Perlecan domain IV (PlnDIV) peptide were employed as 3D mimics of bone marrow [90]. When C4-2B prostate cancer cells were seeded on these fibrous scaffolds, they displayed enhanced adhesion and infiltrated the scaffold recapitulating the metastatic infiltration in vivo. In addition, an increase in focal adhesion kinase (FAK) and its tyrosine 397 activity was noted suggesting that the scaffold activated physiologically relevant signaling pathways, thus providing a platform to study prostate cancer metastasizing to the bone [90].

An emerging area to study metastasis is to create a tissue engineered organ implant which when implanted into mice serves as a humanized metastatic site [91, 92]. Following the implantation, cancer cells are orthotopically injected into mice; which then colonize the implant. Electrospun fibers have been utilized to create tunable tissue engineered organ implants which serve as a humanized metastatic site [91, 92]. Such an approach was used to study breast cancer metastasis to bone. The tissue engineered bone mimics were created using aligned electrospun PCL fibers functionalized with bone morphogenetic protein-7 (BMP-7) [91]. These tissue engineered bone was implanted into mice followed by the orthotopic injection of MDA-MB-231 bone seeking breast cancer cells which colonized the engineered bone in vivo recapitulating breast cancer bone metastasis [91]. In a follow up study, the same strategy was used to study the effect of β1-integrins on breast cancer bone metastasis [92] (Fig. 11).

Overall, electrospun fibers can be successfully utilized to create 3D models for studying cancer cell behaviors in vitro. In particular, electrospun fibers can be engineered to model certain cancer cell behaviors or recapitulate certain aspects of the tumor microenvironment as they provide the necessary 3D context to the cancer cells partially or closely mimicking the in vivo setting. Their ability to capture phenotypic shifts and drug resistance observed in cancer cells makes them suitable for robust anti-cancer drug screening applications.

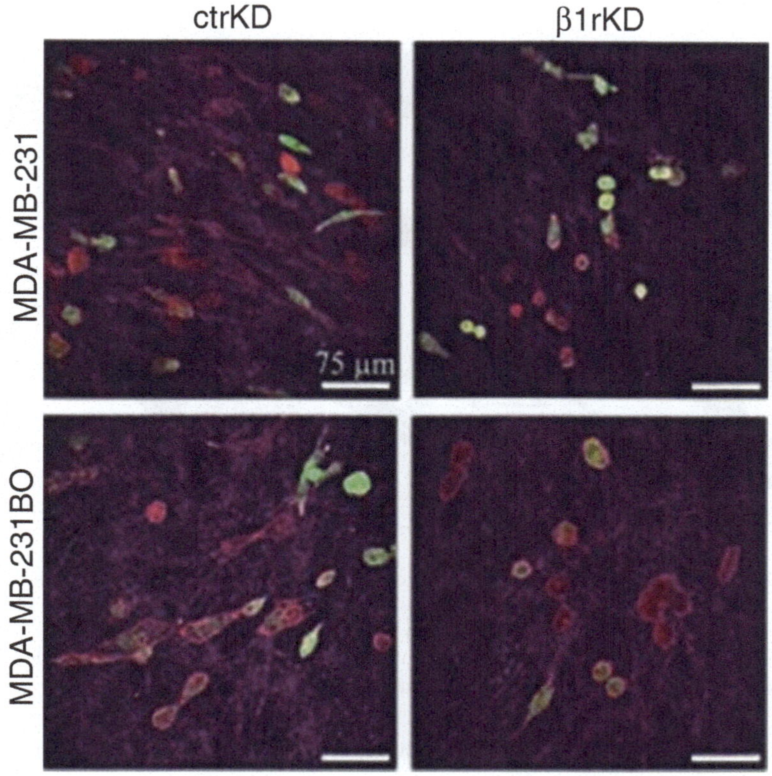

Fig. 11 Electrospun fibers to create tissue engineered humanized bone implants for studying cancer metastasis reveals the role of β1-integrins in cell spreading on tissue engineered bone implants in vitro. Left panel: Control group without the knockdown of β1-integrins shows spreading of MDA-MB-231 and MDA-MB-231 BO (bone seeking) breast cancer cells on the tissue engineered bone implants examined via confocal fluorescence microscopy. Right panel: MDA-MB-231 and MDA-MB-231 BO breast cancer cells take up a rounded morphology on tissue engineered bone implants following the knockdown of β1-integrins (Red: F-actin staining, Green: GFP positive cells) (scale bar = 75 μm) (reprinted with permission from Thibaudeau et al. [92])

5 Electrospun Biomaterials in Cancer Diagnostics

Diagnosis of cancer at later stages and a lack of personalized therapeutic regimen worsen the chances of patient survival. Poor prognosis of the disease is the main factor leading to relapse and metastasis [94]. Therefore, continual efforts are being made to develop biosensor technologies for early detection of cancer and to capture prognostic markers to determine the progression of disease [3]. Owing to precise control over the mechanical and chemical properties, reliability, easy and cost effective processing; electrospun fibers have been widely investigated for their application as biosensors in recent years [3]. Versatility of electrospinning technique also

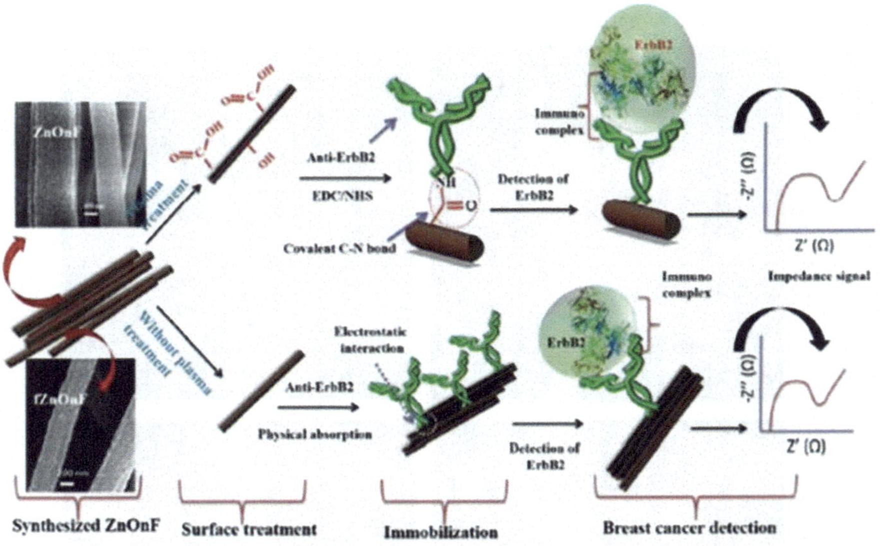

Fig. 12 Process flow diagram for fabricating electrochemical impedance based biosensor using electrospun ZnO nanofibers for early diagnosis of breast cancer. The electrospun ZnO fibers are conjugated to anti-ErbB2 antibody which can detect EGFR expressed on the surface of breast cancer cells over a wide concentration range of 1 fM–0.5 μM and produce a detectable electrochemical signal (reprinted with permission from Ali et al. [95])

allows metals and metal oxides to be spun into fibers as they play a key role in the signal transduction and amplification.

5.1 Electrospun Fibers for Electrochemical Impedance Based Biosensors

Electrospun fibers have been used to develop electrochemical impedance based biosensors to detect surface markers over expressed on cancerous cells and even genetic mutations for early detection of cancer [3]. Recently, an electrochemical impedance based biosensor was developed by taking advantage of such an over expressed surface protein—human epidermal growth factor receptor 2 (ErbB2) for diagnosis of breast cancer [95] (Fig. 12). ErbB2 is particularly over expressed in breast cancers. To create the sensor, mesoporous electrospun zinc oxide (ZnO) nanofibers were oxygen treated for conjugation of anti-ErbB2 antibody on the fiber surface [95]. This antibody can bind the EGFR expressed on the surface of breast cancer cells which generates an electrochemical signal that is detectable. A wide range of concentration (1 fM–0.5 μM) of ErbB2 was detected using this device [95]. In addition to proteins, changes in gene levels could also be detected. For example, an electrochemical biosensor that can sense mutation in K-ras gene and transduce it to a

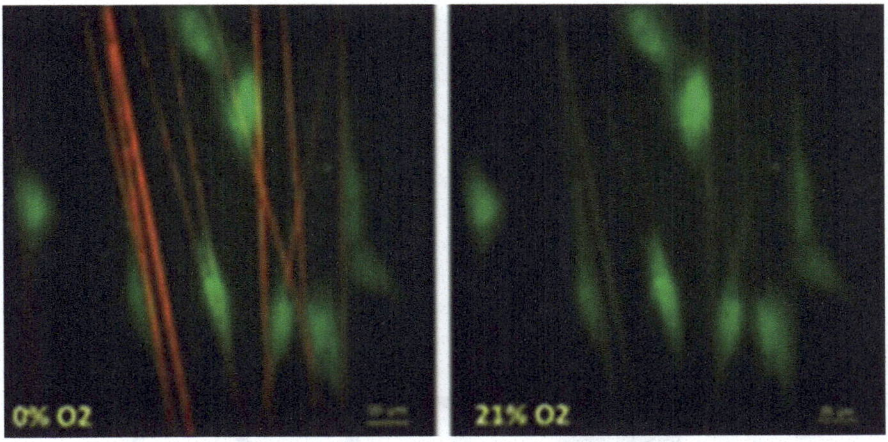

Fig. 13 Luminescence based hypoxia detection in tumor microenvironment using metalloporphyrins doped electrospun fibers. Core-shell electrospun fibers containing metalloporphyrins doped polycarbonate core and PCL shell support U251 GBM cell adhesion and luminesce in presence of U251 GBM cells under hypoxic conditions (0% O_2, left panel) and do not luminesce under normoxic conditions (21% O_2, right panel) (reprinted with permission from Xue et al. [99])

detectable electrical signal has been reported [96]. The biosensor was fabricated out of carboxylated multi-walled carbon nanotube doped nylon-6 electrospun fibers. The biosensor was found to successfully detect the K-ras mutation in SW480 colorectal cancer cells with detection limit as low as 30 fM [96].

5.2 *Electrospun Fibers for Fluorescence/Luminescence Based Biosensors*

Electrospun fibers have also been utilized in developing fluorescence/luminescence based biosensors for early detection and prognosis of cancer. For example, fluorescence based chemo-sensors fabricated using electrospun PCL-metalloporphyrin fibers have been developed recently to detect histamines [97]. An increased level of histamines in the human urine is attributed to a variety of diseases including cancer [3, 97]. Metalloporphyrins were used as probes for sensing histamines [97]. Similarly, electrospun fibers have been utilized to develop luminescence based biosensors to detect hypoxia in the tumor microenvironment [98, 99]. Real time oxygen sensors were fabricated using core-shell electrospun nanofibers containing luminescent organometallic compounds (viz. metalloporphyrins) as oxygen sensing probes and were tested in vitro in presence of U-251 and CNS1 GBM cells [98, 99] (Fig. 13). Hypoxia in the tumor microenvironment has been implicated in radio-/chemo-resistance and metastasis [100]. Therefore, sensing hypoxia can serve as excellent prognostic marker.

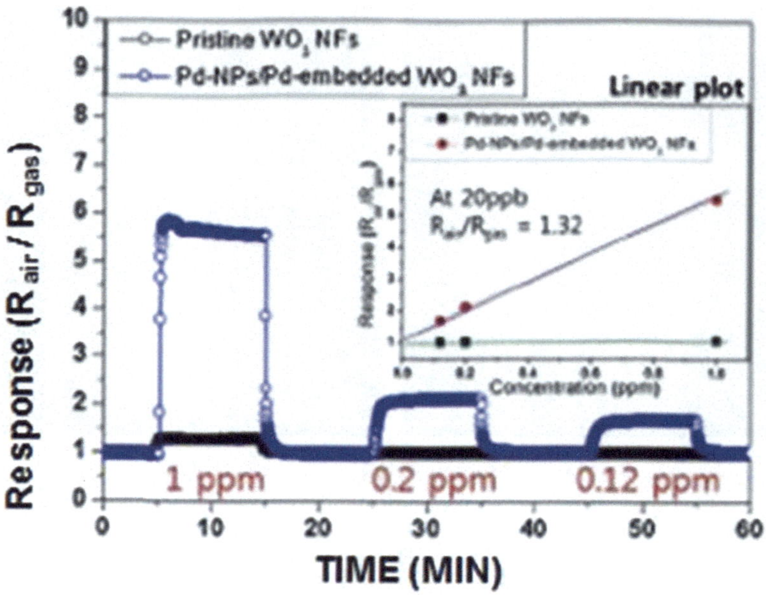

Fig. 14 Chemo-resistive gas sensors based on electrospun fibers for early detection of lung cancer. Cyclic response levels of pristine Tungsten oxide (WO_3) nanofibers and Palladium (Pd)-nanoparticles (NPs)/Pd-embedded WO_3 nanofibers indicating higher toluene sensing ability in presence of palladium catalyst (reprinted with permission from Kim et al. [101])

5.3 *Electrospun Fibers for Chemo-Resistive Gas Sensors*

Exhaled breath contains a variety of volatile organic compounds (VOCs) formed as metabolites [3, 101]. Certain VOCs are present at a higher concentration in the exhaled breath of a diseased individual compared to a healthy individual [101]. Specifically, in the case of lung cancer, 2–3 folds higher concentration (80–100 ppb) of toluene is present in the patient's breath compared to a healthy person [101]. Inspired by this, researchers have successfully designed chemo-resistive gas sensors based on electrospun fibers [3, 101, 102]. Recently, Kim et al., developed a tungsten oxide (WO_3) electrospun fiber based toluene detection biosensor which can sense toluene in the exhaled breath with a high toluene detection sensitivity (response (R_{air}/R_{gas}) = 5.5 @ 1 ppm) [101]. It was observed that by coating the WO_3 fibers with Palladium (Pd) catalyst decreased the activation energy, thus increasing the gas sensing ability of the sensor [101] (Fig. 14).

5.4 Electrospun Fibers for Capturing Circulating Tumor Cells (CTCs)

In addition to their applications in biosensors, electrospun fibers have been used for capturing circulating tumor cells (CTCs). Cancer cells at the primary tumor disengage from the primary tumor and enter the blood circulation. These cancer cells (CTCs) are known to colonize a distant organ forming metastasis. If captured, CTCs can be analyzed for certain biomarkers and genetic variations providing crucial information for therapeutic decision making [3]. Electrospun fibers serve as a 3D template with high surface area and porosity to which the cancer cells can adhere and hence are being investigated for their application in capturing CTCs. However, detection of CTCs faces a major technical challenge because of an extremely low population (up to hundreds per mL of blood) amongst large number of blood cells (10^9 per mL of blood) [103, 104]. To address this issue, Zhang et al. developed a new CTC detection system based on electrospun TiO_2 nanofibers with epithelial cell adhesion molecule antibody (anti-EpCAM) grafted onto the fiber surface for capturing colorectal (HCT116) and gastric cancer cells (BGC823) in vitro [103]. Similarly, another group devised anti-EpCAM grafted MnO_2 nanofibers fabricated into microchannels to trap the MCF-7 breast cancer CTCs with 80% capture efficiency at 0.1 mL h^{-1} flowrate of cell suspension in phosphate buffered saline [104]. Hyaluronic acid is one of the ECM components which interacts with the CD44 receptors overexpressed in the aggressive cancer cells. CD44+ breast cancer cells are known to be highly metastatic [105]. Taking advantage of over expressed CD44 receptors, recently, electrospun PVA/polyethyleneimine nanofibers conjugated with hyaluronic acid were developed to trap HeLa cells as they overexpress CD44 [106].

One of the major limitations hindering analysis of trapped CTCs relates to the contamination caused by white blood cells (WBCs) creating a need for better cell sorting system along with isolation [107]. To address this issue, Zhao et al. devised PLGA nanofiber based setup to efficiently capture prostate CTCs reducing the impurities [107] (Fig. 15). Similarly, Blackstone et al. showed that MCF-7 breast cancer cells can be specifically sorted out of human fibroblasts and keratinocytes by using CF_4 plasma treated electrospun nanofibers based on differences in their adhesion potential [108]. PVA/polyethyleneimine nanofibers can be chemically modified by introducing Folic acid, which renders them hemocompatible and can be further used in cancer cell capture [109]. Integrating the CTC capture with therapeutic aspects, electrospun polystyrene fiber containing iron oxide nano-particles functionalized with collagen were used to capture ovarian (SKOV-3) cancer cells and kill them [110]. The iron oxide nano-particles generate heat when exposed to alternating magnetic field killing the captured cancer cells. The fibers can undergo multiple heating cycles without loss of heating capacity [110].

Overall, electrospun fibers exhibit superior structure dependent properties which make them suitable for developing biosensors sensing various cues ranging from microenvironmental changes to genetic mutations as well as cancer cell capture. These ultra-sensitive electrospun fiber based biosensors can potentially be used to

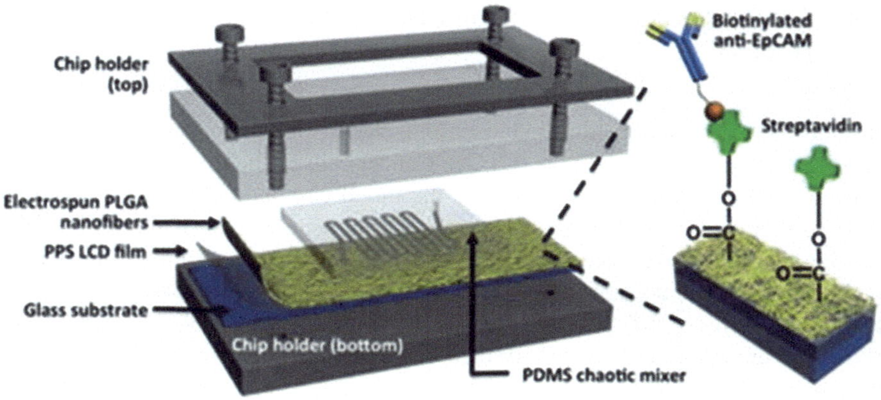

Fig. 15 Electrospun PLGA fiber based device for high purity capture of prostate cancer CTCs for whole exome sequencing. Streptavidin is covalently attached to electrospun PLGA fibers to facilitate the conjugation of biotinylated anti-EpCAM antibody onto the fibers which binds to the EpCAM expressed on the CTCs (reprinted with permission from Zhao et al. [107])

develop point of care cancer detection strategies leading to better prognosis and personalized therapies.

6 Electrospun Biomaterials in Targeted Stem Cell Delivery

Because electrospun fibers can be engineered to mimic the topography of the ECM closely, it makes them ideal for supporting stem cell cultures [111]. Stem cells are known to interact with the microenvironment around them and this interaction is crucial in maintaining stemness and supporting their proliferation. By controlling the structural properties of electrospun fibers, it is possible to closely mimic the stem cell niche evidenced in vivo and provide them with crucial microenvironmental cues [111].

Stem cells have received great attention for their potential in regenerative medicine. The stem cell niche has striking similarities to the tumor microenvironment due to which stem cells exhibit unique tumor-homing properties [12, 112]. This makes stem cells potential candidates for anti-cancer drug and gene carriers. Though this proposition seems promising, retaining the viability and stemness as well as delivering to and retaining them at the target site is not without challenges. Electrospun fibrous scaffolds can immobilize these stem cells at the therapeutic site and simultaneously provide them with essential cues while retaining their stemness and viability. A recent study developed poly-lactic acid (PLA) electrospun nanofibrous scaffolds loaded with bioengineered mesenchymal stem cells specifically for targeting residual GBM cells in the surgical cavity following resection of the primary tumor [12]. The mesenchymal stem cells were bioengineered to secrete tumor-necrosis factor-related apoptosis-inducing ligand (TRAIL) known to induce

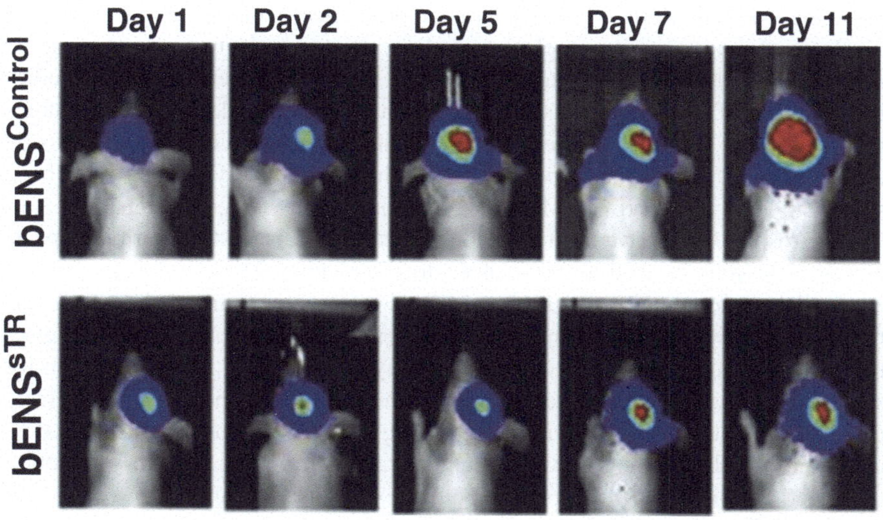

Fig. 16 In vivo therapeutic effect of electrospun PLA scaffold bearing bioengineered mesenchymal stem cells (MSCs) against the residual GBM cells in the post surgical cavity. Delayed relapse of GBM was observed in the mice implanted with electrospun PLA scaffold bearing therapeutic bioengineered MSCs ($bENS^{sTR}$) following surgical resection of GBM tumor compared to control group ($bENS^{control}$) over a period of 11 days (reprinted with permission from Bago et al. [12])

apoptosis in the cancerous cells. The poly-lactic acid (PLA) electrospun nanofibrous scaffolds loaded with bioengineered mesenchymal stem cells were implanted in the post-surgical cavity to induce apoptosis in the residual GBM cells which prevents or lowers the risk of relapse (Fig. 16). In vivo, the bioengineered stem cells seeded onto the fibrous scaffold remained viable for over 3 weeks and induced apoptosis in residual GBM cells following tumor resection. In a murine model, almost 2.3 fold increase in the median survival was noted [12]. The potential of electrospun fibers for targeted stem cell delivery is promising and could be pursued for targeting other types of cancers.

7 Conclusions

Electrospun biomaterials have been applied in multiple areas in cancer research including targeted drug and gene delivery, biomimetic 3D models, drug screening, bio-sensing, as well as stem cell delivery. Advances in electrospinning technique have made it possible to engineer electrospun scaffolds with a range of properties providing a modular platform for cancer research. Because of the high surface area to volume ratio, biocompatibility and easy tailoring, electrospun fibers prove to be excellent templates for delivering bioactive molecules ranging from synthetic drugs to genetic materials. The fibers can be engineered by tailoring the process parameters to yield desired pharmacokinetic properties. Multiple strategies such as co-electrospinning, emulsion electrospinning, co-axial electrospinning, can be applied for loading the therapeutic agents onto the fibers. Whereas, several studies in the context of drug delivery using electrospun fibers exist, relatively fewer studies have been done in context of gene delivery. Future research would likely focus on exploring anti-cancer gene delivery. Most of the drug delivery systems have been studied in vitro which provides further scope for their in vivo evaluation.

Electrospun fibers have been applied as disease models to study various aspects of cancer biology as well as anti-cancer drug screening. These materials provide controlled environments to study cancer cell-matrix interactions as well as the molecular pathways involved in cancer progression. Because of the ability of electrospun fibers to support stem cell culture; they provide an attractive platform for stem cell bioengineering and targeted anti-cancer stem cell therapies. Apart from providing a platform for studying and treating cancer, electrospun fibers have be utilized to develop biosensors which demonstrate their potential in early detection of cancer. Sophisticated strategies are being devised using electrospun scaffolds to capture CTCs to serve as prognostic markers for cancer metastasis. Although a lot of progress has been made, particularly using in vitro studies, additional studies would be needed to validate the performance of these biosensors for clinical applications.

Overall, electrospun biomaterials have enormous potential in the field of cancer research. Future work would likely focus on successful translation of electrospun biomaterial-based drug delivery and bio-sensing techniques into clinical trials. Considering the advent of anti-cancer immunotherapies; investigation into the role of electrospun biomaterials in enhancing the efficacy of these immunotherapies is much awaited. Future studies into adapting the electrospun biomaterial-based disease models to serve as high throughput anti-cancer drug screens could transform current cancer treatment strategies.

References

1. Stewart B, Wild CP (eds) (2014) World cancer report 2014. WHO Press, Geneva
2. Balaji A, Vellayappan MV, John AA et al (2015) An insight on electrospun-nanofibers-inspired modern drug delivery system in the treatment of deadly cancers. RSC Adv 5:57984–58004
3. Chen Z, Chen Z, Zhang A et al (2016) Electrospun nanofibers for cancer diagnosis and therapy. Biomater Sci 4:922–932
4. Aguirre-Ghiso JA (2007) Models, mechanisms and clinical evidence for cancer dormancy. Nat Rev Cancer 7:834–846
5. Gu L, Mooney DJ (2016) Biomaterials and emerging anticancer therapeutics: engineering the microenvironment. Nat Rev Cancer 16:56–66
6. Lee S, Jin G, Jang J-H (2014) Electrospun nanofibers as versatile interfaces for efficient gene delivery. J Biol Eng 8:30
7. Quail DF, Joyce JA (2013) Microenvironmental regulation of tumor progression and metastasis. Nat Med 19:1423–1437

8. Bhowmick NA, Moses HL (2005) Tumor–stroma interactions. Curr Opin Genet Dev 15:97–101
9. Zustiak SP (2015) The role of matrix compliance on cell responses to drugs and toxins: towards predictive drug screening platforms. Macromol Biosci 15:589–599
10. Baker BM, Handorf AM, Ionescu LC et al (2009) New directions in nanofibrous scaffolds for soft tissue engineering and regeneration. Expert Rev Med Devices 6:515–532
11. Feng S, Duan X, Lo PK et al (2013) Expansion of breast cancer stem cells with fibrous scaffolds. Integr Biol (Camb) 5:768–777
12. Bago JR, Pegna GJ, Okolie O et al (2016) Electrospun nanofibrous scaffolds increase the efficacy of stem cell-mediated therapy of surgically resected glioblastoma. Biomaterials 90:116–125
13. Zamani M, Prabhakaran MP, Ramakrishna S (2013) Advances in drug delivery via electrospun and electrosprayed nanomaterials. Int J Nanomed 8:2997–3017
14. Haider A, Haider S, Kang I-K (2015) A comprehensive review summarizing the effect of electrospinning parameters and potential applications of nanofibers in biomedical and biotechnology. Arab J Chem. https://doi.org/10.1016/j.arabjc.2015.11.015
15. Haider S, Al-Zeghayer Y, Ahmed Ali FA et al (2013) Highly aligned narrow diameter chitosan electrospun nanofibers. J Polym Res 20:105
16. Baumgarten PK (1971) Electrostatic spinning of acrylic microfibers. J Colloid Interface Sci 36:71–79
17. Sun B, Long YZ, Zhang HD et al (2014) Advances in three-dimensional nanofibrous macrostructures via electrospinning. Prog Polym Sci 39:862–890
18. Zhang C, Yuan X, Wu L et al (2005) Study on morphology of electrospun poly(vinyl alcohol) mats. Eur Polym J 41:423–432
19. Zong X, Kim K, Fang D et al (2002) Structure and process relationship of electrospun bioabsorbable nanofiber membranes. Polymer 43:4403–4412
20. Sill TJ, von Recum HA (2008) Electrospinning: applications in drug delivery and tissue engineering. Biomaterials 29:1989–2006
21. Lubasova D, Martinova L (2011) Controlled morphology of porous polyvinyl butyral nanofibers. J Nanomater 2011:6
22. Megelski S, Stephens JS, Chase DB et al (2002) Micro-and nanostructured surface morphology on electrospun polymer fibers. Macromolecules 35:8456–8466
23. Pelipenko J, Kristl J, Janković B et al (2013) The impact of relative humidity during electrospinning on the morphology and mechanical properties of nanofibers. Int J Pharm 456:125–134
24. Park J-Y, Lee I-H (2010) Relative humidity effect on the preparation of porous electrospun polystyrene fibers. J Nanosci Nanotechnol 10:3473–3477
25. Bae H-S, Haider A, Selim KMK et al (2013) Fabrication of highly porous PMMA electrospun fibers and their application in the removal of phenol and iodine. J Polym Res 20:158
26. De Vrieze S, Van Camp T, Nelvig A et al (2008) The effect of temperature and humidity on electrospinning. J Mater Sci 44:1357
27. Sridhar R, Ravanan S, Venugopal JR et al (2014) Curcumin-and natural extract-loaded nanofibres for potential treatment of lung and breast cancer: in vitro efficacy evaluation. J Biomater Sci Polym Ed 25:985–998
28. Son YJ, Kim WJ, Yoo HS (2014) Therapeutic applications of electrospun nanofibers for drug delivery systems. Arch Pharm Res 37:69–78
29. Xu X, Chen X, Xu X et al (2006) BCNU-loaded PEG–PLLA ultrafine fibers and their in vitro antitumor activity against Glioma C6 cells. J Control Release 114:307–316
30. Zeng J, Yang L, Liang Q et al (2005) Influence of the drug compatibility with polymer solution on the release kinetics of electrospun fiber formulation. J Control Release 105:43–51
31. Xie J, Wang CH (2006) Electrospun micro- and nanofibers for sustained delivery of paclitaxel to treat C6 glioma in vitro. Pharm Res 23:1817–1826
32. Liu D, Liu S, Jing X et al (2012) Necrosis of cervical carcinoma by dichloroacetate released from electrospun polylactide mats. Biomaterials 33:4362–4369

33. Liu D, Wang F, Yue J et al (2015) Metabolism targeting therapy of dichloroacetate-loaded electrospun mats on colorectal cancer. Drug Deliv 22:136–143
34. Sedghi R, Shaabani A, Mohammadi Z et al (2017) Biocompatible electrospinning chitosan nanofibers: a novel delivery system with superior local cancer therapy. Carbohydr Polym 159:1–10
35. Ignatova MG, Manolova NE, Toshkova RA et al (2010) Electrospun nanofibrous mats containing quaternized chitosan and polylactide with in vitro antitumor activity against HeLa cells. Biomacromolecules 11:1633–1645
36. Toshkova R, Manolova N, Gardeva E et al (2010) Antitumor activity of quaternized chitosan-based electrospun implants against Graffi myeloid tumor. Int J Pharm 400:221–233
37. Sun K, Li Z (2011) Preparations, properties and applications of chitosan based nanofibers fabricated by electrospinning. Express Polym Lett 5:342–361
38. Pakravan M, Heuzey M-C, Ajji A (2011) A fundamental study of chitosan/PEO electrospinning. Polymer 52:4813–4824
39. Xu X, Yang L, Xu X et al (2005) Ultrafine medicated fibers electrospun from W/O emulsions. J Control Release 108:33–42
40. Xu X, Chen X, Ma P et al (2008) The release behavior of doxorubicin hydrochloride from medicated fibers prepared by emulsion-electrospinning. Eur J Pharm Biopharm 70:165–170
41. Xu X, Chen X, Wang Z et al (2009) Ultrafine PEG–PLA fibers loaded with both paclitaxel and doxorubicin hydrochloride and their in vitro cytotoxicity. Eur J Pharm Biopharm 72:18–25
42. Luo X, Xie C, Wang H et al (2012) Antitumor activities of emulsion electrospun fibers with core loading of hydroxycamptothecin via intratumoral implantation. Int J Pharm 425:19–28
43. Huang HH, He CL, Wang HS et al (2009) Preparation of core-shell biodegradable microfibers for long-term drug delivery. J Biomed Mater Res A 90:1243–1251
44. Estrella V, Chen T, Lloyd M et al (2013) Acidity generated by the tumor microenvironment drives local invasion. Cancer Res 73:1524–1535
45. Salehi R, Irani M, Eskandani M et al (2014) Interaction, controlled release, and antitumor activity of doxorubicin hydrochloride from pH-sensitive P(NIPAAm-MAA-VP) nanofibrous scaffolds prepared by green electrospinning. Int J Polym Mater 63:609–619
46. Salehi R, Irani M, Rashidi M-R et al (2013) Stimuli-responsive nanofibers prepared from poly(N-isopropylacrylamide-acrylamide-vinylpyrrolidone) by electrospinning as an anticancer drug delivery. Des Monomers Polym 16:515–527
47. Kim Y-J, Ebara M, Aoyagi T (2013) A smart hyperthermia nanofiber with switchable drug release for inducing cancer apoptosis. Adv Funct Mater 23:5753–5761
48. Ben-Nun Y, Fichman G, Adler-Abramovich L et al (2016) Cathepsin nanofiber substrates as potential agents for targeted drug delivery. J Control Release. https://doi.org/10.1016/j.jconrel.2016.11.028
49. Jain A, Betancur M, Patel GD et al (2014) Guiding intracortical brain tumour cells to an extracortical cytotoxic hydrogel using aligned polymeric nanofibres. Nat Mater 13:308–316
50. Zeng J, Aigner A, Czubayko F et al (2005) Poly(vinyl alcohol) nanofibers by electrospinning as a protein delivery system and the retardation of enzyme release by additional polymer coatings. Biomacromolecules 6:1484–1488
51. Achille C, Sundaresh S, Chu B et al (2012) Cdk2 silencing via a DNA/PCL electrospun scaffold suppresses proliferation and increases death of breast cancer cells. PLoS One 7:e52356
52. Chen M, Gao S, Dong M et al (2012) Chitosan/siRNA nanoparticles encapsulated in PLGA nanofibers for siRNA delivery. ACS Nano 6:4835–4844
53. Che HL, Lee HJ, Uto K et al (2015) Simultaneous drug and gene delivery from the biodegradable poly(epsilon-caprolactone) nanofibers for the treatment of liver cancer. J Nanosci Nanotechnol 15:7971–7975
54. Sukumar UK, Packirisamy G (2015) Bioactive core-shell nanofiber hybrid scaffold for efficient suicide gene transfection and subsequent time resolved delivery of prodrug for anticancer therapy. ACS Appl Mater Interfaces 7:18717–18731

55. Li C, Tzeng SY, Tellier LE et al (2013) (3-aminopropyl)-4-methylpiperazine end-capped poly(1,4-butanediol diacrylate-co-4-amino-1-butanol)-based multilayer films for gene delivery. ACS Appl Mater Interfaces 5:5947–5953
56. Yohe ST, Herrera VL, Colson YL et al (2012) 3D superhydrophobic electrospun meshes as reinforcement materials for sustained local drug delivery against colorectal cancer cells. J Control Release 162:92–101
57. Chen P, Q-S W, Ding Y-P et al (2010) A controlled release system of titanocene dichloride by electrospun fiber and its antitumor activity in vitro. Eur J Pharm Biopharm 76:413–420
58. Xie J, Tan RS, Wang CH (2008) Biodegradable microparticles and fiber fabrics for sustained delivery of cisplatin to treat C6 glioma in vitro. J Biomed Mater Res A 85:897–908
59. Ranganath SH, Wang CH (2008) Biodegradable microfiber implants delivering paclitaxel for post-surgical chemotherapy against malignant glioma. Biomaterials 29:2996–3003
60. Xie C, Li X, Luo X et al (2010) Release modulation and cytotoxicity of hydroxycamptothecin-loaded electrospun fibers with 2-hydroxypropyl-beta-cyclodextrin inoculations. Int J Pharm 391:55–64
61. Llorens E, Ibanez H, Del Valle LJ et al (2015) Biocompatibility and drug release behavior of scaffolds prepared by coaxial electrospinning of poly(butylene succinate) and polyethylene glycol. Mater Sci Eng C Mater Biol Appl 49:472–484
62. Guo G, Fu S, Zhou L et al (2011) Preparation of curcumin loaded poly(ε-caprolactone)-poly(ethylene glycol)-poly(ε-caprolactone) nanofibers and their in vitro antitumor activity against Glioma 9L cells. Nanoscale 3:3825–3832
63. Chen Y, Liu S, Hou Z et al (2015) Multifunctional electrospinning composite fibers for orthotopic cancer treatment in vivo. Nano Res 8:1917–1931
64. Yu Y, Kong L, Li L et al (2015) Antitumor activity of doxorubicin-loaded carbon nanotubes incorporated poly(lactic-co-glycolic acid) electrospun composite nanofibers. Nanoscale Res Lett 10:1044
65. Jain S, Meka SRK, Chatterjee K (2016) Engineering a piperine eluting nanofibrous patch for cancer treatment. ACS Biomater Sci Eng 2:1376–1385
66. Zhang J, Wang X, Liu T et al (2016) Antitumor activity of electrospun polylactide nanofibers loaded with 5-fluorouracil and oxaliplatin against colorectal cancer. Drug Deliv 23:794–800
67. Ma G, Liu Y, Peng C et al (2011) Paclitaxel loaded electrospun porous nanofibers as mat potential application for chemotherapy against prostate cancer. Carbohydr Polym 86:505–512
68. Edmondson R, Broglie JJ, Adcock AF et al (2014) Three-dimensional cell culture systems and their applications in drug discovery and cell-based biosensors. Assay Drug Dev Technol 12:207–218
69. Rao SS, Lannutti JJ, Viapiano MS et al (2014) Toward 3D biomimetic models to understand the behavior of glioblastoma multiforme cells. Tissue Eng B Rev 20:314–327
70. Rape A, Ananthanarayanan B, Kumar S (2014) Engineering strategies to mimic the glioblastoma microenvironment. Adv Drug Del Rev 0:172–183
71. Szot CS, Buchanan CF, Gatenholm P et al (2011) Investigation of cancer cell behavior on nanofibrous scaffolds. Mater Sci Eng C 31:37–42
72. Saha S, Duan X, Wu L et al (2012) Electrospun fibrous scaffolds promote breast cancer cell alignment and epithelial-mesenchymal transition. Langmuir 28:2028–2034
73. Kievit FM, Cooper A, Jana S et al (2013) Aligned chitosan-polycaprolactone polyblend nanofibers promote the migration of glioblastoma cells. Adv Healthc Mater 2:1651–1659
74. Guiro K, Patel SA, Greco SJ et al (2015) Investigating breast cancer cell behavior using tissue engineering scaffolds. PLoS One 10:e0118724
75. Agudelo-Garcia PA, De Jesus JK, Williams SP et al (2011) Glioma cell migration on three-dimensional nanofiber scaffolds is regulated by substrate topography and abolished by inhibition of STAT3 signaling. Neoplasia 13:831–840
76. Nelson MT, Short A, Cole SL et al (2014) Preferential, enhanced breast cancer cell migration on biomimetic electrospun nanofiber 'cell highways'. BMC Cancer 14:825

77. Johnson J, Nowicki MO, Lee CH et al (2009) Quantitative analysis of complex glioma cell migration on electrospun polycaprolactone using time-lapse microscopy. Tissue Eng C Methods 15:531–540
78. Beliveau A, Thomas G, Gong J et al (2016) Aligned nanotopography promotes a migratory state in glioblastoma multiforme tumor cells. Sci Rep 6:26143
79. Rao SS, Nelson MT, Xue R et al (2013) Mimicking white matter tract topography using core-shell electrospun nanofibers to examine migration of malignant brain tumors. Biomaterials 34:5181–5190
80. Cha J, Kang S-G, Kim P (2016) Strategies of mesenchymal invasion of patient-derived brain tumors: microenvironmental adaptation. Sci Rep 6:24912
81. Grodecki J, Short AR, Winter JO et al (2015) Glioma-astrocyte interactions on white matter tract-mimetic aligned electrospun nanofibers. Biotechnol Prog 31:1406–1415
82. Singh A, Settleman J (2010) EMT, cancer stem cells and drug resistance: an emerging axis of evil in the war on cancer. Oncogene 29:4741–4751
83. Imamura Y, Mukohara T, Shimono Y et al (2015) Comparison of 2D- and 3D-culture models as drug-testing platforms in breast cancer. Oncol Rep 33:1837–1843
84. Hartman O, Zhang C, Adams EL et al (2009) Microfabricated electrospun collagen membranes for 3-D cancer models and drug screening applications. Biomacromolecules 10:2019–2032
85. Fong ELS, Lamhamedi-Cherradi S-E, Burdett E et al (2013) Modeling Ewing sarcoma tumors in vitro with 3D scaffolds. Proc Natl Acad Sci U S A 110:6500–6505
86. Kim YJ, Bae HI, Kwon OK et al (2009) Three-dimensional gastric cancer cell culture using nanofiber scaffold for chemosensitivity test. Int J Biol Macromol 45:65–71
87. Girard YK, Wang C, Ravi S et al (2013) A 3D fibrous scaffold inducing tumoroids: a platform for anticancer drug development. PLoS One 8:e75345
88. Bulysheva AA, Bowlin GL, Petrova SP et al (2013) Enhanced chemoresistance of squamous carcinoma cells grown in 3D cryogenic electrospun scaffolds. Biomed Mater 8:055009
89. Sims-Mourtada J, Niamat RA, Samuel S et al (2014) Enrichment of breast cancer stem-like cells by growth on electrospun polycaprolactone-chitosan nanofiber scaffolds. Int J Nanomed 9:995–1003
90. Hartman O, Zhang C, Adams EL et al (2010) Biofunctionalization of electrospun PCL-based scaffolds with perlecan domain IV peptide to create a 3-D pharmacokinetic cancer model. Biomaterials 31:5700–5718
91. Thibaudeau L, Taubenberger AV, Holzapfel BM et al (2014) A tissue-engineered humanized xenograft model of human breast cancer metastasis to bone. Dis Model Mech 7:299–309
92. Thibaudeau L, Taubenberger AV, Theodoropoulos C et al (2015) New mechanistic insights of integrin β1 in breast cancer bone colonization. Oncotarget 6:332–344
93. Kim TE, Kim CG, Kim JS et al (2016) Three-dimensional culture and interaction of cancer cells and dendritic cells in an electrospun nano-submicron hybrid fibrous scaffold. Int J Nanomed 11:823–835
94. Ramaswamy S, Ross KN, Lander ES et al (2003) A molecular signature of metastasis in primary solid tumors. Nat Genet 33:49–54
95. Ali MA, Mondal K, Singh C et al (2015) Anti-epidermal growth factor receptor conjugated mesoporous zinc oxide nanofibers for breast cancer diagnostics. Nanoscale 7:7234–7245
96. Wang X, Shu G, Gao C et al (2014) Electrochemical biosensor based on functional composite nanofibers for detection of K-ras gene via multiple signal amplification strategy. Anal Biochem 466:51–58
97. Seong DY, Choi M-S, Kim Y-J (2012) Fluorescent chemosensor for the detection of histamine based on dendritic porphyrin-incorporated nanofibers. Eur Polym J 48:1988–1996
98. Xue R, Behera P, Xu J et al (2014) Polydimethylsiloxane core–polycaprolactone shell nanofibers as biocompatible, real-time oxygen sensors. Sens Actuators B Chem 192:697–707
99. Xue R, Nelson MT, Teixeira SA et al (2016) Cancer cell aggregate hypoxia visualized in vitro via biocompatible fiber sensors. Biomaterials 76:208–217

100. Teicher BA (1994) Hypoxia and drug resistance. Cancer Metast Rev 13:139–168
101. Kim N-H, Choi S-J, Yang D-J et al (2014) Highly sensitive and selective hydrogen sulfide and toluene sensors using Pd functionalized WO3 nanofibers for potential diagnosis of halitosis and lung cancer. Sens Actuators B Chem 193:574–581
102. Shin J, Choi SJ, Lee I et al (2013) Thin-wall assembled SnO2 fibers functionalized by catalytic Pt nanoparticles and their superior exhaled-breath-sensing properties for the diagnosis of diabetes. Adv Funct Mater 23:2357–2367
103. Zhang N, Deng Y, Tai Q et al (2012) Electrospun TiO2 nanofiber-based cell capture assay for detecting circulating tumor cells from colorectal and gastric cancer patients. Adv Mater 24:2756–2760
104. H-q L, X-l Y, Cai B et al (2015) Capture and release of cancer cells using electrospun etchable MnO2 nanofibers integrated in microchannels. Appl Phys Lett 106:093703
105. Sheridan C, Kishimoto H, Fuchs RK et al (2006) CD44+/CD24- breast cancer cells exhibit enhanced invasive properties: an early step necessary for metastasis. Breast Cancer Res 8:R59
106. Zhao Y, Fan Z, Shen M et al (2015) Hyaluronic acid-functionalized electrospun polyvinyl alcohol/polyethyleneimine nanofibers for cancer cell capture applications. Adv Mater Interfaces 2:1–9
107. Zhao L, YT L, Li F et al (2013) High-purity prostate circulating tumor cell isolation by a polymer nanofiber-embedded microchip for whole exome sequencing. Adv Mater 25:2897–2902
108. Blackstone BN, Willard JJ, Lee CH et al (2012) Plasma surface modification of electrospun fibers for adhesion-based cancer cell sorting. Integr Biol (Camb) 4:1112–1121
109. Z-y F, Y-l Z, Luo Y et al (2016) Folic acid modified electrospun poly (vinyl alcohol)/polyethyleneimine nanofibers for cancer cell capture applications. Chin J Polym Sci 34:755–765
110. Huang C, Soenen SJ, Rejman J et al (2012) Magnetic electrospun fibers for cancer therapy. Adv Funct Mater 22:2479–2486
111. Lim SH, Mao H-Q (2009) Electrospun scaffolds for stem cell engineering. Adv Drug Del Rev 61:1084–1096
112. Li L, Neaves WB (2006) Normal stem cells and cancer stem cells: the niche matters. Cancer Res 66:4553–4557

Electrospun Nanofibrous Nerve Conduits

Jeong In Kim, Tae In Hwang, Joshua Lee, Chan Hee Park, and Cheol Sang Kim

Abstract An injury to the human nervous system, which plays a major role in our daily lives by being involved in our thought and action processes, has been one of the greatest issues in the medical field. Social costs are considerably high because of these injuries and the many ongoing studies searching for cures to nervous system injuries. As a result of these efforts, electrospinning technology has been found to a suitable alternative to fabricating scaffolds for nerve regeneration. The electrospun nanofibrous scaffold can provide the regenerating nervous system with cell-friendly environments that have sufficient porosity, mechanical strength, guidance cues, etc. First, the anatomies of the central and peripheral nervous systems and their regeneration mechanisms are introduced and compared to each other. Second, the mechanisms, requirements, and favored properties are discussed. Finally, various fabrication methods and the current evolving concept of electrospun nerve conduits with functionalization strategies such as cell loading, neurotrophic biomolecule or nanoparticle immobilization, and conductive polymer use are discussed.

Keywords Electrospinning • Aligned nanofibers • Nanofibrous scaffolds • Nanofibrous nerve conduits • Biomimicry

J.I. Kim • J. Lee
Department of Bionanosystem Engineering, Graduate School, Chonbuk National University, Jeonju, Republic of Korea

T.I. Hwang
Department of Medical Practicing, Woori Convalescent Hospital, Jeonju, Republic of Korea

Department of Bionanosystem Engineering, Graduate School, Chonbuk National University, Jeonju, Republic of Korea

C.H. Park • C.S. Kim (✉)
Department of Bionanosystem Engineering, Graduate School, Chonbuk National University, Jeonju, Republic of Korea

Division of Mechanical Design Engineering, College of Engineering, Chonbuk National University, Jeonju, Republic of Korea
e-mail: biochan@jbnu.ac.kr; chskim@jbnu.ac.kr

J. Almodovar (ed.), *Electrospun Biomaterials and Related Technologies*,
DOI 10.1007/978-3-319-70049-6_7

List of Abbreviations

a-FGF	Acidic fibroblast growth factor
ASIA	American Spinal Injury Association
ATF-3	Activating transcription factor-3
BDNF	Brain-derived neurotrophic factor
CNS	Central nervous system
CNTF	Ciliary neurotrophic factor
DRG	Dorsal root ganglia
ECM	Extracellular matrix
EMSCs	Ectomesenchymal stem cells
GAP-43	Growth-associated protein-43
GDNF	Glial cell line-derived neurotrophic factor
GVHD	Graft-versus-host disease
MAIs	Myelin-associated inhibitors
MSCs	Mesenchymal stem cells
MWCNT	Multi-walled carbon nanotubes
NGFs	Nerve growth factors
NPCs	Neural progenitor cells
NPs	Nanoparticles
NRG1	Neuregulin 1
NSCs	Neural stem cells
NT-3	Neurotrophin-3
NTFs	Neurotrophic factors
OECs	Olfactory ensheathing cells
OEG	Olfactory ensheathing glia
PAN	Polyacrylonitrile
PCL	Polycaprolactone
PE	Polyethylene
PHB	Poly-β-hydroxybutyrate
PLGA	Poly(lactide-co-glycolide)
PNS	Peripheral nervous system
PPy	Polypyrrole
PVC	Polyvinyl chloride
RAGs	Regeneration-associated genes
rNSCs	Rat neural stem cells
SCI	Spinal cord injuries
siRNA	Small-interfering RNA
Sox11	SRY-box containing gene 11
SPRR1A	Small proline-repeat protein 1A
SWCNT	Single-walled carbon nanotube
TF-MSNs	Transferrin-modified mesoporous silica nanoparticles
VEGF	Vascular endothelial growth factor
α1-GP	Alpha-1 glycoprotein

1 Introduction

The human nervous system is responsible for the thought processes and action control that occurs in our bodies. The nervous system consists of two main parts: the central nervous system (CNS) and peripheral nervous system (PNS), which are based on location and functions. The PNS collects and integrates somatic and autonomic information from different sensory nerves and organs throughout the trunk and extremities. The CNS determines the responses based on the collected information and sends commands throughout our body via the PNS. These processes are conducted literally millions of times each minute of our daily lives [1].

Despite being connected systems and having similarities in functions and gross anatomy, the CNS and PNS show distinct differences. For instance, the healing capacities of PNS injuries are considerably greater than those of CNS lesions owing to the difference in the intrinsic properties of neurons and extrinsic cellular environments.

With the nervous system playing such a major role, an injury to the system can cause not only a socio-economic problem but a decline in an individual's quality of life. For example, spinal cord injuries (SCI) can cause various clinical manifestations including loss of motor control and sensation, pain, numbness in related areas, and rarely death. The one-year survival rate of SCIs in the U.S. (2015, NSCISC annual statistical report) was reported to be higher than 95.5%, due in part to the fact that damage to the spinal cord below the cervical level rarely leads to death. However, many American Spinal Injury Association (ASIA) Classification "complete" injury cases result in permanent disabilities because neuronal cells generally fail to regenerate in a CNS injury [2–4]. The lifetime healthcare costs for a 25-year-old patient with cervical level tetraplegia are expected to be more than $4 million. Even though there are various biological, pharmacological, mechanical, and surgical treatments to overcome these problems, there is no established treatment strategy yet [5–7].

Even if these therapeutic strategies do not guarantee full recovery from spinal injury, by facilitating functional recovery in the early stages of post-trauma, overall lifetime expenses are reduced and quality of life is increased. On the other hand, peripheral nerve injuries can be healed and can recover to some degree from neurotmesis, although complete functionality is not recovered and some complications may arise even after recovery [8].

Physicians decide how to go about treatment by taking into account many different factors, but the deciding factor is the size of the defective gap between the two stumps. Surgical approaches, tensioned direct repair, and grafting techniques have been the conventional treatments for nerve injuries in the past. The autograft transplantation has become the most popular surgical technique since Berger and Millesi demonstrated its superiority over direct tension repairs if the injured nerve gap is narrow enough to avoid tension [9]. The autograft is an instantly available source in most cases and also offers the following advantages: a peripheral nerve-friendly environment, guidance cues, perineurium scaffolding, and the support of Schwann cells. Although autograft transplantation has been the treatment of choice for the past few decades, many complications and limitations still remain [10].

The reported complications from autograft nerve repair include tender neuroma formation, dysesthesia, paresthesia, paralysis, contracture, etc. excluding iatrogenic complications. They are more frequently found in wounds with inadequate tension and vascularization. Also, the donor site from where the autograft tissue is harvested suffers from a permanent nerve injury and potential postoperative complications from the required additional incision. More recently, alternative sutureless nerve repair techniques by coaptation using fibrin glue and laser have come into the limelight offering competitive benefits [11]. Sutureless techniques are arguably more efficient than conventional techniques, eliminate the tension caused by suturing, and improve the alignment of fascicles.

Another emerging technology, the bioartificial nerve conduit, is a supporting tube-like scaffold proposed as a promising new alternative or complementary therapeutic technology to autografts. With extensive researches and advantages, nerve conduit techniques are considered to be the current gold standard for nerve repair [12]. This scaffolding technique focuses on entubulation, the guiding of axonal regrowth through an enclosed tubal structure. One of the significant advantages of the entubulating nerve conduit is that it is able to be used to upgrade other conventional suture techniques and sutureless techniques or even as a stand-alone procedure.

Various artificial nerve conduit researches are being carried out to enhance functionality and improve properties such as biomimetics, nanotopography, material selection, and enhancement by the addition of biomolecules [13]. The biodegradable aligned electrospun nanofibrous nerve conduit is currently the option with the best nanotopographic properties.

The aligned electrospun nanofibrous nerve conduit is fabricated by electrospinning, which is one of the most suitable techniques for producing nerve conduits. Electrospinning is also one of the simplest methods that can produce aligned topography with tunable porosity from a large variety of materials and additive substances. The tailored surface topography and porosity of aligned electrospun nanofibrous nerve conduits foster nerve regeneration by providing guidance cues, permissiveness, and a cell-friendly environment [14, 15]. Also, the polymer composition for nerve conduit and doping substances also influences nerve regeneration with controllable drug loading capacity of electrospun nanofibers. Although nerve conduit technology is commonly used to facilitate nerve regeneration in the PNS, it is also expected to take part in CNS injury treatment combined with other therapeutic strategies such as cell transplantation, neurotrophic factors, and nanoparticles [16, 17].

We will discuss in brief nerve regeneration physiology, electrospinning methods for obtaining aligned nanofibers, and the role of aligned nanofibers in cell guidance in vitro and in vivo, by comparing aligned fibers with randomly oriented nanofibrous fibers.

1.1 Anatomy of a Nerve and Neuroregeneration

The human nervous system has two main components: the central nervous system and the peripheral nervous system. The CNS is composed of the brain and spinal cord while the PNS can be categorized into spinal nerves and cranial nerves according to the initiation lesion/level or three groups according to the direction of signal conduction: afferent nerves, efferent nerves, and mixed nerves. A nerve is a cord-like axonal bundle in the PNS that delivers electric and chemical signals from the CNS to the innervated organ and vice versa. Axons are covered by the endoneurium, an outer layer of connective tissue, and bundle up into fascicles, which are wrapped in the perineurium. These fascicles are bundled together to finally form the hierarchical structure of the nerve covered by the epineurium, the outermost layer of connective tissue as seen in Fig. 1 [1].

As seen in Fig. 2, the two nervous systems show several big differences in their functions, structures and, more importantly, in their regeneration processes. The difference in regenerative capacities between the two systems is explained mainly by their intrinsic regenerative potentials and different glial cells, also known as signal transduction assistants, and their different reactions to a damaged neuron [19, 20].

Slender peripheral nerves are usually located at mechanically vulnerable positions surrounded by tissues. They can be easily injured by various traumatic events including cuts and compressions. Once a nerve is damaged, the remnant distal nerve tissues undergo Wallerian degeneration, the fragmentation and disintegration of the axon, to start the regenerative process in which Schwann cells, the basal lamina, and the neurilemma around the damaged sites begin to construct a primitive regeneration tube [22–24]. Upregulation of regeneration-associated genes (RAGs) with nerve growth factors (NGFs) also highly contributes to this process. Once the regeneration tube is completed, the nerves grow guided by the tube to reach its destination [25]. However, this entire regenerative process takes several months and incomplete functional restoration often occurs when the regenerative processes are not able to complete their tasks. Regeneration in the CNS used to be regarded as an unachievable goal due to the limited intrinsic capacity for axonal growth and inhibitory milieu. The inhibitory environmental cues in the CNS were experimentally demonstrated by David and Aguayo, transplanting peripheral nerve graft to the CNS and vice versa. The transplanted PNS nerve graft vastly promoted CNS neuron regeneration but the CNS grafts dampened the regeneration capabilities of the PNS neuron [26]. The incapability of the CNS neuron to recover postpones prompt initiation of axonal regeneration. Moreover, the injury site is further beset by secondary reactive events.

First, the glial cellular function of oligodendrocytes as a scavenger is generally unsuccessful. As a result, they fail to rejuvenate and to make matters worse, the astrocyte-supported oligodendrocytes secrete inhibitory substances instead of helpful trophic factors. Secretory inhibitors including myelin-associated inhibitors (MAIs) and chondroitin sulfate proteoglycans (CSPGs) coincide the upregulation of RAGs such as c-Jun, activating transcription factor-3 (ATF-3), SRY-box containing

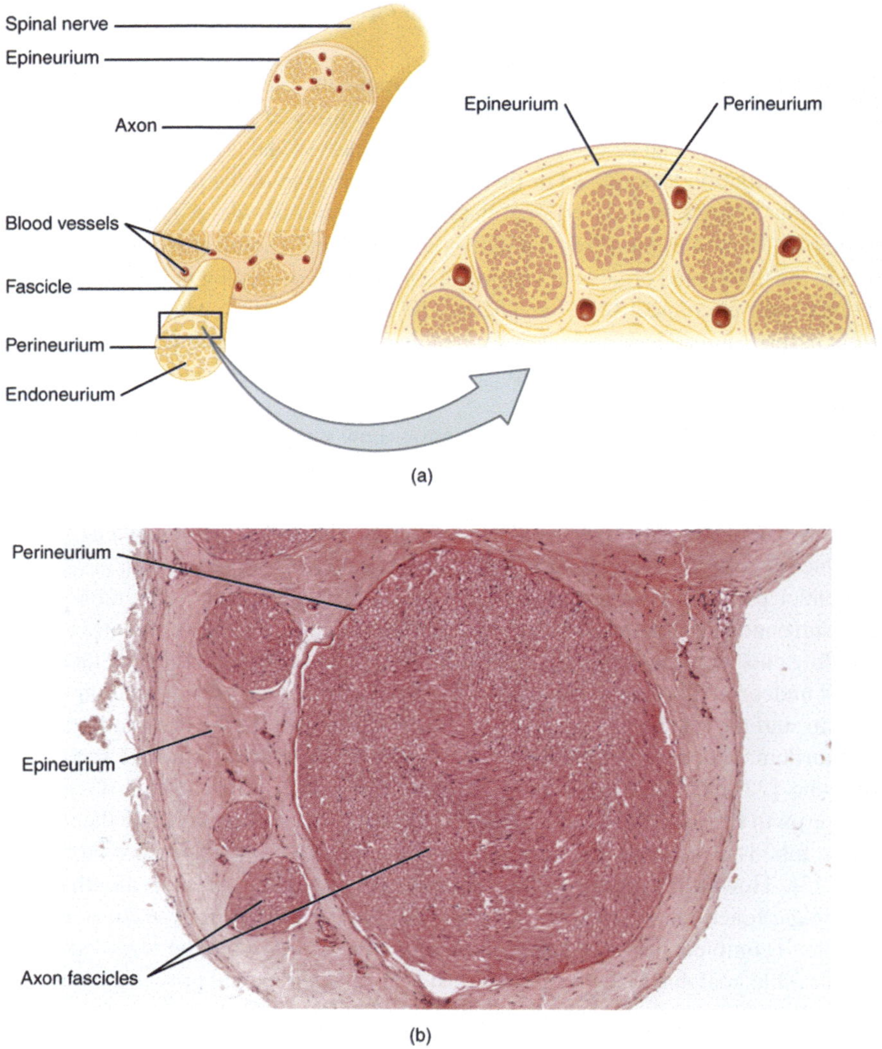

Fig. 1 The schematic of layered-structure of a nerve (**a**), histologic picture of a nerve (**b**) (tissue source: simian). LM × 40. Image by OpenStax College, licensed under CC BY 3.0 [18]

gene 11 (Sox11), small proline-repeat protein 1A (SPRR1A), growth-associated protein-43 (GAP-43) and CAP-23 [27–31]. Finally, they initiate fibrosis and a dense scar formation, or sometimes a cyst formation, which acts as a natural obstacle that prevents the nerve stumps from reuniting by blocking neural cell migration. Figure 3 is a graphical representation of the regeneration process of the axon.

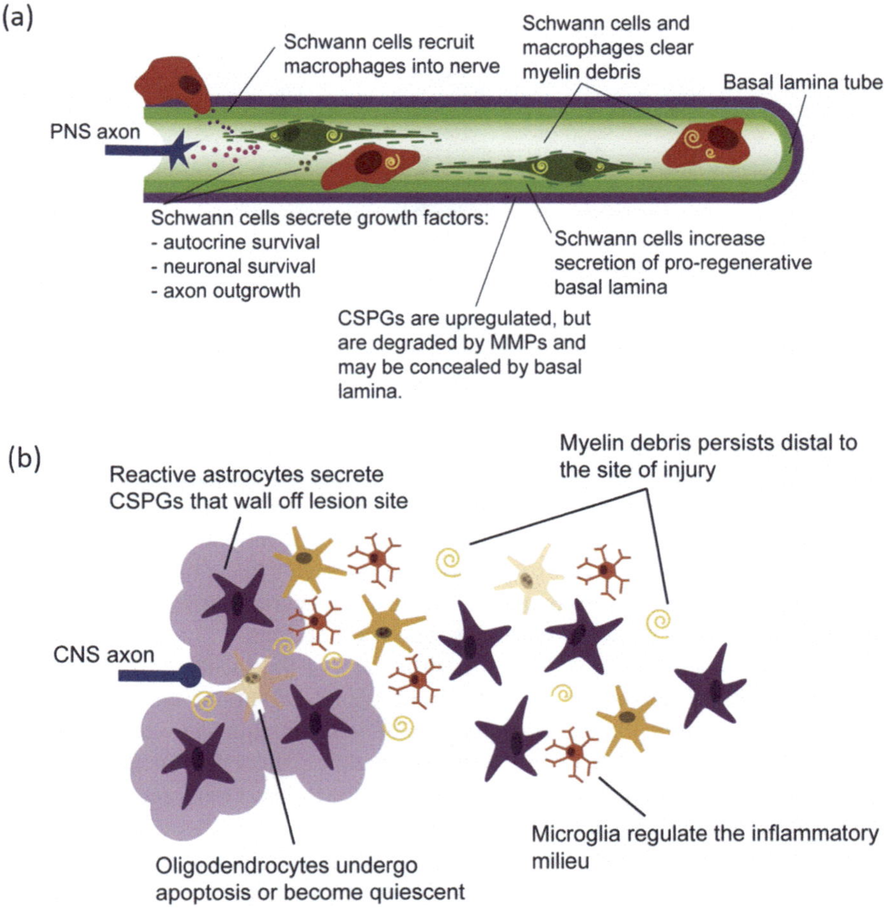

Fig. 2 Schematics of the nervous system responses to axon injury; (**a**) response in the PNS and (**b**) response in the CNS. Adapted from Ref. [21] with permission from Elsevier [21]

1.2 Nerve Guide Conduits

1.2.1 Requirements of Nerve Conduit

The history of manufactured nerve conduits is traced back to the tube-shaped decalcified bone made by Gluck and his colleagues for the connection of transected nerve ends in the 1880s before the microsurgery [32]. This trial for the treatment of neurotmesis injuries by an encapsulation strategy continued until Dahlin and Lundborg's silicon-based nerve conduit. Dahlin and Lundborg characterized the mechanism of nerve regeneration in a tube-structured nerve conduit. The nerve conduit encases both nerve stumps within the lumen of the tube to provide a gross alignment and fluid accumulation between the stumps for nerve regeneration. The fluid accumulated

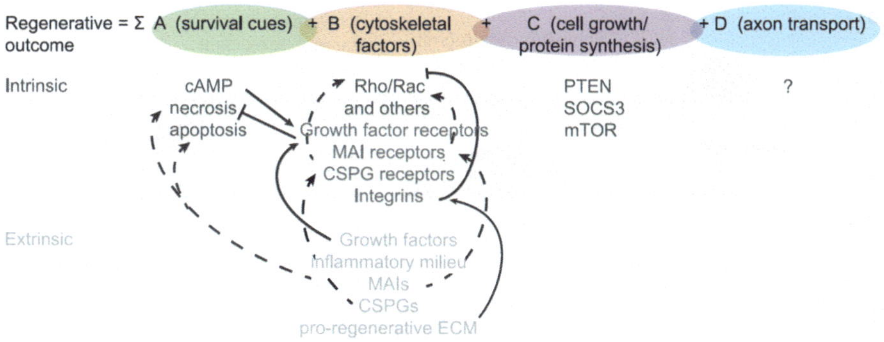

Fig. 3 The axon regeneration equation. Pro-regenerative and anti-regenerative factors in axonal regeneration process. Reprinted from Ref. [21] with permission from Elsevier [21]

inside the tube's inner chamber initiates a rudimentary fibrin matrix formation that connects the two nerve stumps. When the accumulation reaches a sufficient amount, this primitive fibrin matrix can serve as a bridge for cell migration [23]. Once the cells have migrated, connecting structures called Büngner's linear band forms within the disordered rudimentary fibrin matrix. The juvenile neurites grow along these linear bands surrounded by newly-created pseudo nerve sheaths.

This process depends heavily on the sufficiency of fluid leakage volume from both nerve ends. If the volume of fluid is insufficient to fill up the inner lumen, the newly formed fibrin matrix and neo nerve can often be too thin and weak due to the mechanical contraction. The thinner nerve regeneration alters the functional regain because the axonal regeneration is proportional to the thinnest cross sectional diameter of the cable [33]. Also, there has been another issue that can cause neuroma formation by tightly holding neurites to prevent them from potential escape. Even with these limitations, the nerve encapsulation strategy demonstrates apparent benefits. The axoplasm and milieu-containing nerve conduits provide treatment to the injured nerve tissue by keeping it away from the inflammation of the wound bed.

A series of mixed nerve repairs in the forearm were researched by the Lundborg group and a 5-year follow-up study reported that the peripheral nervous tubal scaffolding strategy is highly comparable to direct sutures. They showed a greater sensory recovery in less than 5 mm gaps even with non-permissive, permanent silicon tubes [23].

After a series of researches with efforts to improve the nerve conduit, some requirements have been explored and are being optimized. As a tissue scaffold, a nerve conduit must be composed of biocompatible materials to avoid rejection and further wound inflammation. The porosity and permeability of a scaffold are also important for invasion and migration of nerve cells, including Schwann cells and neural stem cells, and also neurological biomolecules (such as laminin-1, NGF, & BDNF). The porosity of a scaffold should be kept in a proper range to avoid complications. If the pores are too big, the fluids and cells needed to form a fibrin matrix in the early stages escape before the cells can attach, leading to failed neuroregeneration in the later stages. On the other hand, if pores are too narrow, the supportive cells cannot penetrate through to their destinations [14]. Also, the biodegradability of the nerve conduit also supports the dynamic neuroregeneration condition. When the active neuroregeneration process ends, a permanent nerve conduit that remains surrounding the neo-tissue may generate unwanted compression as an unnecessary structure.

Recently, the microscopic alignment of the scaffold surface and the electrical conductivity of the scaffold have emerged as new supporting strategies [34, 35]. Nanotopography is considered to provide nerve cells and neurites with guidance cues since the nerve cells need to directionally regenerate. Biodegradable aligned electrospun nanofibers have been attracting attention for their ability to fulfill the described requirements as nerve conduit materials.

2 Electrospun Nerve Conduits

2.1 Nanotopography of Nerve Conduit and Alignments

The nervous system comprises central and peripheral branches and functions to deliver information to all parts of the body. An extensive network of neurons and glia support the communication process. Nerve injury could result in painful neuropathies because of reduction in sensory perception and motor function depending on the location of the injury. There are many limitations in microsurgery techniques for the treatment of serious peripheral nerve injuries. NGCs are limited to treating nerve gaps of less than 4 cm in length, and sometimes the postoperative nerve is not well connected after implanting [36]. For this reason, recent developments in biomaterials and tissue engineering approaches seek to overcome the limitations associated with these methods of treatment. The incorporation of topographical guidance features and intraluminal structures have been studied to induce Schwann cell migration and regrowth of the axon toward their distal target. Several similar studies have been performed using various combinations of intraluminal guide structures and external conduit materials. One approach to this intraluminal guidance structure, including gels, sponges, films, filaments, and fibers, taken alone or with multiple support factors, is to add nanoscale guidance cues to micrometric in situ guiding structures [35, 37, 38].

Nanofiber-based scaffolds are most commonly used to add nanoscale guidance functionality to micrometric in situ guidance structures. Nanofibers are generally fabricated using three methods: self-assembly, phase separation, and electrospinning. Of these three, electrospinning is more widely used because of the excellent tunability of the nanotopography and diameter of the nanofibers [39]. Many studies have demonstrated the superiority of nanofibrous scaffolds in terms of cell activity. Also, nanofibrous scaffolds with specific patterning exhibit excellent mechanical strength as well as significant advantages in terms of cell proliferation processes associated with cell proliferation such as cell attachment, migration, and orientation. For example, highly aligned nanofibers, when compared to randomly distributed nanofibers, play an important role in neurite outgrowth in the case of neurons [40]. The reason is that the extracellular matrix of neural cells and neurites tend to grow parallel to the nanofibers aligned along the nanofiber array [14].

The environment at the single-cell level is considered to be key in elucidating the fundamental mechanisms of tissue regeneration [41]. While the extracellular matrix (ECM) has been a main target to reproduce or mimic, it has been revealed that the various properties of ECM including not only the soluble chemical factors, but also the physical characteristics, regulate cellular processes such as attachment, migration, proliferation, and differentiation. The architecture (structural and morphological properties) of the ECM is also a major determinant of the fate of both stem cells and differentiated cells in its vicinity, aiding them to inherit the characteristics of the original tissue [42].

The cell's behaviors and functions heavily depend on the cell polarity and shape [43]. A single abnormal polarity among many different cell types can cause an organ malformation during fetal development and is also closely related to the pathophysiology of various human diseases, including cancer metastasis. There are reported cellular regulators of polarity such as Par (partitioning defective) complex and associated Rho GTPase signaling involved in various cellular activities. Although the physiological mechanism of the architecture of native ECM in situ is yet unclear, the influence of topography on cell polarity in various cell types has been researched and proven by employing an experimental process known as contact guidance [44].

The polarizations of the affected cells were investigated with nano-patterned surfaces that had various modifications such as gradients and isotropic and anisotropic nanotopographies. The many different cells grown on the nano-patterned substrate were noted to prefer elongation and parallel alignment to the patterned nanogrooves. For instance, the PNS neurons are also polarized along the fabricated nanogrooves as they regenerate in neurite bundles. The cell proliferation rate is greatly dependent on and sensitive to the size of the nanostructures and cell type. For example, neural stem cells cultured on electrospun nanofibrous meshes with larger nanostructures proliferated much less than on flat surfaces while proliferation increased with decreased fiber size [45]. In contrast, the cell proliferation rate of mouse osteoblasts was enhanced on hollow 100 nm diameter nanotubes.

Even though the underlying molecular mechanisms of cellular response remain elusive, some clues can be found from investigating the integrin family of adhesion molecules which play a key role in adhesion and mechanical signal transduction and its related intrinsic pathways. Since the nerve acts as an electric wire, neural cell orientation and polarization in the correct direction are crucial for nerve regeneration [46]. Aligned patterned nano-sized scaffolds can support nerve regeneration not only at the macroscopic level but also at the nano level by providing neuronal cells with anchors in the form of nanogrooves.

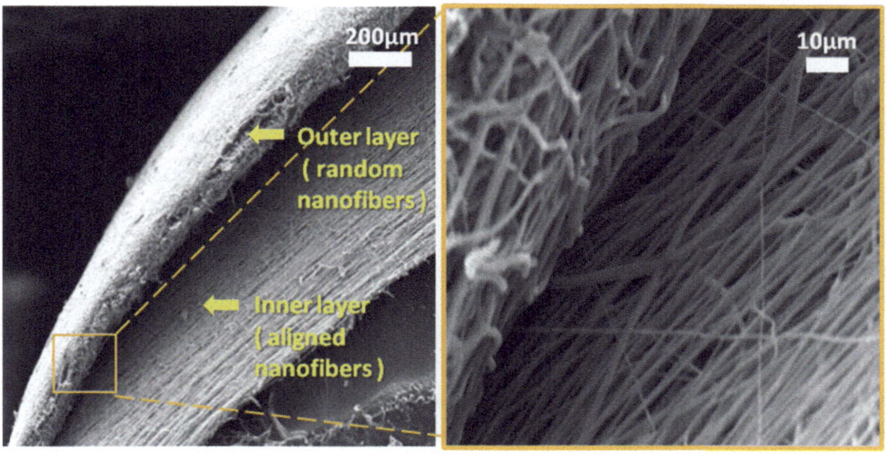

Fig. 4 (Left) Cross-sectional SEM image of a bilayer membrane consisting of a randomly oriented outer layer and an aligned inner layer. (Right) Closer view of the outer and inner layer interface. From Ref. [14] by Kim et al., licensed under CC BY 4.0 [14]

2.2 *Aligned Electrospun Nanofibers*

Aligned nanofibers provide topographic cues for nerve regeneration. Randomly oriented nanofibers are generally fabricated without preferential direction by a typical electrospinning set-up because the polymer jet that travels from nozzle tip to collector is disordered [14, 47]. It is challenging to produce a conduit from a neat aligned nanofibrous mat. In addition, aligned nanofibers have not been used for surgical applications as a nerve guide conduit due to their insufficient mechanical strength. For these reasons, a conduit with a highly aligned electrospun mat is produced with double coating using randomly oriented nanofibers by a modified electrospinning method. This double coated conduit favorable features like selective permeability and good hydrophilicity were made as nerve guide conduits [14, 48]. The inner part of the nerve guide conduit is covered with ordered nanofibers for enhancement of the proliferation of neural cells and the outer part of the conduit is double-coated with random nanofibers over ordered nanofibers for strengthening the mechanical properties of the inner part of the aligned nanofibrous conduit as shown in Fig. 4.

2.3 *Fabrication Methods*

A nerve conduit made of aligned nanofibers is favorable for nerve regeneration because of their superior nerve cell proliferation and attachment. However, it is challenging to fabricate a neat mat form with aligned nanofibers for biological applications as a nerve guide conduit because of their insufficient tensile strength.

For this reason, extensive efforts have focused on producing aligned nanofibers and controlling the orientation of the fibers to meet requirements for medical applications. The most common fabrication technique for aligned nanofibers is the introduction and the modification of the collectors like a rotating drum, cone and disk as shown in Fig. 5.

Many researchers have used high-speed rotating drums to collect ordered nanofibers which are parallel to each other along a common axis. Increasing the speed of the rotating collector results in highly aligned nanofibers. Edwards et al. reported that the structures of aligned polycaprolactone (PCL) nanofibers are influenced by the speed of the rotating collector [40, 49]. Theron et al. investigated an electrospinning set-up with a thin rotating disk that has a tapered edge to fabricate continuous aligned nanofibers. Afifi et al. proposed a modified rotating collector covered with insulating materials and fixed with conductive fins to produce aligned nanofibers. By setting up a gap between grounded conductive materials, one can obtain ordered nanofibers between the conductive materials. Li et al. reported the density of collected aligned nanofibers can differ according to the width of the gap between conductive materials. Sun et al. investigated that a rotating grooved collector could achieve the results of fabricating ordered nanofibrous mat in the electrospinning [50]. The charged nanofibers are stretched and spun across a gap between grooved collectors and form axially aligned nanofibers. Kim et al. demonstrated a modification of the electrospinning process in order to collect both aligned and random nanofibers on a mat via a single step electrospinning process. The copper wires were attached to a rotating collector and a semi-conductive mat was attached to the fixed copper wires in the horizontal and vertical axes for fabrication of aligned nanofibers. In this process of fabricating aligned nanofibers, the collector base was rotated at a rate of about 1000 rpm to get a neat mat with aligned nanofibers as shown in Fig. 6.

Aligned nanofibers can achieve a better cell viability and migration compared to randomly oriented nanofibers and act as a guide for neurite growth (Fig. 7). These processing methods for ordered nanofibers can be promising candidates for manufacturing scaffolds for neural tissue engineering.

2.4 *Improvement of Electrospun Nerve Conduit*

2.4.1 Cell-Seeded Constructs

Therapeutic cell transplantation technology has taken its place in SCI treatment in the last few decades as the importance of the supportive cellular environment in neuroregeneration is highlighted. Several approaches to understanding more efficient cell delivery routes also have been under investigation [21, 51]. Reports have observed that the treatment outcome of cell transplantation is highly influenced by the volumetric cell density, type, and delivery route [52, 53].

While the cell density and type are adjustable, the optimal delivery routes and methods for cell transplantation vary from direct injection to scaffolds—seeding

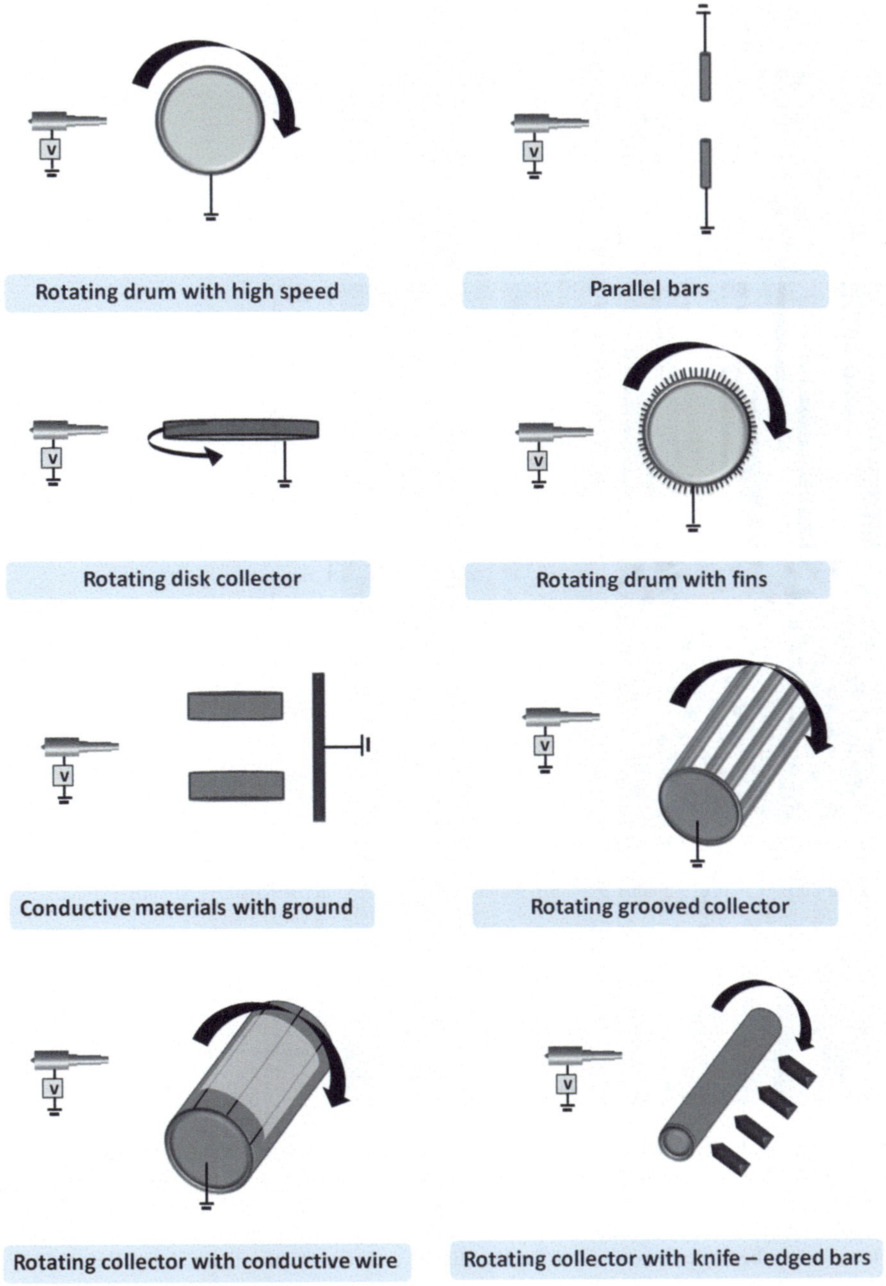

Fig. 5 Schematics of electrospinning methods for fabricating ordered nanofibers

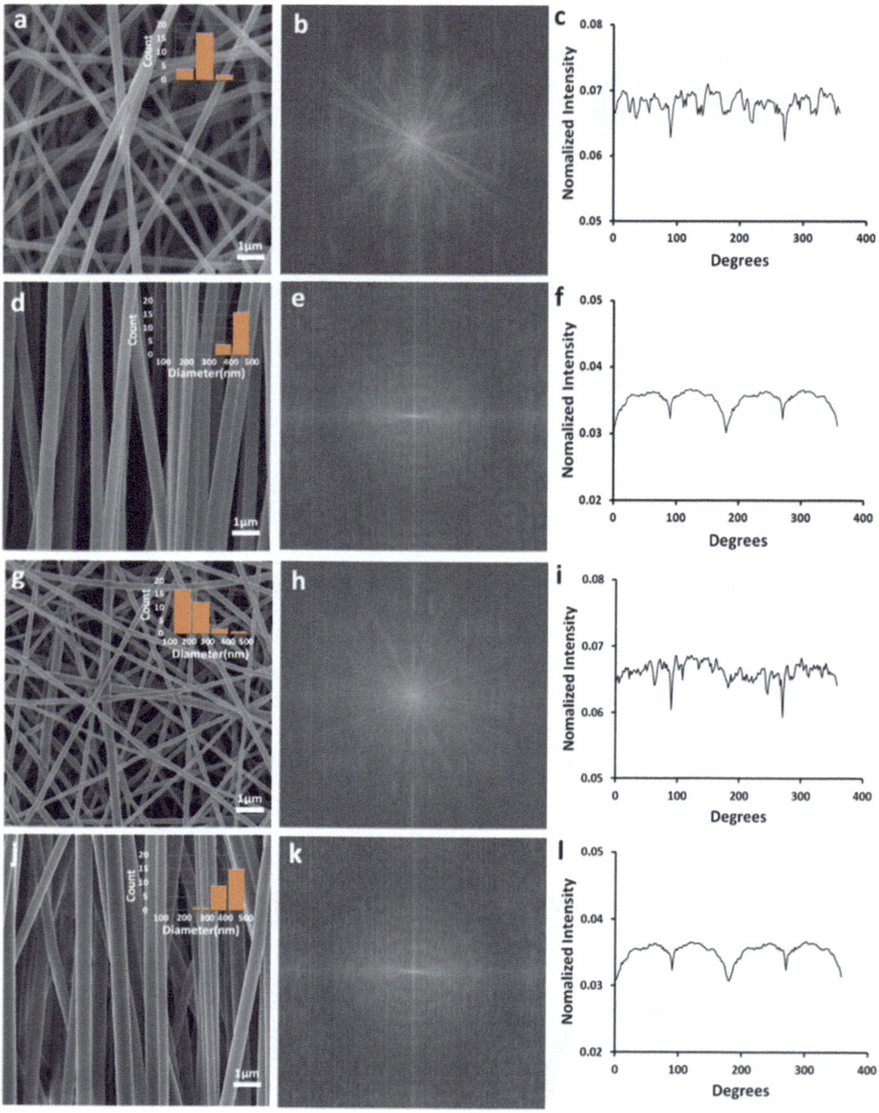

Fig. 6 Morphological analysis of PU and PLGA nanofibrous mat with diameter graph (**a**) SEM image of randomly oriented PU nanofibers (**d**) SEM images of aligned PU nanofibers at a speed of 1000 rpm. (**g**) SEM image of randomly oriented PLGA nanofibers (**j**) SEM images of aligned PLGA nanofibers at a speed of 1000 rpm (**b**, **e**, **h**, **k**) FFT output images (**c**, **f**, **i**, **l**) Pixel intensity plots against the angle of acquisition for the aligned and random nanofibers. From Ref. [14] by Kim et al., licensed under CC BY 4.0 [14]

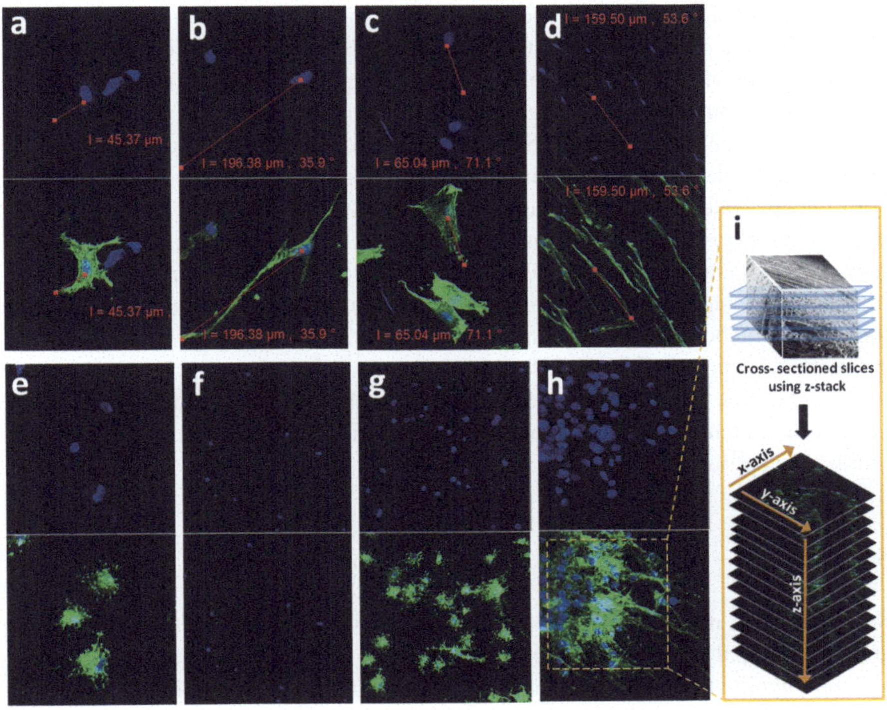

Fig. 7 (**a**, **b**) Confocal microscopy images of PC12 cells attached after 1 day of culture on a randomly oriented and aligned nanofibrous PLGA scaffold (**c**, **d**) Confocal microscopy images of PC12 cells attached after 5 day of culture on a random and aligned nanofibrous PLGA scaffold (**e**, **f**) Confocal microscopy images of S42 cells attached after 1 day of culture on a random and aligned nanofibrous PLGA scaffold (**g**, **h**) Confocal microscopy images of S42 cells attached after 5 day of culture on a random and aligned nanofibrous PLGA scaffold. Actin Green 488 (green) was applied for actin filament and DAPI (blue) for staining nuclei (**i**) Z stack from a PLGA scaffold with aligned nanofibers in which S42 cells. Images were collected at 0.37 μm intervals using the 488 laser. From Ref. [14] by Kim et al., licensed under CC BY 4.0 [14]

according to the target site. For SCI cell transplantation, injection through lumbar puncture is less invasive, cost-effective, and readily available [52, 54]. The preliminary cell transplantation therapeutic outcomes of the epicenter, rostral, and caudal injection sites at 1–2 weeks after SCI appeared to have similar levels of functional restoration. However, the result may imply the injection requires a higher population of therapeutic cells to reach a favorable cell density for filling up the spinal cavity.

As a combinatorial treatment, a cell-seeded nerve conduit or a nerve conduit-mediated cell transplantation along with additional mechanical supports and guidance cues can be a promising alternative in providing the injured tissue with the appropriate cell density needed in order to efficiently utilize the limited supply of cells.

Among the many types of cells available Schwann cells, neural stem cells (NSCs), neural progenitor cells (NPCs), ectomesenchymal stem cells (EMSCs), mesenchymal cells (MSCs) and olfactory ensheathing cells (OECs) are considered to be the most promising in bringing out the best outcome from cell-seeded nerve conduit treatments. Particularly, Schwann cells are most studied neurotrophic cell type and have demonstrated the most superior effect in treatments for both spinal cord and peripheral nerve injuries. For example, Schwann cells are reported to improve propriospinal axons around the injury site and also enhance the axonal regrowth of dorsal root ganglia (DRG). However, Schwann cells alone are incapable of helping axonal sprouts penetrate into the proximal destination so Schwann cell therapy should be aided by a secondary scaffold such as aligned nerve conduits with a directional topology for better therapeutic results [53].

To determine the best option and understand the behaviors and environmental cues of the potential cells, the types and characteristics of a few neurotrophic cells will be briefly discussed.

Schwann Cells

Schwann cells, named after physiologist Theodor Schwann, or neurolemmocytes, are the principal glial cells of the PNS. They can be categorized into myelinating and nonmyelinating Schwann cells depending on the myelin sheath wrapping around the axons of neurons in the PNS. All glial cells including Schwann cells functionally support neurons by supporting the conduction of electric signals and synaptic activity. However, different glial cells play roles in the development and regeneration of nerves, extracellular matrix synthesis, and immunologic monocytic antigen presentation.

One of the many roles of glial cells and Schwann cells, in general, include rejuvenating the distal portion of the damaged area and making it permissive for further neuronal regrowth and re-innervation. Following this cellular debridement, the Schwann cells layer upon on each other to form what are known as bands of Büngner, tunnels that guide axon regeneration toward the destination. Axons begin to regenerate at a rate of approximately 1 mm per day under favorable cellular milieu [55]. Schwann cells contribute to axonal regeneration by providing direction and the synthesis of ECM and neurotrophic biomolecules such as NGF, BDNF, and cell adhesion molecules. Once nerve fibers in the PNS get injured, the genes related to Schwann cell formation are quickly upregulated. The most well-known genes that contribute to Schwann cell formation and maintenance are SOX10 and Neuregulin 1. SOX10 is known as a determining transcription factor for glial cell generation from trunk crest cells and Neuregulin 1 (NRG1) promotes the formation of Schwann cells and support the survival of immature Schwann cells [56, 57].

Owing to their neurotrophic properties, Schwann cells are popularly used in attempts to treat neuronal injuries in a variety of ways. Novikova et al. seeded Schwann cells on poly-β-hydroxybutyrate (PHB) nerve conduit and implanted the nerve scaffold to cervical SCI rat models. They reported the nerve conduit enhanced axonal regeneration by supporting cell attachment and proliferation in the vicinity of the injury [58]. Xu et al. evaluated the influence of Schwann cell-seeded PAN/PVC mini-channel on a SCI rat model. Significant axonal regrowth in both directions, myelination, and improved vasculature were observed only in the cell-seeded group [59]. Blits et al. also loaded Schwann cells to their PAN/polyvinyl chloride (PVC)-based nerve conduit with other substances such as fibrinogen, gentamycin, and aprotinin to promote neuroregeneration. Two adenovirus-associated vectors were additionally administrated to two groups separately to evaluate the roles of BDNF and NT-3 and significantly improved functional gains of the hind-limb in both groups were observed [60].

Meanwhile, through various assessments such as histologic findings, electrophysiologic analyses and walking track analyses, the neuroregenerative effect of Schwann cells seeded on nerve conduits were confirmed by Keeley et al. [61]. The supportive cellular mechanisms of Schwann cells are also gradually being revealed. Williams et al. reported for the first time that the supportive roles of Schwann cells in nerve regeneration are performed mainly during the early stages [54]. The guidance cue provided by Schwann cells was emphasized by Brayan et al. through their experiment on Schwann cell-seeded poly-L-lysine precoated polyethylene (PE) nerve conduit implanted in a 20 mm neurotmesis rat model [53, 62].

To recapitulate briefly, once PNS nerve injury occurs, Schwann cells convert to a cell phenotype that is specialized to promote nerve repair and rapidly divide, migrate, express an appropriate set of genes, and facilitate axonal regrowth in the early phase of nerve regeneration. Because of their many supportive properties, Schwann cells have been extensively studied to be exploited in various neural injury treatments [3]. Schwann cells are the most studied and potentially most favorable candidate for cell loading nervous scaffold use, however, their application is restricted by low availability owing to insufficient nerve donors and the time-consuming culture processes for cell expansion.

Neural Stem Cells (NSCs) & Neural Progenitor Cells (NPCs)

Neural stem cells (NSCs) refer to multipotent cells closely involved in the embryonic development of the nervous system. More specifically, the neurons and glia of the animal nervous system are generated in embryonic development by NSCs and some NSCs will remain in the adult brain and continue to produce neurons [63, 64]. Neural stem cells can be divided into two daughter cells through asymmetric or symmetric cell division. When NSCs undergo asymmetric cell division, one of the daughter cells is differentiated into another cell type, primarily astrocytes, neurons, or oligodendrocytes, while the other daughter cell remains multipotent [65]. NSCs can also improve axonal regeneration by releasing metalloprotease-2 and multiple neurotrophic factors [66].

Olson et al. reported the supportive influence of NSCs and Schwann cells on neuroregeneration. In the study, Schwann cells and NSCs were seeded separately on a multi-channel PLGA scaffold which was implanted in an SCI-injury rat model.

Increased axonal regeneration was observed in both the Schwann cell and NSC seeded-scaffolds, although the number of newly developed axons appeared to be slightly higher in the Schwann cell scaffold [67]. Another group, Lee et al., confirmed the positive effects of NSCs in a canine SCI model instead of the conventional rodent model.

Stem cell transplantation always carries with it concerns of toxicity and graft-versus-host disease (GVHD) along with post-transplantation fever. Therefore, regimens for stem cell transplantation using preventive strategies such as alloimmunization and upregulation of Regulatory T cells, have been extensively studied [68]. Neural progenitor cells (NPCs) are also multipotent as differentiated into hierarchically lower cell types similarly to NSCs. The differences of NPCs from NSCs can be found in the more specific differentiating potentials and the finite replication capabilities of the cells [69]. However, these two types of cells are sometimes considered to be equal and the concept of NPCs is still evolving.

Other Cells

Ectomesenchymal Stem Cells (EMSCs) are capable of differentiating into Schwann cells and also supporting neuroregeneration. Nie et al. reported the effect of EMSCs-loaded PLGA nerve conduit model in a rat sciatic neurotmesis model. The EMSCs-loaded nerve conduit showed statistically similar results as the autograft transplant group in a 3-month follow-up study using the sciatic functional index (SFI) while the result of the cell-free PLGA group was substantially inferior to the other groups [70].

Mesenchymal Stem Cells (MSCs), usually referred to as bone marrow stromal cells, are also available for cell-loaded nerve conduit technology. Although MSCs do not differentiate into nerve cells, many studies have reported improved neurogenesis with MSCs transplantation. Through a lumbar puncture, human BMSCs transplanted into a SCI rat model by Pal et al. appeared to guide axonal growth and also significantly improved nerve functions. The transplanted hBMSCs survived in the spinal cavity for at least 1 month and the functional regain was found to be dependent on the volume of the transplanted cells [71]. Pereira Lopes et al. reported the supportive effect of BMSCs as seeded on biodegradable collagen tube implanted in a sciatic neurotmesis rat model with a 3 mm nerve gap. Compared to the control group, a significantly greater number of regenerating clusters including both myelinated and non-myelinated fibers were observed in the BMSC-loaded nerve conduit after 6 weeks. They also confirmed the secretion of two neural growth factors, BDNF and NGF β [72].

Olfactory ensheathing cells (OECs) also known as olfactory ensheathing glia (OEG), are a type of glial cells distributed throughout the olfactory epithelium, olfactory nerve, and also olfactory bulb. By phagocytosing pathogens, they contribute to the immunoprotection of the olfactory nerve which lies under a mucosal layer of the upper nasal cavity. The human olfactory system keeps regenerating its neurons even in adulthood because the exposed nerve ends must degenerate to maintain the function [73]. During the regeneration processes, OECs play an important role in functionalization of new olfactory receptor neurons by cleaning up debris and providing damaged neurons with a favorable environment for neuroregeneration. Since the 1990s, the therapeutic possibility of transplantation of OECs to SCI to promote axonal regeneration and neurogenesis began to be reported possibly owing to the similar function of OECs in the olfactory system. In 2014, a Polish patient suffering from SCI-induced paraplegia regained mobility via therapeutic OEG transplantation. It was the first report of paraplegia recovery by SCI treatment [74]. The supportive mechanisms of OECs are still under investigation, but widely thought to be due to the upregulation of NGF receptors rather than the release of neurotrophic factors.

Neurotrophic Factors

Neurotrophic factors (NTFs) are various biomolecules related to neuronal growth, survival, and differentiation. This family of biomolecules is widely studied to understand the mechanisms of neuronal regeneration in both the CNS and PNS. NTFs are categorized into three main groups: the neurotrophin family, the CNTF family, and the GDNF family, based on the neurotrophic mechanisms at the cellular level [75]. Neurotrophin-3 (NT-3) of the neurotrophin family, is known for its significant role in neuronal survival in the PNS, and improves neuroregeneration in the CNS, specifically in the corticospinal tracts. Fan et al. reported that the addition of NT-3 onto PLGA nerve conduits promotes neural regrowth and motor function [76].

Another neurotrophin family member, brain-derived neurotrophic factor (BDNF) is one of the most studied biomolecules in various different aspects among neurotrophic factors. BDNF shares structural similarities to other family members such as NGF, NT-3 and NT-4/5 but BDNF is more closely related to the survival of neurons, particularly of dorsal root ganglion neurons, compared to other factors although the mechanism is still unclear [77]. Liang et al. suggested that collagen-bound BDNF enables sufficient BDNF delivery to the injured tissue [78].

Other neurotrophic factors, acidic fibroblast growth factor (a-FGF) and basic fibroblast growth factor also demonstrate neuroprotective effects and increase proliferation of NSCs and NPCs [79]. There are also other potential neurotrophic growth factors reported and used to support neuronal regeneration.

Vascular endothelial growth factor (VEGF) produced by cells is another supporting biomolecule that promotes neuronal reconstruction. Excluding traumatic neural injury in the CNS, these angiogenetic factors have demonstrated clinical significances in prognosis and treatment strategies of many other different diseases such as breast cancer, rheumatoid arthritis, diabetic retinopathy, age-related macular degeneration, and angiosarcoma. The members of the VEGF family are activated through tyrosine kinase receptors, similar to the cellular activation mechanism of most NTFs [80]. After the spinal cord is injured, VEGF represses apoptosis of nerve cells in order to spare neural tissues. Improved behavior, increased vascularization, increased spare tissue and decreased apoptosis levels were observed in rat SCI models with VEGF injection [81]. VEGF administration to SCI rat model also resulted in attenuated cavity formation and the production of a more permissive tissue environment for axonal ingrowth as reported by Sundberg et al. [82]. However, the clinical VEGF therapy for acute SCI is controversial because other studies exhibit few adverse effects including chronic pain after VEGF injection and acute exacerbation to injured neural tissue possibly due to VEGF-induced microvascular permeability.

Many ongoing researches on immobilization techniques and a number of potential substances including glial cell line-derived neurotrophic factor (GDNF), nerve growth factor (NGF), ciliary neurotrophic factor (CNTF), and alpha-1 glycoprotein (α1-GP) are expected to achieve great improvements in the treatment of neural injury.

2.4.2 Nanoparticles

Nerve guide conduits have benefited from the many recent advances in nanotechnology, in particular aligned nanofiber membranes and nanoparticles (NPs). Various materials from gold, silica, poly(lactide-co-glycolide) (PLGA), and titanium dioxide have been used to make nanoparticles for the delivery of drugs and growth factors to the injured site [83–85]. Other materials such as carbon nanotubes were utilized to promote and direct the growth of neuronal cells [86].

Nanoparticles have the great advantage of being able to penetrate the blood-brain barrier for more efficient and effective treatment of the CNS. To be used as drug carriers to the brain, NPs must be smaller than 100 nm, stable in blood, avoid platelet aggregation as well as meet many other conditions [87]. Besides being exceptional drug carriers, NPs have also been reported to promote the growth of neuronal cells and elongate neurites uniaxially for enhanced recovery rates. A property of metallic nanoparticles that is particularly attractive is their distinctive optical properties. When metallic NPs are illuminated by external light, they generate an oscillation that is called localized surface plasmon resonance. Depending on the wavelength of the absorbed light the localized surface plasmon resonance can match the therapeutic window of biological tissues to help with regeneration. For example, silica-coated Au nanorods irradiated with near infrared light can stimulate electrical activity in auditory neurons by inducing temperature increases between 0.5 and 6 °C [88]. Metallic NPs have also been used to deliver small-interfering RNA (siRNA) into neural stem cells for controlling their differentiation as depicted in Fig. 8.

Polymer NPs have the advantage of being biocompatible and biodegradable, making them a prime choice in neuroprotective therapeutic strategy and drug release. Rittchen et al. reported the delivery of pro-remyelinating factors to the CNS using PLGA nanoparticles and targeting antibodies to induce oligodendrocyte precursor cell maturation and improve remyelination. PLGA, like other FDA-approved polymers, degrades to biocompatible agents in the body, eliminating the need for a surgery to recover the particles. Silica NPs have large surface areas for protein binding, but they also can be easily functionalized for targeted delivery of cargos to neuronal cells. One study utilized transferrin-modified mesoporous silica nanopar-

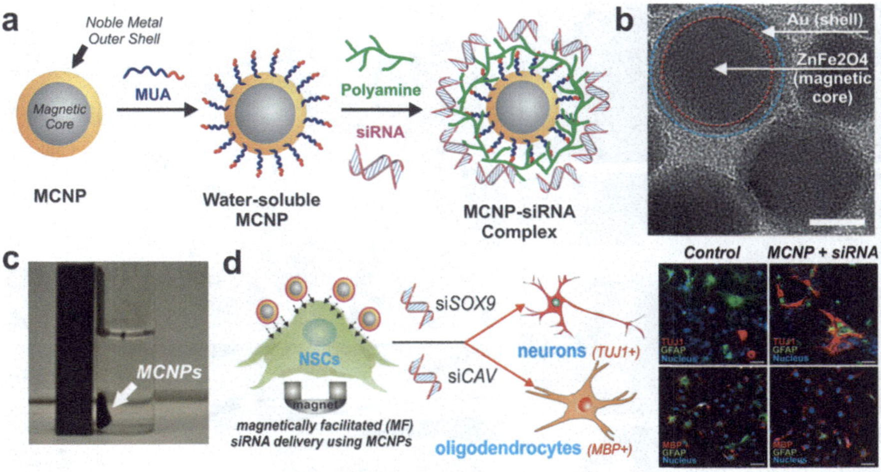

Fig. 8 Magnetic core-shell nanoparticles (MCNPs) for the delivery of small-interfering RNA (siRNA) into rat neural stem cells (rNSCs). (**a**) MCNPs functionalized with siRNA. (**b**) Transmission electron microscopy image of MCNPs. Scale bar: 10 nm. (**c**) MCNPs dispersed in water attracted to a magnet. (**d**) Schematic of magnetically facilitated delivery of siRNA to induce neural differentiation of rNSCs using MCNPs. Fluorescence images of neuronal (top) and oligodendrocyte (bottom) differentiation after siSOX9 and siCAV delivery, respectively. Adapted with permission from Ref. [85]. Copyright 2016 American Chemical Society [85]

ticles (TF-MSNs) conjugated with HI-6 to prevent brain damage caused by soman poisoning [89]. These NPs were chosen because of their mesoporous structure for storage and rapid release, important for fighting toxic nerve agents.

Carbon nanotubes (CNTs) are cylindrical nanostructures comprised of graphene sheets that are wrapped onto themselves. The most frequently used CNTs are single (SWCNT) and multi-walled (MWCNT) carbon nanotubes, which are made of one layer of graphene and several concentric graphene cylinders, respectively. CNTs have many important applications in neuroscience currently and of these is acting as a platform to promote neuronal growth and performance. They have had unexpected and exciting impacts on neuronal signaling and behavior.

2.4.3 Conductive Polymers

There are many different types of polymers that have unique properties and characteristics such as biocompatibility and conductivity. Polymers that are conductive can conduct charge because electrons can jump within and between the chains of the polymer with ease. The polymers contain a conjugated backbone, meaning that single and double bonds alternate along the polymer chain. Both single and double bonds contain chemically strong localized σ-bonds that hold the atoms together, while double bonds also have a weaker π-bond that allow electrons to be more easily delocalized and move freely. Moreover, doping conducting polymers can

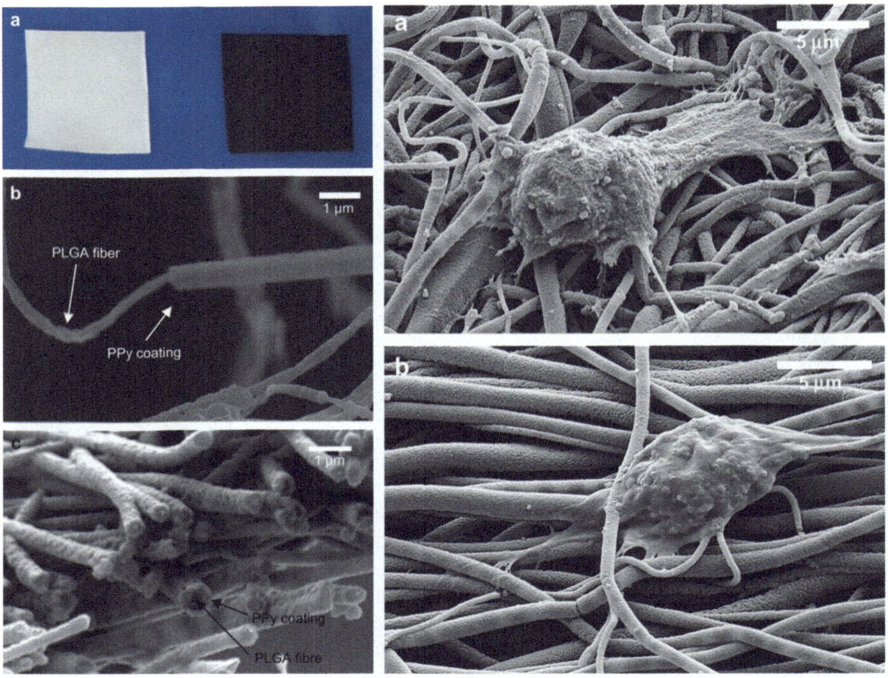

Fig. 9 Left column. (**a**) Uncoated PLGA mesh (left) and PPy-PLGA mesh (right). (**b**) SEM of single strands of PPy-PLGA fibers. (**c**) SEM image of section of the PPy-PLGA meshes. Right column. (**a**) PC12 cells cultured on PPy-random fibers and (**b**) PPy-aligned fibers for 2 days. Black arrows indicated neurites. Reprinted from Ref. [94] with permission from Elsevier [94]

increase their conductivity even further by introducing oxidizing or reducing agents and is dependent on the type and molecular size of the dopant.

Conductive polymers typically allow excellent control of the electrical stimulus, possess very good electrical and optical properties, and can be made biocompatible and biodegradable [34]. Many conductive polymers such as polypyrrole (PPy) are not inherently biodegradable, but there are ways to make them be so. One method is to prepare a composite containing both the conducting polymer with a biodegradable polymer. However, a downside to this method is that the conductive polymers are not degraded in the body. A second route is to modify the polymer structure itself. Studies have reported that the addition of ionizable or hydrolysable side groups to the backbone of PPy have successfully made the PPy degradable [90].

Considering the electroactive nature of our nervous system, conductive polymers such as PPy and PEDOT are attractive for use in neural engineering applications. Recent studies have used these conductive polymers as neural electrodes and scaffolds for nerve regeneration in neural-tissue engineering [91, 92]. Conductive polymers allow for the electrical stimulation of cells cultured on the polymers, as shown by Schmidt et al. [93]. PC12 cells were cultured on PPy and subjected to electrical stimulation, resulting in a significant increase in neurite lengths when compared to the passive control group. Another group explored the possibilities of combining electrical stimulation and nanotopography as shown in Fig. 9 [94]. Several studies have also combined neural growth factors with conductive polymers for enhanced neurite outgrowth and obtained positive results.

3 Conclusion

Since the introduction of electrospinning technology, a number of studies on collecting various types of polymers into nanofibrous architectures have been reported. The nanofiber producing technique is also useful for tissue engineering applications because it creates scaffolds with a fibrous structure that is similar in structure to that of natural extracellular matrix. The biomimetic structure of electrospun nanofibers is favorable for cell viability, attachment, growth, migration, and division. In particular, uniaxially aligned nanofibers have been reported to aid in neuronal lining growth. Aligned nanofibrous nerve conduits meet the requirements of porosity, biocompatibility, biodegradability, and mechanical strength that scaffolds need while providing guidance cues. However, in vivo transplantation of the nerve conduit itself requires further studies such as a development of optimal surgical access to the spinal cord, long-term adverse effects, and graft-versus-host disease. Various methods have been studied to develop nanofibrous nerve conduits, and the incorporation of Schwann cells, neurotrophic factors, CNTs, and conductive polymers have been reported to showing remarkable results.

References

1. Hall JE, Guyton AC (2011) Guyton and Hall textbook of medical physiology, 12th edn. Saunders/Elsevier, Philadelphia, PA
2. Kwon BK, Okon E, Hillyer J, Mann C, Baptiste D, Weaver LC, Fehlings MG, Tetzlaff W (2011) A systematic review of non-invasive pharmacologic neuroprotective treatments for acute spinal cord injury. J Neurotrauma 28(8):1545–1588. https://doi.org/10.1089/neu.2009.1149
3. Han D, Cheung KC (2011) Biodegradable cell-seeded nanofiber scaffolds for neural repair. Polymers-Basel 3(4):1684–1733. https://doi.org/10.3390/polym3041684
4. National Spinal Cord Injury Statistical Center (NSCISC). Available online: https://www.nscisc.uab.edu/. Accessed 10 Jan 2017 (2015)
5. Kwon BK, Okon EB, Plunet W, Baptiste D, Fouad K, Hillyer J, Weaver LC, Fehlings MG, Tetzlaff W (2011) A systematic review of directly applied biologic therapies for acute spinal cord injury. J Neurotrauma 28(8):1589–1610. https://doi.org/10.1089/neu.2009.1150
6. Sykova E, Homola A, Mazanec R, Lachmann H, Konradova SL, Kobylka P, Padr R, Neuwirth J, Komrska V, Vavra V, Stulik J, Bojar M (2006) Autologous bone marrow transplantation in patients with subacute and chronic spinal cord injury. Cell Transplant 15(8–9):675–687
7. Geron C (2009) World's first clinical trial of human embryonic stem cell therapy cleared. Regen Med 4(2):161
8. Ruijs AC, Jaquet JB, Kalmijn S, Giele H, Hovius SE (2005) Median and ulnar nerve injuries: a meta-analysis of predictors of motor and sensory recovery after modern microsurgical nerve repair. Plast Reconstr Surg 116(2):484–494; discussion 495-486

9. Berger A, Millesi H (1978) Nerve grafting. Clin Orthop Relat Res 133:49–55
10. Mikami Y, Nagano A, Ochiai N, Yamamoto S (1997) Results of nerve grafting for injuries of the axillary and suprascapular nerves. J Bone Joint Surg Br 79(4):527–531
11. Barton MJ, Morley JW, Stoodley MA, Lauto A, Mahns DA (2014) Nerve repair: toward a sutureless approach. Neurosurg Rev 37(4):585–595. https://doi.org/10.1007/s10143-014-0559-1
12. Phillips JB, Bunting SC, Hall SM, Brown RA (2005) Neural tissue engineering: a self-organizing collagen guidance conduit. Tissue Eng 11(9–10):1611–1617. https://doi.org/10.1089/ten.2005.11.1611
13. de Ruiter GC, Malessy MJ, Yaszemski MJ, Windebank AJ, Spinner RJ (2009) Designing ideal conduits for peripheral nerve repair. Neurosurg Focus 26(2):E5. https://doi.org/10.3171/FOC.2009.26.2.E5
14. Kim JI, Hwang TI, Aguilar LE, Park CH, Kim CS (2016) A controlled design of aligned and random nanofibers for 3D bi-functionalized nerve conduits fabricated via a novel electrospinning set-up. Sci Rep 6:23761. https://doi.org/10.1038/srep23761
15. Yang F, Murugan R, Wang S, Ramakrishna S (2005) Electrospinning of nano/micro scale poly(L-lactic acid) aligned fibers and their potential in neural tissue engineering. Biomaterials 26(15):2603–2610. https://doi.org/10.1016/j.biomaterials.2004.06.051
16. Koh HS, Yong T, Chan CK, Ramakrishna S (2008) Enhancement of neurite outgrowth using nano-structured scaffolds coupled with laminin. Biomaterials 29(26):3574–3582. https://doi.org/10.1016/j.biomaterials.2008.05.014
17. Ahmed I, Liu HY, Mamiya PC, Ponery AS, Babu AN, Weik T, Schindler M, Meiners S (2006) Three-dimensional nanofibrillar surfaces covalently modified with tenascin-C-derived peptides enhance neuronal growth in vitro. J Biomed Mater Res A 76a(4):851–860. https://doi.org/10.1002/jbm.a.30587
18. College O (2013) Illustration from anatomy & physiology. https://commons.wikimedia.org/wiki/File:1319_Nerve_StructureN.jpg
19. Kandel ER, Schwartz JH (1981) Principles of neural science. Elsevier, North Holland, New York
20. Brosius Lutz A, Barres BA (2014) Contrasting the glial response to axon injury in the central and peripheral nervous systems. Dev Cell 28(1):7–17. https://doi.org/10.1016/j.devcel.2013.12.002
21. Lutz AB, Barres BA (2014) Contrasting the glial response to axon injury in the central and peripheral nervous systems. Dev Cell 28(1):7–17. https://doi.org/10.1016/j.devcel.2013.12.002
22. Huebner EA, Strittmatter SM (2009) Axon regeneration in the peripheral and central nervous systems. Results Probl Cell Differ 48:339–351. https://doi.org/10.1007/400_2009_19
23. Dahlin LB, Lundborg G (2001) Use of tubes in peripheral nerve repair. Neurosurg Clin N Am 12(2):341
24. Moore AM, Kasukurthi R, Magill CK, Farhadi HF, Borschel GH, Mackinnon SE (2009) Limitations of conduits in peripheral nerve repairs. Hand (N Y) 4(2):180–186. https://doi.org/10.1007/s11552-008-9158-3
25. Ma TC, Willis DE (2015) What makes a RAG regeneration associated? Front Mol Neurosci 8:43. https://doi.org/10.3389/fnmol.2015.00043
26. David S, Aguayo AJ (1981) Axonal elongation into peripheral nervous-system bridges after central nervous-system injury in adult-rats. Science 214(4523):931–933. https://doi.org/10.1126/science.6171034
27. Jankowski MP, McIlwrath SL, Jing X, Cornuet PK, Salerno KM, Koerber HR, Albers KM (2009) Sox11 transcription factor modulates peripheral nerve regeneration in adult mice. Brain Res 1256:43–54. https://doi.org/10.1016/j.brainres.2008.12.032
28. Seijffers R, Allchorne AJ, Woolf CJ (2006) The transcription factor ATF-3 promotes neurite outgrowth. Mol Cell Neurosci 32(1-2):143–154. https://doi.org/10.1016/j.mcn.2006.03.005
29. GrandPre T, Nakamura F, Vartanian T, Strittmatter SM (2000) Identification of the Nogo inhibitor of axon regeneration as a Reticulon protein. Nature 403(6768):439–444
30. Bomze HM, Bulsara KR, Iskandar BJ, Caroni P, Skene JHP (2001) Spinal axon regeneration evoked by replacing two growth cone proteins in adult neurons. Nat Neurosci 4(1):38–43

31. Bonilla IE, Tanabe K, Strittmatter SM (2002) Small proline-rich repeat protein 1A is expressed by axotomized neurons and promotes axonal outgrowth. J Neurosci 22(4):1303–1315
32. Safa B, Buncke G (2016) Autograft substitutes: conduits and processed nerve allografts. Hand Clin 32(2):127–140. https://doi.org/10.1016/j.hcl.2015.12.012
33. Chen ZL, Yu WM, Strickland S (2007) Peripheral regeneration. Annu Rev Neurosci 30:209–233. https://doi.org/10.1146/annurev.neuro.30.051606.094337
34. Balint R, Cassidy NJ, Cartmell SH (2014) Conductive polymers: towards a smart biomaterial for tissue engineering. Acta Biomater 10(6):2341–2353. https://doi.org/10.1016/j.actbio.2014.02.015
35. Ghasemi-Mobarakeh L, Prabhakaran MP, Morshed M, Nasr-Esfahani MH, Ramakrishna S (2009) Electrical stimulation of nerve cells using conductive nanofibrous scaffolds for nerve tissue engineering. Tissue Eng A 15(11):3605–3619. https://doi.org/10.1089/ten.tea.2008.0689
36. Oh SH, Kim JH, Song KS, Jeon BH, Yoon JH, Seo TB, Namgung U, Lee IW, Lee JH (2008) Peripheral nerve regeneration within an asymmetrically porous PLGA/Pluronic F127 nerve guide conduit. Biomaterials 29(11):1601–1609. https://doi.org/10.1016/j.biomaterials.2007.11.036
37. Assmann U, Szentivanyi A, Stark Y, Scheper T, Berski S, Drager G, Schuster RH (2010) Fiber scaffolds of polysialic acid via electrospinning for peripheral nerve regeneration. J Mater Sci Mater Med 21(7):2115–2124. https://doi.org/10.1007/s10856-010-4072-y
38. Dinis TM, Elia R, Vidal G, Dermigny Q, Denoeud C, Kaplan DL, Egles C, Marin F (2015) 3D multi-channel bi-functionalized silk electrospun conduits for peripheral nerve regeneration. J Mech Behav Biomed Mater 41:43–55. https://doi.org/10.1016/j.jmbbm.2014.09.029
39. Goh YF, Shakir I, Hussain R (2013) Electrospun fibers for tissue engineering, drug delivery, and wound dressing. J Mater Sci 48(8):3027–3054. https://doi.org/10.1007/s10853-013-7145-8
40. Jenkins PM, Laughter MR, Lee DJ, Lee YM, Freed CR, Park D (2015) A nerve guidance conduit with topographical and biochemical cues: potential application using human neural stem cells. Nanoscale Res Lett 10(1):972. https://doi.org/10.1186/s11671-015-0972-6
41. Gattazzo F, Urciuolo A, Bonaldo P (2014) Extracellular matrix: a dynamic microenvironment for stem cell niche. BBA-Gen Subjects 1840(8):2506–2519. https://doi.org/10.1016/j.bbagen.2014.01.010
42. Lutolf MP, Blau HM (2009) Artificial stem cell niches. Adv Mater 21(32–33):3255–3268. https://doi.org/10.1002/adma.200802582
43. Kim DH, Provenzano PP, Smith CL, Levchenko A (2012) Matrix nanotopography as a regulator of cell function. J Cell Biol 197(3):351–360. https://doi.org/10.1083/jcb.201108062
44. Kim MH, Sawada Y, Taya M, Kino-oka M (2014) Influence of surface topography on the human epithelial cell response to micropatterned substrates with convex and concave architectures. J Biol Eng 8:13. https://doi.org/10.1186/1754-1611-8-13
45. Beachley V, Wen XJ (2010) Polymer nanofibrous structures: fabrication, biofunctionalization, and cell interactions. Prog Polym Sci 35(7):868–892. https://doi.org/10.1016/j.progpolymsci.2010.03.003
46. Hoffman-Kim D, Mitchel JA, Bellamkonda RV (2010) Topography, cell response, and nerve regeneration. Annu Rev Biomed Eng 12(12):203–231. https://doi.org/10.1146/annurev-bioeng-070909-105351
47. Jha BS, Colello RJ, Bowman JR, Sell SA, Lee KD, Bigbee JW, Bowlin GL, Chow WN, Mathern BE, Simpson DG (2011) Two pole air gap electrospinning: fabrication of highly aligned, three-dimensional scaffolds for nerve reconstruction. Acta Biomater 7(1):203–215. https://doi.org/10.1016/j.actbio.2010.08.004
48. Gupta D, Venugopal J, Prabhakaran MP, Dev VR, Low S, Choon AT, Ramakrishna S (2009) Aligned and random nanofibrous substrate for the in vitro culture of Schwann cells for neural tissue engineering. Acta Biomater 5(7):2560–2569. https://doi.org/10.1016/j.actbio.2009.01.039
49. Bhutto MA, Wu T, Sun B, Ei-Hamshary H, Al-Deyab SS, Mo X (2016) Fabrication and characterization of vitamin B5 loaded poly (l-lactide-co-caprolactone)/silk fiber aligned electrospun nanofibers for schwann cell proliferation. Colloids Surf B: Biointerfaces 144:108–117. https://doi.org/10.1016/j.colsurfb.2016.04.013

50. Sun B, Jiang XJ, Zhang SC, Zhang JC, Li YF, You QZ, Long YZ (2015) Electrospun anisotropic architectures and porous structures for tissue engineering. J Mater Chem B 3(27):5389–5410. https://doi.org/10.1039/c5tb00472a
51. Rowan A (2006) Nerve regeneration - a strain on regeneration. Nat Rev Neurosci 7(8):596–597. https://doi.org/10.1038/nrn1972
52. Tetzlaff W, Okon EB, Karimi-Abdolrezaee S, Hill CE, Sparling JS, Plemel JR, Plunet WT, Tsai EC, Baptiste D, Smithson LJ, Kawaja MD, Fehlings MG, Kwon BK (2011) A systematic review of cellular transplantation therapies for spinal cord injury. J Neurotraum 28(8):1611–1682. https://doi.org/10.1089/neu.2009.1177
53. Bryan DJ, Wang KK, ChakalisHaley DP (1996) Effect of Schwann cells in the enhancement of peripheral-nerve regeneration. J Reconstr Microsurg 12(7):439–446. https://doi.org/10.1055/s-2007-1006616
54. Williams LR, Longo FM, Powell HC, Lundborg G, Varon S (1983) Spatial-temporal progress of peripheral-nerve regeneration within a silicone chamber - parameters for a bioassay. J Comp Neurol 218(4):460–470. https://doi.org/10.1002/cne.902180409
55. Bunge RP (1994) The role of the Schwann-cell in trophic support and regeneration. J Neurol 242(1):S19–S21. https://doi.org/10.1007/Bf00939235
56. Jankowski MP, Cornuet PK, McILwrath S, Koerber HR, Albers KM (2006) SRY-box containing gene 11 (Sox11) transcription factor is required for neuron survival and neurite growth. Neuroscience 143(2):501–514. https://doi.org/10.1016/j.neuroscience.2006.09.010
57. Barres BA, Raff MC (1999) Axonal control of oligodendrocyte development. J Cell Biol 147(6):1123–1128. https://doi.org/10.1083/jcb.147.6.1123
58. Novikova LN, Pettersson J, Brohlin M, Wiberg M, Novikov LN (2008) Biodegradable poly-beta-hydroxybutyrate scaffold seeded with Schwann cells to promote spinal cord repair. Biomaterials 29(9):1198–1206. https://doi.org/10.1016/j.biomaterials.2007.11.033
59. Xu XM, Zhang SX, Li HY, Aebischer P, Bunge MB (1999) Regrowth of axons into the distal spinal cord through a Schwann-cell-seeded mini-channel implanted into hemisected adult rat spinal cord. Eur J Neurosci 11(5):1723–1740. https://doi.org/10.1046/j.1460-9568.1999.00591.x
60. Blitts B, Oudega M, Boer GJ, Bunge MB, Verhaagen J (2003) Adeno-associated viral vector-mediated neurotrophin gene transfer in the injured adult rat spinal cord improves hind-limb function. Neuroscience 118(1):271–281. https://doi.org/10.1016/S0306-4522(02)00970-3
61. Keeley R, Atagi T, Sabelman E, Padilla J, Kadlcik P, Agras J, Eng L, Wiedman TW, Nguyen K, Sudekum A, Rosen J (1994) Synthetic nerve graft containing collagen and synthetic Schwann-cells improves functional, electrophysiological, and histological parameters of peripheral-nerve regeneration (Vol 5, Pg 353, 1993). Restor Neurol Neurosci 6(2):161
62. Lohmeyer JA, Shen ZL, Walter GF, Berger A (2007) Bridging extended nerve defects with an artificial nerve graft containing Schwann cells pre-seeded on polyglactin filaments. Int J Artif Organs 30(1):64–74
63. Nie X, Zhang YJ, Tian WD, Jiang M, Dong R, Chen JW, Jin Y (2007) Improvement of peripheral nerve regeneration by a tissue-engineered nerve filled with ectomesenchymal stem cells. Int J Oral Maxillofac Surg 36(1):32–38. https://doi.org/10.1016/j.ijom.2006.06.005
64. Clarke DL, Johansson CB, Wilbertz J, Veress B, Nilsson E, Karlstrom H, Lendahl U, Frisen J (2000) Generalized potential of adult neural stem cells. Science 288(5471):1660–1663
65. Temple S (1989) Division and differentiation of isolated CNS blast cells in microculture. Nature 340(6233):471–473. https://doi.org/10.1038/340471a0
66. Heine W, Conant K, Griffin JW, Hoke A (2004) Transplanted neural stem cells promote axonal regeneration through chronically denervated peripheral nerves. Exp Neurol 189(2):231–240. https://doi.org/10.1016/j.expneurol.2004.06.014
67. Olson HE, Rooney GE, Gross L, Nesbitt JJ, Galvin KE, Knight A, Chen B, Yaszemski MJ, Windebank AJ (2009) Neural stem cell- and Schwann cell-loaded biodegradable polymer scaffolds support axonal regeneration in the transected spinal cord. Tissue Eng A 15(7):1797–1805. https://doi.org/10.1089/ten.tea.2008.0364
68. Lee SH, Chung YN, Kim YH, Kim YJ, Park JP, Kwon DK, Kwon OS, Heo JH, Kim YH, Ryu S, Kang HJ, Paek SH, Wang KC, Kim SU, Yoon BW (2009) Effects of human neural stem cell

transplantation in canine spinal cord hemisection. Neurol Res 31(9):996–1002. https://doi.org/10.1179/174313209x385626
69. Zahir T, Nomura H, Guo XD, Kim H, Tator C, Morshead C, Shoichet M (2008) Bioengineering neural stem/progenitor cell-coated tubes for spinal cord injury repair. Cell Transplant 17(3): 245–254
70. Nie X, Zhang YJ, Tian WD, Jiang M, Dong R, Chen JW, Jin Y (2007) Improvement of peripheral nerve regeneration by a tissue-engineered nerve filled with ectomesenchymal stem cells. Int J Oral Max Surg 36(1):32–38. https://doi.org/10.1016/j.ijom.2006.06.005
71. Pal R, Gopinath C, Rao NM, Banerjee P, Krishnamoorthy V, Venkataramana NK, Totey S (2010) Functional recovery after transplantation of bone marrow-derived human mesenchymal stromal cells in a rat model of spinal cord injury. Cytotherapy 12(6):792–806. https://doi.org/10.3109/14653249.2010.487899
72. Lopes FRP, Campos LCD, Correa JD, Balduino A, Lora S, Langone F, Borojevic R, Martinez AMB (2006) Bone marrow stromal cells and resorbable collagen guidance tubes enhance sciatic nerve regeneration in mice. Exp Neurol 198(2):457–468. https://doi.org/10.1016/j.expneurol.2005.12.019
73. Su ZD, He C (2010) Olfactory ensheathing cells: biology in neural development and regeneration. Prog Neurobiol 92(4):517–532. https://doi.org/10.1016/j.pneurobio.2010.08.008
74. http://www.livemint.com/Politics/L800Xj31ulYJVBUaDrwaiL/Paralysed-man-walks-again-after-breakthrough-treatment.html (2014)
75. Hempstead BL (2006) Dissecting the diverse actions of pro- and mature neurotrophins. Curr Alzheimer Res 3(1):19–24
76. Fan J, Zhang H, He J, Xiao Z, Chen B, Xiaodan J, Dai J, Xu R (2011) Neural regrowth induced by PLGA nerve conduits and neurotrophin-3 in rats with complete spinal cord transection. J Biomed Mater Res B Appl Biomater 97(2):271–277. https://doi.org/10.1002/jbm.b.31810
77. Joosten EA, Houweling DA (2004) Local acute application of BDNF in the lesioned spinal cord anti-inflammatory and anti-oxidant effects. Neuroreport 15(7):1163–1166
78. Liang W, Han Q, Jin W, Xiao Z, Huang J, Ni H, Chen B, Kong J, Wu J, Dai J (2010) The promotion of neurological recovery in the rat spinal cord crushed injury model by collagen-binding BDNF. Biomaterials 31(33):8634–8641. https://doi.org/10.1016/j.biomaterials.2010.07.084
79. Cheng H, Cao Y, Olson L (1996) Spinal cord repair in adult paraplegic rats: partial restoration of hind limb function. Science 273(5274):510–513
80. De Laporte L, des Rieux A, Tuinstra HM, Zelivyanskaya ML, De Clerck NM, Postnov AA, Preat V, Shea LD (2011) Vascular endothelial growth factor and fibroblast growth factor 2 delivery from spinal cord bridges to enhance angiogenesis following injury. J Biomed Mater Res A 98(3):372–382. https://doi.org/10.1002/jbm.a.33112
81. Widenfalk J, Lipson A, Jubran M, Hofstetter C, Ebendal T, Cao Y, Olson L (2003) Vascular endothelial growth factor improves functional outcome and decreases secondary degeneration in experimental spinal cord contusion injury. Neuroscience 120(4):951–960
82. Sundberg LM, Herrera JJ, Narayana PA (2011) Effect of vascular endothelial growth factor treatment in experimental traumatic spinal cord injury: in vivo longitudinal assessment. J Neurotrauma 28(4):565–578. https://doi.org/10.1089/neu.2010.1533
83. Kim MS, El-Fiqi A, Kim JW, Ahn HS, Kim H, Son YJ, Kim HW, Hyun JK (2016) Nanotherapeutics of PTEN inhibitor with mesoporous silica nanocarrier effective for axonal outgrowth of adult neurons. ACS Appl Mater Inter 8(29):18741–18753. https://doi.org/10.1021/acsami.6b06889
84. Sun BB, Taing A, Liu HY, Nie GC, Wang JY, Fang YL, Liu L, Xue Y, Shi J, Liao YP, Ku J, Xia T, Liu Y (2016) Nerve growth factor-conjugated mesoporous silica nanoparticles promote neuron-like PC12 cell proliferation and neurite growth. J Nanosci Nanotechnol 16(3):2390–2393. https://doi.org/10.1166/jnn.2016.10958
85. Shah S, Solanki A, Lee KB (2016) Nanotechnology-based approaches for guiding neural regeneration. Acc Chem Res 49(1):17–26. https://doi.org/10.1021/acs.accounts.5b00345
86. Fabbro A, Prato M, Ballerini L (2013) Carbon nanotubes in neuroregeneration and repair. Adv Drug Deliver Rev 65(15):2034–2044. https://doi.org/10.1016/j.addr.2013.07.002

87. Gilmore JL, Yi X, Quan L, Kabanov AV (2008) Novel nanomaterials for clinical neuroscience. J Neuroimmune Pharm 3(2):83–94. https://doi.org/10.1007/s11481-007-9099-6
88. Yong J, Needham K, Brown WGA, Nayagam BA, McArthur SL, Yu AM, Stoddart PR (2014) Gold-nanorod-assisted near-infrared stimulation of primary auditory neurons. Adv Healthc Mater 3(11):1862–1868. https://doi.org/10.1002/adhm.201400027
89. Yang J, Fan LX, Wang FJ, Luo Y, Sui X, Li WH, Zhang XH, Wang YG (2016) Rapid-releasing of HI-6 via brain-targeted mesoporous silica nanoparticles for nerve agent detoxification. Nanoscale 8(18):9537–9547. https://doi.org/10.1039/c5nr06658a
90. Zelikin AN, Lynn DM, Farhadi J, Martin I, Shastri V, Langer R (2002) Erodible conducting polymers for potential biomedical applications. Angew Chem Int Ed 41(1):141–144. https://doi.org/10.1002/1521-3773(20020104)41:1<141::Aid-Anie141>3.0.Co;2-V
91. Sun BB, Wu T, Wang J, Li DW, Wang J, Gao Q, Bhutto MA, El-Hamshary H, Al-Deyab SS, Mo XM (2016) Polypyrrole-coated poly(L-lactic acid-co-epsilon-caprolactone)/silk fibroin nanofibrous membranes promoting neural cell proliferation and differentiation with electrical stimulation. J Mater Chem B 4(41):6670–6679. https://doi.org/10.1039/c6tb01710j
92. Yan L, Zhao BX, Liu XH, Li X, Zeng C, Shi HY, Xu XX, Lin T, Dai LM, Liu Y (2016) Aligned nanofibers from polypyrrole/graphene as electrodes for regeneration of optic nerve via electrical stimulation. ACS Appl Mater Inter 8(11):6834–6840. https://doi.org/10.1021/acsami.5b12843
93. Schmidt CE, Shastri VR, Vacanti JP, Langer R (1997) Stimulation of neurite outgrowth using an electrically conducting polymer. Proc Natl Acad Sci U S A 94(17):8948–8953. https://doi.org/10.1073/pnas.94.17.8948
94. Lee JY, Bashur CA, Goldstein AS, Schmidt CE (2009) Polypyrrole-coated electrospun PLGA nanofibers for neural tissue applications. Biomaterials 30(26):4325–4335. https://doi.org/10.1016/j.biomaterials.2009.04.042

Electrospun Scaffold for Retinal Tissue Engineering

Samad Nadri and Ghasem Barati

Abstract Many treatment strategies have been planned to slow down the progression of retinal diseases. Although cell based therapies showed promising results for retinal diseases, but the injection of cell suspensions into the subretinal space may result in inadequate cell survival, organization and differentiation. For efficient cell therapy, cell populations need to survive and function suitably inside the transplanted environment. With the emergence of tissue engineering, regenerative medicine is being advanced and offers great promise for treatment retinal diseases. There are some innovative and developing approaches for retinal tissue engineering. To date, different types of cellular sources and electrospun scaffolds have been considered to define the best donor cell and delivery system to set the degenerating retina free. It seems that fibrous scaffolds, as a substrate, may hold many advantages for retinal stem cell transplantation. In this book chapter, the advantages of electrospun fibrous scaffold and challenges in front of retinal tissue engineering have been discussed.

Keywords Electrospun Scaffold • Retinal • Tissue Engineering • Cell Delivery • Drug and Growth Factor

1 Introduction

Retina, a light-sensitive tissue, lines the back of the eye and converts focused light into neural signals. Nerve fibers send these signals into the brain, which then interprets them as visual images [1].

The retinal tissue is made up of numerous layers including pigment epithelium, rods and cones cells, Müller cells, bipolar cells, amacrine cells, ganglion cells, inner fiber layers and inner limiting membranes. Three layers of nerve cell bodies and two layers of synapses exist in all vertebrate retinas. The outer nuclear layer contains

S. Nadri (✉) • G. Barati
Department of Medical Biotechnology and Nanotechnology, School of Medicine, Zanjan University of Medical Science, Zanjan, Iran
e-mail: nadri_s@ZUMS.ac.ir

J. Almodovar (ed.), *Electrospun Biomaterials and Related Technologies*, DOI 10.1007/978-3-319-70049-6_8

cell bodies of the rods and cones, the inner nuclear layer contains cell bodies of the bipolar, horizontal and amacrine cells and the ganglion cell layer contains cell bodies of ganglion cells and displaced amacrine cells. Central retina near the fovea is significantly thicker than peripheral retina due to the increased density of photoreceptors, and their associated bipolar and ganglion cells in central retina.

Degenerative diseases of the retina include glaucoma (17% of world blindness), macular degeneration (the third leading cause of world blindness), diabetic retinopathy (17% of world blindness) and retinitis pigmentosa. Macular degeneration, retinitis pigmentosa and glaucoma are categorized by the degeneration of the photoreceptors, retinal pigmented epithelium (RPE) and retinal ganglion cells (RGCs), respectively [2, 3]. These degenerations can lead to significant vision difficulties and blindness if left untreated [3].

Existing treatments for glaucoma include a surgical procedure named trabeculectomy. The trabeculectomy can lead to problems such as the control of intraocular pressure, invasive procedure, inflammation [4] and infection.

Many treatment strategies have been designed to decelerate the progression of macular degeneration or diabetic maculopathy, such as administration of growth factors along with gene therapy, neuroprotective treatment, anti-angiogenic therapy and intravitreal injections containing ocular drug treatments [5–9]. However, the injections are invasive, require repeated treatments and can lead to endophthalmitis, a potentially blinding infection [10]. To date, no commercially available treatments for retinitis pigmentosa have been reported.

Therefore, there is a need for more effective treatments for retinal diseases. Cell based therapies showed promising results for retinal diseases. However, the injection of cell suspensions into the subretinal space may result in inadequate cell survival, integration, organization/alignment and differentiation [7, 11–13]. Furthermore, overcoming the problem of immune rejection and the guidance of axonal growth to create valuable networks are challenges with aspects of cell delivery and integration of regenerative materials to the retinal tissue [14]. For efficient cell therapy, cell populations need to survive and function suitably inside the transplanted environment [15]. With the emergence of tissue engineering, where cell populations are delivered via a material scaffold, regenerative medicine is being advanced and offers great promise for treatment retinal diseases [7].

2 Retinal Tissue Engineering

Tissue engineering should be able to overcome problems such as donor shortages, graft rejections, and inflammatory responses following transplantation. There are two main strategies in tissue engineering: (1) cell-based and, (2) scaffold-based. In retina, both cell and substrate-based tissue engineering approaches have been examined mainly in animal models. There are some innovative and developing approaches for retinal tissue engineering [16]. Cell transplantation with a bio-compatible polymer scaffold enhances survival [17], differentiation [18] and integration of cells in host retinal tissue [18, 19]. In retinal tissue-engineered approach, the aim is to integrate cell populations into the retina, differentiate and organize them into effective cells, and provide a promising treatment to restore vision in patients with retinal degeneration.

3 Cells

To date, different types of cellular sources have been considered to define the best donor cell to set the degenerating retina free (Table 1).

Among these, Retinal Progenitor Cells (RPCs), photoreceptor precursor cells, pluripotent (ES and iPSc) cell lines and Retinal Pigment Epithelium (RPE) cells have been widely investigated [63]. RPCs and RPE have shown the capability of self-renewal and differentiation into the different retinal cell types including photoreceptors in vitro. They have revealed ability as a source of cell transplantation in experimental models of retinal degeneration. RPE cells are hexagonal shaped and have a complex structural and functional polarity which helps them to perform numerous special functions [64]. Photoreceptor precursors (Late-stage) isolated from the retina cannot be expanded in vitro but can be used for retinal regeneration with the best functional outcome [65]. Pluripotent cell-derived photoreceptors (ES and iPSc) have suitable self-renewal properties; however, they exhibit no effective differentiation capability for retinal regeneration [24].

Despite the fact that different cell sources are considered to be well-accepted for retinal tissue engineering, the clinical use of these cells have presented problems including ethical issues, risk of teratoma formation, painful aspiration, low photoreceptor efficiency, complication of microsurgical procedures for extraction [66], limited availability of pluripotent retinal stem cells, formation of an abnormal cell orientation (rosette cells) [67, 68] and rejection by the host immune system in either allogeneic or xenogenic host species recipients [69, 70]. This has led many researchers to investigate alternate sources for retinal tissue engineering. It has been observed that mesenchymal stem cells (MSCs) can differentiate into photoreceptor cells [66].

Bone marrow mesenchymal stem cells (BMMSCs) were differentiated within 2 weeks using a mixture of growth factors in vitro. However, BMMSC is not an appropriate candidate for retinal therapy as researchers have never detected any expression of main photoreceptor markers [71].

Previous studies have demonstrated that merely 20–30% of CD90 positive MSCs, express photoreceptor-specific markers (rhodopsin and recoverin) [66, 72, 73]. Due to down regulation of some of the genes involved in cell proliferation and cell cycle progression during cell isolation by immunodepletion method, CD90 positive MSCs cannot be expanded very well in vitro [74, 75].

Recently, conjunctiva mesenchymal stem cells (CJMSCs) and trabecular meshwork mesenchymal stem cells (TMMSCs), isolated from conjunctiva and trabecular meshwork, have shown a good proliferation capability and differentiation into mature photoreceptor cells in vitro. Surprisingly, the photoreceptor differentiation rate of CJMSCs and TMMSCs are higher than other sources of MSCs.

Table 1 Source of cells used in different retinal cell differentiation and treatment of retinal disease

No.	Cell type	Advantages	Disadvantages	References
1	Pluripotent stem cells (ESCs/iPS)	Successfully differentiate into Retinal progenitor cells (RPCs) or photoreceptors up to 80% efficiency in vitro using growth factors Successfully integrate into retinal layers	Risk of teratoma formation Ethical issue (in the case of ESCs)	[20–36]
2	BM-MSCs NSCs	Protect retinal cells by releasing paracrine factors Successfully integrate into retinal layers	Painful aspiration and contamination with non-MSCs Differentiation into retinal cells with low efficiency	[37–47]
3	DP-MSCs	Neuroprotect retinal ganglion cells (RGCs) survival and axon regeneration after optic nerve injury	Do not differentiate into retinal cells Fails to integrate into retinal layers	[48]
4	AD-MSCs	Conserve the retinal vasculature in mouse models of diabetic retinopathy by differentiation into pericytes	Differentiate into retinal cells with low efficiency Fails to integrate into retinal layers	[49, 50]
5	Corneal-derived MSCs (TMMSCs, CJMSCs)	High yield differentiation into photoreceptor-like cells in vitro and in vivo (data not published)	Required person skilled at cell culture	[51, 52]
6	RPCs	High capacity for differentiation into retinal-like cells and photoreceptor Capability of integration into retinal layers	Complication of surgical procedures	[5, 53, 54]
7	RPE/IPE	Do not require synaptic connection to perform their role, Improve recurrence rate of CNV Prevent further loss of vision	Ethical issue, Immune rejection (Allogenic PREs) Slow integration and phagocytosis properties	[55–62]

ESCs embryonic stem cells, *iPS* induced pluripotent stem cells, *BM-MSCs* bone marrow-derived mesenchymal stem cells, *NSCs* neural stem cells, *AD-MSCs* adipocyte-derived mesenchymal stem cells, *DP-MSCs* dental pulp mesenchymal stem cells, *RPCs* retinal progenitor cells, *RPE* retinal pigment epithelium, *IPE* iris pigment epithelium, *CNV* choroidal neovascularization. Corneal-derived MSCs include trabecular meshwork mesenchymal stem cells (TMMSCs) and conjunctiva mesenchymal stem cells (CJMSCs)

4 Electrospun Scaffold

Scaffolds can be categorized into three different types: cylindrical, hydrogel and fibrous. Due to their mimicking of the extracellular matrix (ECM), Fibrous scaffolds have become very common in tissue engineering (Fig. 1). The fibers offer a large surface area for cell attachment and create interconnected pores for transport of nutrients and gases within scaffold [76]. While a wide range of techniques have been investigated for fabrication of fibrous scaffolds in tissue engineering and subretinal space transplantation, one of the most common is through electrospinning [77–79]. Electrospinning technique uses high electrostatic forces to create polymer fibers [80]. Electrospun scaffolds in retinal tissue engineering are established to study the cellular response to nano/microtopological cues [81].

In recent years, polymeric scaffolds have been employed to increase the number of viable cells delivered to the degenerated retina. Suitable polymer substrate design is critical to provide a three-dimensional environment (mimic the fibrous nature of natural extracellular matrix (ECM)) that can facilitate cell adhesion, proliferation and post transplantation migration into the host environment [19].

The ideal scaffold for retinal tissue engineering should have criteria including, porosity, biodegradability, elasticity, and permeability. Scaffold material must be biocompatible, does not induce immunogenic responses [82], and contains the correct Young's modulus and capability for drug and protein loading.

Due to the small size of the subretinal chamber, fabricated scaffold must be extremely thin (thinner than 50 μm) [83] and should be implantable by a minimally invasive methods [84, 85]. Thick scaffolds formed large retinal detachments and caused retinal damage to the retina during implantation. Scaffolds, less than 50 μm in thickness, permit the interaction of seeded cells on scaffold with retinal cells in vivo. To hold out the necessary surgical manipulation and to escape damages to the surrounding tissues, the scaffold materials must be mechanically strong and

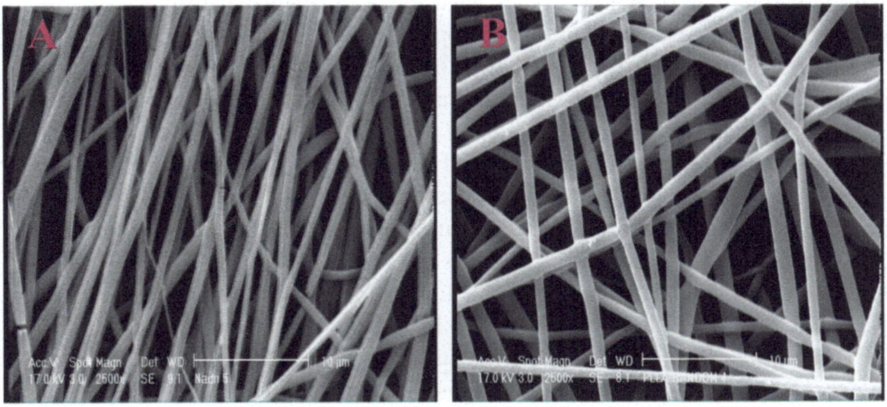

Fig. 1 SEM micrographs of aligned (**a**) and random (**b**) nanofibrous scaffolds fabricated using electrospinning methods

Table 2 Electrospun scaffolds used in retinal tissue engineering

No.	Polymer	Cell	Description	Reference
1.	PLGA	RPCs	MMP-2 control-released of electrospun fibers increased cell migration into retina (three to fourfold) and PRC differentiation into photoreceptors (10–15-fold)	[90]
2.	PCL	RPCs	Transplanted cells onto retinal explants migrated into control and Rho−/− retinas and only differentiated into glial cells	[88]
3.	CS-PCL/PCL	RPCs	Better proliferation and differentiation of RPCs into photoreceptors and glial cells	[89]
4.	SF:PLCL	RPCs	SF: PLCL yielded the best RPC growth, proliferation and preferentially differentiation toward retinal neurons (photoreceptors)	[91]
5.	PLLA	CJMSCs	High yield of photoreceptor-like cells (rod photoreceptor genes) on randomly-oriented compared to aligned nanofibers	[51]
6.	PEG/PCL	CJMSCs	Core/shell PEG/PCL fibrous scaffold incorporated taurine could increase the expression of rhodopsin genes in vitro	[92]

PLGA poly(lactic-co-glycolic acid), *PCL* polycaprolactone, *CS-PCL/PCL* cationic chitosan-graft-poly(ε-caprolactone)/polycaprolactone, *SF:PLCL* silk fibroin and poly(L-lactic acid-co-ε-caprolactone), *PLLA* poly-L-lactic acid, *PEG* polyethylene glycol, *RPCs* retinal progenitor cells, *MMP-2* matrix metallopeptidase 2, *CJMSCs* conjunctiva mesenchymal stem cells

flexible [86]. To limit the damage during insertion into the subretinal space, scaffolds were scrolled, which was probable due to the higher elasticity of materials (such as PGS) [6]. Furthermore, it should provide sufficient signals for proper cell attachment and alignment. To improve cell binding, these scaffolds were coated with extracellular matrix component (such as laminin, Fibronectin etc.).

Numerous polymers fulfill this criteria including poly(lactic-co-glycolic acid) (PLGA), poly(lactic acid) (PLLA), poly(caprolactone) (PCL) and poly(glycerol sebacate) (PGS) [87]. A few studies have revealed promising results using these polymers for retinal tissue engineering applications.

RPCs grown on electrospun PCL scaffolds coated with laminin [88] were transplanted onto retinal explants, which migrated into Rho−/− retinas and differentiation into glial cells.

Fibrous electrospun scaffolds produced from a blended PCL and chitosan, which cause a reduction in hydrophobicity of scaffold, not only influenced the capability of the RPCs to attach on the scaffold but also improved the proliferation rate of cultivated RPCs [89]. Retinal Progenitors cells (RPCs) cultivated on fibrous scaffolds have revealed limited ability to photoreceptor differentiation following transplantation. RPCs seeded on electrospun PCL were differentiated into glial cells (GFAP-positive) [88]. When cells were cultured on chitosan/PCL composite scaffolds, expression of recoverin and rhodopsin (as a main photoreceptor marker) and GFAP were upregulated [81, 89] (Table 2).

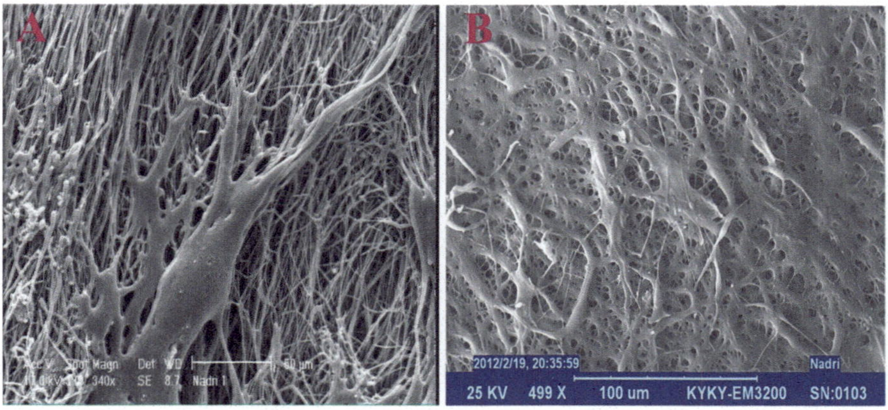

Fig. 2 SEM micrographs of CJMSCs differentiated on aligned (**a**) and random (**b**) PLLA nanofibrous scaffolds

SF/PLCL nanofibrous scaffolds have shown good cytocompatibility due to no up-regulation in expression levels of IL-6 (a pro-inflammatory cytokine) [93], MCP-1 (key gene in the migration of immune-competent cells) [94], and Caspase 3 (a major terminal cleavage ribozyme in the apoptosis process) [95] in RPCs cultured on SF/PLCL. SF: PLCL scaffolds with a suitable pore diameter (less than 2 μm) would offer larger surface areas, which mimic the topographic structures of the ECM contributing to increases in attachment, proliferation and differentiation of cells into photoreceptors [96, 97]. Furthermore, electrospinning of SF (with its powerful hydrophilicity) along with PLCL will improve the hydrophilicity of PLCL and the introduction of biological functional groups, such as $-NH_2$ and –COOH, via SF in SF/PLCL may increase the proliferation and photoreceptor differentiation of RPCs, respectively [76].

Recently, CJMSCs were cultured on PLLA fiber matrices made of electrospun nanofibers, and subsequently, differentiation into photoreceptor cells was examined. Electrospun scaffolds could enhance (recoverin) and diminish (CRX, GFAP and Nestin) the expression of some specific photoreceptor genes. Taken together, CJMSCs on scaffolds are more likely to differentiate toward photoreceptors and retinal neural cells, while those cultured on tissue culture plates (TCPS) preferentially differentiate into glial cells.

Aligned fibers with various degrees of order and fiber directions for two- and three-dimensional assemblies have been developed using electrospinning. Electrospun aligned fibers are usually achieved using a pair of parallel conducting electrodes to produce an electric field or a rotating drum at a speed of 1000–2500 rpm to collect aligned fibers [51]. Fiber orientation strongly influences growth and other related functions in cells [80]. On aligned fibrous scaffolds, the CJMSCs direction of elongation and its neurite outgrowth were parallel to PLLA fiber direction (Fig. 2). The significant changes were observed on the photoreceptor gene expression with respect to the fiber orientation. CJMSCs cultured on aligned nanofibrous scaffolds demonstrated a higher rate in expressing photoreceptor and retinal neuronal genes including PKC and nestin compared to random nanofibers. Aligned nanofiber scaffolds could discourage the differentiation of CJMSC cells into retinal neuronal cells, which is much desirable in therapies targeting retinal diseases. CJMSCs on random nanofibrous scaffolds demonstrated a higher expression rate of rhodopsin gene than on aligned nanofibrous scaffolds [51]. By blending PLGA with PLLA, the flexibility profile of PLGA can be increased (low modulus) [98]. Good mechanical properties of a scaffold are helpful for successful cell transplantation. Retina is a soft and flexible tissue with elastic modulus of 0.1 MPa. PLGA has an elastic modulus of 1.4–2.8 GPa while SF: PLCL has an elastic modulus of 105.3 MPa, which is closer to that of retinal tissue [99]. Such scaffolds may be mainly related to subretinal applications. The excellent mechanical properties (stiffness ~0.056 to 1.5 MPa after curing), high flexibility, biodegradability and minimal swelling of PGS are fine matched to the subretinal space. Although a wide range of natural and synthetic polymers, such as PCL, PLLA, PLGA, collagen and gelatin, have been fabricated by electrospinning, PGS crosslinks into an elastomer cannot be dissolved in organic solvents and so could not be electrospun. However, the modification of PGS with rigid polymers such as PMMA and formation of PGS-PMMA leads to the fabrication of nanofibrous scaffold [100].

5 Cell Delivery

Retinal tissue engineering displays huge potential in treatment of retinal degeneration by using scaffold-based delivery systems of cells into the subretinal space. Cell transplantation with a scaffold has been defined as an effective approach to improve survival [17, 101], differentiation [18] and integration of cells in host retinal tissue [19]. The formation of synapses and right tight junction with each other are essential for rescue of vision after cell transplantation and differentiation into mature retinal cell types [102, 103]. Scaffolds can also direct the organization of retinal cell populations.

Building on these advantages, proliferation and integration dynamics of mouse RPC (mRPC) were examined on fabricated ultra-thin (15 μm-thick) electrospun PCL scaffold [104]. In strategy of electrospun PCL scaffold, a middle porosity percentage (the diameter of mRPC is 10–20 μm) and scaffold thickness of 15 μm (the thickness of mouse retina is 200–300 μm), with the goal of upholding nutrient, gas exchange, three-dimensional cellular infiltration is selected to reduce disorder to retina while protects scaffold rigidity for transplantation purposes [105–107]. After Ex vivo transplantation, healthy migration of mRPCs from the scaffolds into mice retinal tissue was observed. Most of mRPCs took the round shape characteristic of proliferating progenitor cells, but the integrated cells showed retinal neuronal morphology [108]. However, ultra-thin electrospun PCL scaffold is capable of delivering high numbers of localized cells to mouse retinal explants, with strong migration, integration and differentiation.

6 Drug and Growth Factor Delivery

Local delivery of bioactive molecules by an implantable device could decrease the amount of required drug dose as well as non-target site toxicities. Recently, electrospinning methods have been used to improve drug or protein-loaded nanofibrous scaffold for use as a novel implantable delivery system.

Coaxial electrospinning, as an advanced extension of electrospinning, could be used to fabricate unique core-shell or hollow structures due to multichannel spinneret consisting of core and shell channels (Fig. 3) [109]. As those seen in the conventional one, under an adequate electrical field, the core-shell jets become ejected and after an electrospinning process, the core-shell fibers will be collected on the collector. Core-shell structures are aiming to deliver therapeutic molecules such as proteins, growth factors, and drugs. The growth factors such as Vascular endothelial growth factor (**VEGF**), recombinant human bone morphogenetic protein 2 (rhBMP-2) incorporated into core-shell nanofibers have been used for the repair and regeneration of tissues [110, 111]. The macromolecules could be released gradually from the core-shell nanofibrous scaffolds. The shell layer acts as an obstacle to avoid the early release of the core contents. Promising work has been reported with coaxially electrospun fiber incorporated with growth factors or bioactive agents for use in tissue engineering. The potential of the coaxial electrospinning methods for controlled release of taurine for retinal tissue engineering has been reported. PEG/PCL fiber matrices incorporated taurine could increase the expression of photoreceptor genes such as rhodopsin compared to CJMSCs differentiated on TCPS dishes [92].

In another study, to further enhance integration of RPCs in the retina, the enzyme Matrix Metallopeptidase-2 (MMP-2) was incorporated at the core of the PLGA electrospun fibers [90, 112, 113]. The encapsulation of MMP-2 to the scaffold fibers increased the total number of cells able to migrate into retinas and significantly

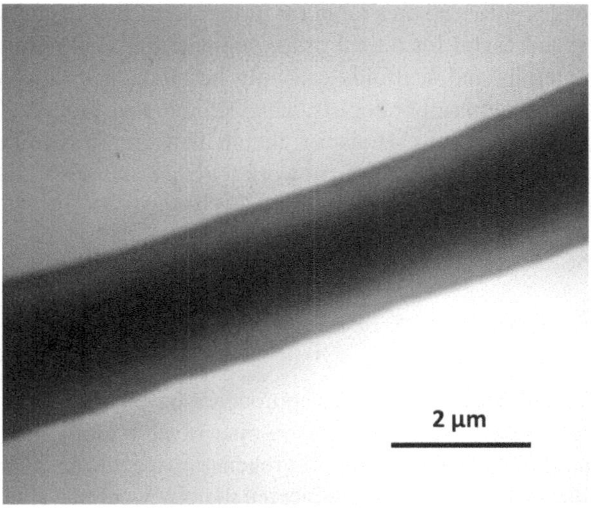

Fig. 3 TEM images of PEG/PCL with distinct core-sheath structure

increased the efficacy of RPC differentiated into photoreceptors (10–15-fold) when compared with electrospun scaffolds without MMP-2 [90]. Also, when transplanted on PLGA nanoscaffolds, an increase migration of RPCs (three to fourfold) in animal model (Rho−/− retinas) has been documented within the retina. In a similar study, a significant overexpression of recoverin and rhodopsin was gotten using a PLGA nanofiber scaffold in the presence of MMP-2 compared with PLGA scaffolds [90].

7 Challenge and Limitation

The main limitation of retinal tissue engineering study is the use of cell lines as an alternative of primary cells or stem cells which are identified to be diverse in terms of their behavior.

Photoreceptor precursors (late-stage) isolated from the retina cannot be expanded in vitro and unsuitable differentiation of pluripotent cell-derived photoreceptors limits the potential of these sources for retinal regeneration [114].

Thickness, surface topography, mechanical properties and degradation characteristics as critical challenges, should be considered during fabrication of the scaffold to avoid damages to the retina and to let interactions between the retina and post-transplant cell population. Although electrospinning is a quick and simple method to fabrication of nanofibrous scaffolds, it is still a big challenge to produce scaffolds with complex structures for retinal tissue engineering. Toxicity of solvents, inhomogeneous distribution and insufficient size, and problems in obtaining 3D structures, are main disadvantages of electrospinning.

Cell manufacture, delivery, survival, physiological behavior, immune response, and a potential risk of cancer development are chief challenges opposing clinical application of cell-scaffold therapy for degenerative retinal disease. Technical challenges still keep on a room for further study before cell-scaffold based therapies can be successful for retinal tissue engineering. The improvement of such techniques, materials and scaffolds can only be attained by means of a collaborative effort among stem cell biologists, nanotechnologies, biomaterial experts, and skilled ophthalmic surgeons. However, the challenge here is to be able to get the scientists, engineers, and clinicians to work together in order to solve challenges and give our patients the best possible treatment.

8 Conclusion

Retinal tissue engineering approaches have been shown to be serious to increasing our understanding of the nano-microscale biology of stem cells, and to supplying valid mechanisms to support regenerative cell transplantation for retinal degeneration. A positive tissue-engineered therapy would be able to restore vision to patients with progressive RPE and photoreceptor loss. Cell-substrate interactions play a critical role in cellular behavior and may demonstrate to be an active device in managing stem/progenitor cells to a differentiated cell fate.

It seems that fibrous scaffolds, as a substrate, may hold many advantages for retinal stem cell transplantation. However, more study is required to expand the properties of the electrospun scaffolds in order to better match to those of retinal tissue. This, however, remains to be seen.

References

1. Livesey FJ, Cepko CL (2001) Vertebrate neural cell-fate determination: lessons from the retina. Nat Rev Neurosci 2:109–118
2. Strauss O (2005) The retinal pigment epithelium in visual function. Physiol Rev 85:845–881
3. Hartong DT, Berson EL, Dryja TP (2006) Retinitis pigmentosa. Lancet 368:1795–1809
4. Hong CH, Arosemena A, Zurakowski D, Ayyala RS (2005) Glaucoma drainage devices: a systematic literature review and current controversies. Surv Ophthalmol 50:48–60
5. Klassen HJ, Ng TF, Kurimoto Y, Kirov I, Shatos M, Coffey P et al (2004) Multipotent retinal progenitors express developmental markers, differentiate into retinal neurons, and preserve light-mediated behavior. Invest Ophthalmol Vis Sci 45:4167–4173
6. Redenti S, Neeley WL, Rompani S, Saigal S, Yang J, Klassen H et al (2009) Engineering retinal progenitor cell and scrollable poly(glycerol-sebacate) composites for expansion and subretinal transplantation. Biomaterials 30:3405–3414
7. Zarbin MA, Montemagno C, Leary JF, Ritch R (2012) Regenerative nanomedicine and the treatment of degenerative retinal diseases. Wiley Interdiscip Rev Nanomed Nanobiotechnol 4:113–137
8. Stout JT, Francis PJ (2011) Surgical approaches to gene and stem cell therapy for retinal disease. Hum Gene Ther 22(5):531
9. Zarbin MA, Montemagno C, Leary JF, Ritch R (2010) Nanomedicine in ophthalmology: the new frontier. Am J Ophthalmol 150:144–62 e2
10. Falavarjani KG, Nguyen QD (2013) Adverse events and complications associated with intravitreal injection of anti-VEGF agents: a review of literature. Eye (Lond) 27:787–794
11. Redenti S, Tao S, Yang J, Gu P, Klassen H, Saigal S et al (2008) Retinal tissue engineering using mouse retinal progenitor cells and a novel biodegradable, thin-film poly(e-caprolactone) nanowire scaffold. J Ocul Biol Dis Infor 1:19–29
12. MacLaren RE, Pearson RA, MacNeil A, Douglas RH, Salt TE, Akimoto M et al (2006) Retinal repair by transplantation of photoreceptor precursors. Nature 444:203–207
13. Tomita M, Lavik E, Klassen H, Zahir T, Langer R, Young MJ (2005) Biodegradable polymer composite grafts promote the survival and differentiation of retinal progenitor cells. Stem Cells 23:1579–1588
14. Ramsden CM, Powner MB, Carr AJ, Smart MJ, da Cruz L, Coffey PJ (2013) Stem cells in retinal regeneration: past, present and future. Development 140:2576–2585
15. West EL, Pearson RA, MacLaren RE, Sowden JC, Ali RR (2009) Cell transplantation strategies for retinal repair. Prog Brain Res 175:3–21
16. Araki M (2007) Regeneration of the amphibian retina: role of tissue interaction and related signaling molecules on RPE transdifferentiation. Develop Growth Differ 49:109–120
17. Tezel TH, Del Priore LV (1997) Reattachment to a substrate prevents apoptosis of human retinal pigment epithelium. Graefes Arch Clin Exp Ophthalmol 235:41–47
18. Neeley WL, Redenti S, Klassen H, Tao S, Desai T, Young MJ et al (2008) A microfabricated scaffold for retinal progenitor cell grafting. Biomaterials 29:418–426

19. Lavik EB, Klassen H, Warfvinge K, Langer R, Young MJ (2005) Fabrication of degradable polymer scaffolds to direct the integration and differentiation of retinal progenitors. Biomaterials 26:3187–3196
20. Lamb TM, Knecht AK, Smith WC, Stachel SE, Economides AN, Stahl N et al (1993) Neural induction by the secreted polypeptide noggin. Science (New York then Washington) 262:713
21. Pera EM, Wessely O, Li S-Y, De Robertis E (2001) Neural and head induction by insulin-like growth factor signals. Dev Cell 1:655–665
22. Ikeda H, Osakada F, Watanabe K, Mizuseki K, Haraguchi T, Miyoshi H et al (2005) Generation of Rx+/Pax6+ neural retinal precursors from embryonic stem cells. Proc Natl Acad Sci U S A 102:11331–11336
23. Lamba DA, Karl MO, Ware CB, Reh TA (2006) Efficient generation of retinal progenitor cells from human embryonic stem cells. Proc Natl Acad Sci 103:12769–12774
24. Tucker BA, Park I-H, Qi SD, Klassen HJ, Jiang C, Yao J et al (2011) Transplantation of adult mouse iPS cell-derived photoreceptor precursors restores retinal structure and function in degenerative mice. PLoS One 6:e18992
25. Gonzalez-Cordero A, West EL, Pearson RA, Duran Y, Carvalho LS, Chu CJ et al (2013) Photoreceptor precursors derived from three-dimensional embryonic stem cell cultures integrate and mature within adult degenerate retina. Nat Biotechnol 31:741–747
26. Hambright D, Park K-Y, Brooks M, McKay R, Swaroop A, Nasonkin IO (2012) Long-term survival and differentiation of retinal neurons derived from human embryonic stem cell lines in un-immunosuppressed mouse retina. Mol Vis 18:920
27. Lamba DA, Gust J, Reh TA (2009) Transplantation of human embryonic stem cell-derived photoreceptors restores some visual function in Crx-deficient mice. Cell Stem Cell 4:73–79
28. Meyer JS, Shearer RL, Capowski EE, Wright LS, Wallace KA, McMillan EL et al (2009) Modeling early retinal development with human embryonic and induced pluripotent stem cells. Proc Natl Acad Sci 106:16698–16703
29. Osakada F, Ikeda H, Sasai Y, Takahashi M (2009) Stepwise differentiation of pluripotent stem cells into retinal cells. Nat Protoc 4:811–824
30. Bharti K, Miller SS, Arnheiter H (2011) The new paradigm: retinal pigment epithelium cells generated from embryonic or induced pluripotent stem cells. Pigment Cell Melanoma Res 24:21–34
31. Jiang Y, Ding Q, Xie X, Libby RT, Lefebvre V, Gan L (2013) Transcription factors SOX4 and SOX11 function redundantly to regulate the development of mouse retinal ganglion cells. J Biol Chem 288:18429–18438
32. Buchholz DE, Hikita ST, Rowland TJ, Friedrich AM, Hinman CR, Johnson LV et al (2009) Derivation of functional retinal pigmented epithelium from induced pluripotent stem cells. Stem Cells 27:2427–2434
33. Chen M, Chen Q, Sun X, Shen W, Liu B, Zhong X et al (2010) Generation of retinal ganglion–like cells from reprogrammed mouse fibroblasts. Invest Ophthalmol Vis Sci 51:5970–5978
34. Decembrini S, Cananzi M, Gualdoni S, Battersby A, Allen N, Pearson RA et al (2010) Comparative analysis of the retinal potential of embryonic stem cells and amniotic fluid-derived stem cells. Stem Cells Dev 20:851–863
35. Zhong X, Gutierrez C, Xue T, Hampton C, Vergara MN, Cao L-H et al (2014) Generation of three-dimensional retinal tissue with functional photoreceptors from human iPSCs. Nat Commun 5:4047
36. Meyer JS, Katz ML, Maruniak JA, Kirk MD (2006) Embryonic stem cell-derived neural progenitors incorporate into degenerating retina and enhance survival of host photoreceptors. Stem Cells 24:274–283
37. Grozdanic SD, Ast AM, Lazic T, Kwon YH, Kardon RH, Sonea IM et al (2006) Morphological integration and functional assessment of transplanted neural progenitor cells in healthy and acute ischemic rat eyes. Exp Eye Res 82:597–607
38. Mellough CB, Cui Q, Spalding KL, Symons NA, Pollett MA, Snyder EY et al (2004) Fate of multipotent neural precursor cells transplanted into mouse retina selectively depleted of retinal ganglion cells. Exp Neurol 186:6–19

39. McGill TJ, Cottam B, Lu B, Wang S, Girman S, Tian C et al (2012) Transplantation of human central nervous system stem cells–neuroprotection in retinal degeneration. Eur J Neurosci 35:468–477
40. Jung G, Sun J, Petrowitz B, Riecken K, Kruszewski K, Jankowiak W et al (2013) Genetically modified neural stem cells for a local and sustained delivery of neuroprotective factors to the dystrophic mouse retina. Trials 4:7–9
41. Yu S, Tanabe T, Dezawa M, Ishikawa H, Yoshimura N (2006) Effects of bone marrow stromal cell injection in an experimental glaucoma model. Biochem Biophys Res Commun 344:1071–1079
42. Levkovitch-Verbin H, Sadan O, Vander S, Rosner M, Barhum Y, Melamed E et al (2010) Intravitreal injections of neurotrophic factors secreting mesenchymal stem cells are neuroprotective in rat eyes following optic nerve transection. Invest Ophthalmol Vis Sci 51:6394–6400
43. Johnson TV, Bull ND, Hunt DP, Marina N, Tomarev SI, Martin KR (2010) Neuroprotective effects of intravitreal mesenchymal stem cell transplantation in experimental glaucoma. Invest Ophthalmol Vis Sci 51:2051–2059
44. Arnhold S, Absenger Y, Klein H, Addicks K, Schraermeyer U (2007) Transplantation of bone marrow-derived mesenchymal stem cells rescue photoreceptor cells in the dystrophic retina of the rhodopsin knockout mouse. Graefes Arch Clin Exp Ophthalmol 245:414–422
45. Lu B, Wang S, Girman S, McGill T, Ragaglia V, Lund R (2010) Human adult bone marrow-derived somatic cells rescue vision in a rodent model of retinal degeneration. Exp Eye Res 91:449–455
46. Zhang Y, Wang W (2010) Effects of bone marrow mesenchymal stem cell transplantation on light-damaged retina. Invest Ophthalmol Vis Sci 51:3742–3748
47. Tzameret A, Sher I, Belkin M, Treves AJ, Meir A, Nagler A et al (2014) Transplantation of human bone marrow mesenchymal stem cells as a thin subretinal layer ameliorates retinal degeneration in a rat model of retinal dystrophy. Exp Eye Res 118:135–144
48. Mead B, Logan A, Berry M, Leadbeater W, Scheven BA (2013) Intravitreally transplanted dental pulp stem cells promote neuroprotection and axon regeneration of retinal ganglion cells after optic nerve injurydental pulp stem cells for optic nerve injury. Invest Ophthalmol Vis Sci 54:7544–7556
49. Haddad-Mashadrizeh A, Bahrami AR, Matin MM, Edalatmanesh MA, Zomorodipour A, Gardaneh M et al (2013) Human adipose-derived mesenchymal stem cells can survive and integrate into the adult rat eye following xenotransplantation. Xenotransplantation 20:165–176
50. Mendel TA, Clabough EB, Kao DS, Demidova-Rice TN, Durham JT, Zotter BC et al (2013) Pericytes derived from adipose-derived stem cells protect against retinal vasculopathy. PLoS One 8:e65691
51. Nadri S, Kazemi B, Eslaminejad MB, Yazdani S, Soleimani M (2013) High yield of cells committed to the photoreceptor-like cells from conjunctiva mesenchymal stem cells on nanofibrous scaffolds. Mol Biol Rep 40:3883–3890
52. Nadri S, Yazdani S, Arefian E, Gohari Z, Eslaminejad MB, Kazemi B et al (2013) Mesenchymal stem cells from trabecular meshwork become photoreceptor-like cells on amniotic membrane. Neurosci Lett 541:43–48
53. Aftab U, Jiang C, Tucker B, Kim JY, Klassen H, Miljan E et al (2009) Growth kinetics and transplantation of human retinal progenitor cells. Exp Eye Res 89:301–310
54. Merhi-Soussi F, Angenieux B, Canola K, Kostic C, Tekaya M, Hornfeld D et al (2006) High yield of cells committed to the photoreceptor fate from expanded mouse retinal stem cells. Stem Cells 24:2060–2070
55. Algvere PV, Berglin L, Gouras P, Sheng Y (1994) Transplantation of fetal retinal pigment epithelium in age-related macular degeneration with subfoveal neovascularization. Graefes Arch Clin Exp Ophthalmol 232:707–716
56. Algvere P, Berglin L, Gouras P, Sheng Y (1997) Transplantation of fetal retinal pigment epithelium in age-related macular degeneration with subfoveal neovascularization. Retina 17:468
57. Binder S, Stolba U, Krebs I, Kellner L, Jahn C, Feichtinger H et al (2002) Transplantation of autologous retinal pigment epithelium in eyes with foveal neovascularization resulting from age-related macular degeneration: a pilot study. Am J Ophthalmol 133:215–225

58. Algvere P, Gouras P, Dafgård KE (1998) Long-term outcome of RPE allografts in non-immunosuppressed patients with AMD. Eur J Ophthalmol 9:217–230
59. Tamai M, Abe T, Tomita H (1999) Autologous iris pigment transplantation in age related macular degeneration. Retina, Current Praxis and Future Trends, Hyderabad:151–161.
60. Thumann G, Aisenbrey S, Schraermeyer U, Lafaut B, Esser P, Walter P et al (2000) Transplantation of autologous iris pigment epithelium after removal of choroidal neovascular membranes. Arch Ophthalmol 118:1350–1355
61. Girman S, Wang S, Lund R (2003) Cortical visual functions can be preserved by subretinal RPE cell grafting in RCS rats. Vis Res 43:1817–1827
62. Thumann G, Viethen A, Gaebler A, Walter P, Kaempf S, Johnen S et al (2009) The in vitro and in vivo behaviour of retinal pigment epithelial cells cultured on ultrathin collagen membranes. Biomaterials 30:287–294
63. Gualdoni S, Baron M, Lakowski J, Decembrini S, Smith AJ, Pearson RA et al (2010) Adult ciliary epithelial cells, previously identified as retinal stem cells with potential for retinal repair, fail to differentiate into new rod photoreceptors. Stem Cells 28:1048–1059
64. Lu L, Yaszemski MJ, Mikos AG (2001) Retinal pigment epithelium engineering using synthetic biodegradable polymers. Biomaterials 22:3345–3355
65. Pearson RA, Barber AC, Rizzi M, Hippert C, Xue T, West EL et al (2012) Restoration of vision after transplantation of photoreceptors. Nature 485:99–103
66. Kicic A, Shen WY, Wilson AS, Constable IJ, Robertson T, Rakoczy PE (2003) Differentiation of marrow stromal cells into photoreceptors in the rat eye. J Neurosci 23:7742–7749
67. Aramant RB, Seiler MJ, Ball SL (1999) Successful cotransplantation of intact sheets of fetal retina with retinal pigment epithelium. Invest Ophthalmol Vis Sci 40:1557–1564
68. Seiler MJ, Aramant RB, Bergstrom A (1995) Co-transplantation of embryonic retina and retinal pigment epithelial cells to rabbit retina. Curr Eye Res 14:199–207
69. Gouras P, Flood MT, Kjedbye H, Bilek MK, Eggers H (1985) Transplantation of cultured human retinal epithelium to Bruch's membrane of the owl monkey's eye. Curr Eye Res 4:253–265
70. Lopez R, Gouras P, Kjeldbye H, Sullivan B, Reppucci V, Brittis M et al (1989) Transplanted retinal pigment epithelium modifies the retinal degeneration in the RCS rat. Invest Ophthalmol Vis Sci 30:586–588
71. Tomita M, Mori T, Maruyama K, Zahir T, Ward M, Umezawa A et al (2006) A comparison of neural differentiation and retinal transplantation with bone marrow-derived cells and retinal progenitor cells. Stem Cells 24:2270–2278
72. Milam AH, Dacey DM, Dizhoor AM (1993) Recoverin immunoreactivity in mammalian cone bipolar cells. Vis Neurosci 10:1–12
73. Gu P, Yang J, Wang J, Young MJ, Klassen H (2009) Sequential changes in the gene expression profile of murine retinal progenitor cells during the induction of differentiation. Mol Vis 15:2111–2122
74. Tropel P, Noel D, Platet N, Legrand P, Benabid AL, Berger F (2004) Isolation and characterisation of mesenchymal stem cells from adult mouse bone marrow. Exp Cell Res 295:395–406
75. Nadri S, Soleimani M (2007) Isolation murine mesenchymal stem cells by positive selection. In Vitro Cell Dev Biol Anim 43:276–282
76. Ma PX, Zhang R (1999) Synthetic nano-scale fibrous extracellular matrix. J Biomed Mater Res 46:60–72
77. Steedman MR, Tao SL, Klassen H, Desai TA (2010) Enhanced differentiation of retinal progenitor cells using microfabricated topographical cues. Biomed Microdevices 12:363–369
78. Mano JF, Silva GA, Azevedo HS, Malafaya PB, Sousa RA, Silva SS et al (2007) Natural origin biodegradable systems in tissue engineering and regenerative medicine: present status and some moving trends. J R Soc Interface 4:999–1030
79. Dosunmu OO, Chase GG, Kataphinan W, Reneker DH (2006) Electrospinning of polymer nanofibres from multiple jets on a porous tubular surface. Nanotechnology 17:1123–1127
80. CY X, Inai R, Kotaki M, Ramakrishna S (2004) Aligned biodegradable nanofibrous structure: a potential scaffold for blood vessel engineering. Biomaterials 25:877–886

81. Chen H, Huang J, Yu J, Liu S, Electrospun GP (2011) chitosan-graft-poly (epsilon -caprolactone)/poly (epsilon-caprolactone) cationic nanofibrous mats as potential scaffolds for skin tissue engineering. Int J Biol Macromol 48:13–19
82. Bellamkonda R, Ranieri JP, Bouche N, Aebischer P (1995) Hydrogel-based three-dimensional matrix for neural cells. J Biomed Mater Res 29:663–671
83. Sodha S, Wall K, Redenti S, Klassen H, Young MJ, Tao SL (2011) Microfabrication of a three-dimensional polycaprolactone thin-film scaffold for retinal progenitor cell encapsulation. J Biomater Sci Polym Ed 22:443–456
84. Fitzpatrick SD, Jafar Mazumder MA, Muirhead B, Sheardown H (2012) Development of injectable, resorbable drug-releasing copolymer scaffolds for minimally invasive sustained ophthalmic therapeutics. Acta Biomater 8:2517–2528
85. Mawad D, Boughton EA, Boughton P, Lauto A (2012) Advances in hydrogels applied to degenerative diseases. Curr Pharm Des 18:2558–2575
86. Aizawa Y, Shoichet MS (2012) The role of endothelial cells in the retinal stem and progenitor cell niche within a 3D engineered hydrogel matrix. Biomaterials 33:5198–5205
87. Langer R (2000) Biomaterials in drug delivery and tissue engineering: one laboratory's experience. Acc Chem Res 33:94–101
88. Cai S, Smith ME, Redenti SM, Wnek GE, Young MJ (2012) Mouse retinal progenitor cell dynamics on electrospun poly (ε-caprolactone). J Biomater Sci Polym Ed 23:1451–1465
89. Chen H, Fan X, Xia J, Chen P, Zhou X, Huang J et al (2011) Electrospun chitosan-graft-poly (ε-caprolactone)/poly (ε-caprolactone) nanofibrous scaffolds for retinal tissue engineering. Int J Nanomedicine 6:453–461
90. Tucker BA, Redenti SM, Jiang C, Swift JS, Klassen HJ, Smith ME et al (2010) The use of progenitor cell/biodegradable MMP2-PLGA polymer constructs to enhance cellular integration and retinal repopulation. Biomaterials 31:9–19
91. Zhang D, Ni N, Chen J, Yao Q, Shen B, Zhang Y et al (2015) Electrospun SF/PLCL nanofibrous membrane: a potential scaffold for retinal progenitor cell proliferation and differentiation. Sci Rep 5:14326
92. Nadri S, Nasehi F, Barati G (2017) Effect of parameters on the quality of core-shell fibrous scaffold for retinal differentiation of conjunctiva mesenchymal stem cells. J Biomed Mater Res A 105:189–197
93. Osakada F, Jin ZB, Hirami Y, Ikeda H, Danjyo T, Watanabe K et al (2009) In vitro differentiation of retinal cells from human pluripotent stem cells by small-molecule induction. J Cell Sci 122:3169–3179
94. Sakaguchi DS, Van Hoffelen SJ, Theusch E, Parker E, Orasky J, Harper MM et al (2004) Transplantation of neural progenitor cells into the developing retina of the Brazilian opossum: an in vivo system for studying stem/progenitor cell plasticity. Dev Neurosci 26:336–345
95. Bhang SH, Lim JS, Choi CY, Kwon YK, Kim BS (2007) The behavior of neural stem cells on biodegradable synthetic polymers. J Biomater Sci Polym Ed 18:223–239
96. Saha K, Keung AJ, Irwin EF, Li Y, Little L, Schaffer DV et al (2008) Substrate modulus directs neural stem cell behavior. Biophys J 95:4426–4438
97. Banerjee A, Arha M, Choudhary S, Ashton RS, Bhatia SR, Schaffer DV et al (2009) The influence of hydrogel modulus on the proliferation and differentiation of encapsulated neural stem cells. Biomaterials 30:4695–4699
98. Lim JM, Byun S, Chung S, Park TH, Seo JM, Joo CK et al (2004) Retinal pigment epithelial cell behavior is modulated by alterations in focal cell-substrate contacts. Invest Ophthalmol Vis Sci 45:4210–4216
99. Arras MM, Grasl C, Bergmeister H, Schima H (2012) Electrospinning of aligned fibers with adjustable orientation using auxiliary electrodes. Sci Technol Adv Mater 13:035008
100. Hu J, Kai D, Ye H, Tian L, Ding X, Ramakrishna S et al (2017) Electrospinning of poly(glycerol sebacate)-based nanofibers for nerve tissue engineering. Mater Sci Eng C Mater Biol Appl 70:1089–1094
101. Tao S, Young C, Redenti S, Zhang Y, Klassen H, Desai T et al (2007) Survival, migration and differentiation of retinal progenitor cells transplanted on micro-machined poly(methyl methacrylate) scaffolds to the subretinal space. Lab Chip 7:695–701

102. Lee E, MacLaren RE (2011) Sources of retinal pigment epithelium (RPE) for replacement therapy. Br J Ophthalmol 95:445–449
103. Pillai RG (2011) Stem cells for ocular tissue engineering and regeneration. Curr Top Med Chem 11:1606–1620
104. Linh NT, Lee KH, Lee BT (2013) Functional nanofiber mat of polyvinyl alcohol/gelatin containing nanoparticles of biphasic calcium phosphate for bone regeneration in rat calvaria defects. J Biomed Mater Res A 101:2412–2423
105. Gnavi S, Fornasari BE, Tonda-Turo C, Laurano R, Zanetti M, Ciardelli G et al (2015) The effect of electrospun gelatin fibers alignment on schwann cell and axon behavior and organization in the perspective of artificial nerve design. Int J Mol Sci 16:12925–12942
106. Yan J, Qiang L, Gao Y, Cui X, Zhou H, Zhong S et al (2012) Effect of fiber alignment in electrospun scaffolds on keratocytes and corneal epithelial cells behavior. J Biomed Mater Res A 100:527–535
107. Lai JY (2010) Biocompatibility of chemically cross-linked gelatin hydrogels for ophthalmic use. J Mater Sci Mater Med 21:1899–1911
108. Chou SF, Luo LJ, Lai JY, Ma DH (2017) Role of solvent-mediated carbodiimide cross-linking in fabrication of electrospun gelatin nanofibrous membranes as ophthalmic biomaterials. Mater Sci Eng C Mater Biol Appl 71:1145–1155
109. Loscertales IG, Barrero A, Guerrero I, Cortijo R, Marquez M, Ganan-Calvo AM (2002) Micro/nano encapsulation via electrified coaxial liquid jets. Science 295:1695–1698
110. Zhu H, Yu D, Zhou Y, Wang C, Gao M, Jiang H et al (2013) Biological activity of a nanofibrous barrier membrane containing bone morphogenetic protein formed by core-shell electrospinning as a sustained delivery vehicle. J Biomed Mater Res B Appl Biomater 101:541–552
111. Seyednejad H, Ji W, Yang F, van Nostrum CF, Vermonden T, van den Beucken JJ et al (2012) Coaxially electrospun scaffolds based on hydroxyl-functionalized poly(epsilon-caprolactone) and loaded with VEGF for tissue engineering applications. Biomacromolecules 13:3650–3660
112. Zhang Y, Klassen HJ, Tucker BA, Perez MT, Young MJ (2007) CNS progenitor cells promote a permissive environment for neurite outgrowth via a matrix metalloproteinase-2-dependent mechanism. J Neurosci 27:4499–4506
113. Yao J, Tucker BA, Zhang X, Checa-Casalengua P, Herrero-Vanrell R, Young MJ (2011) Robust cell integration from co-transplantation of biodegradable MMP2-PLGA microspheres with retinal progenitor cells. Biomaterials 32:1041–1050
114. Tucker BA, Park IH, Qi SD, Klassen HJ, Jiang C, Yao J et al (2015) Correction: transplantation of adult mouse iPS cell-derived photoreceptor precursors restores retinal structure and function in degenerative mice. PLoS One 10:e0125947

Electrospun Polymeric Smart Materials for Tissue Engineering Applications

S. Ribeiro, D.M. Correia, C. Ribeiro, and S. Lanceros-Méndez

Abstract Smart materials are increasingly being implemented in different areas and, particularly in tissue engineering applications. In tissue engineering, cells and materials interplay a central role, the scaffold strongly affecting cell behavior through specific physico-chemical interactions. In this context, smart materials show strong advantages with respect to conventional scaffolds providing additional clues to the cell and supporting the development of suitable microenvironments. Thus, smart and functional materials are being developed to mimic the response capabilities of natural living systems through suitable electromechanical or thermomechanical responses, among others. Among the different shapes and forms in which those materials can be processed for tissue engineering applications, electrospun mats stand out as one of the most promising substrates, allowing to tailor fiber size and orientation as well as degree of porosity, among others, allowing to mimic micro and nanoscale properties and structural characteristics of native extracellular matrix for tissue engineering applications.

The present chapter will provide an insight on the main smart materials interesting for biomedical applications. Further, it will focus on those smart materials that have been processed in the form of electrospun fibers and summarize the main obtained results in tissue engineering applications. Thus, it will be shown that elec-

S. Ribeiro and D.M. Correia contributed equally to this work.

S. Ribeiro
Centro de Física, Universidade do Minho, Braga, Portugal

D.M. Correia
Departamento de Química, Universidade de Trás-os-Montes e Alto Douro, Vila Real, Portugal

BCMaterials, Parque Cientifico y Tecnologico de Bizkaia, Derio, Spain

C. Ribeiro
Centro de Física, Universidade do Minho, Braga, Portugal

CEB – Centre of Biological Engineering, Universidade do Minho, Braga, Portugal

S. Lanceros-Méndez (✉)
BCMaterials, Parque Cientifico y Tecnologico de Bizkaia, Derio, Spain

IKERBASQUE, Basque Foundation for Science, Bilbao, Spain
e-mail: senentxu.lanceros@bcmaterials.net

J. Almodovar (ed.), *Electrospun Biomaterials and Related Technologies*,
DOI 10.1007/978-3-319-70049-6_9

trospinning of smart materials represents an increasingly growing and interesting that hold great promise for finally achieved suitable and implementable tissue engineering strategies.

Keywords Smart materials • Tissue engineering • Electrospinning • Microenvironments • Functional materials • Active materials • Cell response

We cannot live only for ourselves. A thousand fibers connect us with our fellow men; and among those fibers, as sympathetic threads, our actions run as causes, and they come back to us as effects.

Herman Mellville (1819–1891)

1 Introduction

Smart or field responsive materials are playing an increasing role in different areas, such as sensors, energy, aerospace, communications and biomedical. The large interest in this topic is related to the possibility to assemble smarter structures and achieve higher performance systems with potential for recognition, discrimination and adjustability, by the use of smart materials [1].

In particular, smart materials can be used as innovative active biomimetic template for the regeneration of damaged tissue [2]. The three main components of tissue engineering strategies are: cells, biomaterials, and growth factors [3]. Among these, biomaterials play an important role on triggering cell behavior, including adhesion, morphology, proliferation and differentiation, and can act as an artificial extracellular matrix (ECM) [1, 4]. In this way, smart biomaterials can be tailored not only to be passively tolerated by the organism, but also to provide the appropriate environment to assist specific cell responses, such as biological, physical and mechanical functions of native ECM found in the body tissues [5, 6]. For example, smart piezoelectric scaffolds can be used not only passively as support for bone cell culture, but can be also actively used for the mechano-electrical stimulation of such cells, optimizing cell growth [5]. This added value of smart scaffolds is related to the fact that cells from living systems have the capability to modify their environmental conditions with the objective to respond to external stimulus. Aiming to mimic these situations, polymer-based smart materials emerge as the class of materials ideal for such approach. Polymer-based smart materials, particularly shape-changing, electrically-responsive and metamorphic materials exhibit attractive properties appropriate for tissue engineering applications.

The properties of those engineered biomaterials are the main concerns in the search to fabricate ultimate scaffolds for tissue engineering applications, being the design and selection of the biomaterial a critical step in the tissue engineering process. The ideal biomaterial must be able to control the function and the structure of the tissue in a preplanned manner by interacting with host cells or/and the transplanted cells.

In a general way, the perfect biomaterial should be biocompatible, promote specific cellular interaction and tissue development, and possess proper mechanical and physical properties [1].

Polymer-based smart biomaterials are particularly attractive for the development of smart scaffolds due to the ability to be processed by different techniques in order to tailor their shape for permanent or temporary cell supports [7]. Those advantages are well demonstrated by the large number of cells and tissues grown in polymer-based smart materials, including skin, cartilage, liver, heart valves and arteries, bladder, pancreas, nerves, corneas, and various other soft tissues [8].

Among all the different biomaterials forms used in tissue engineering applications, polymer electrospun mats show to be a promising substrate as they can mimic the micro and nanoscale properties and structural characteristics of native ECM [9]. Polymer electrospun fibers have been used for several tissue repair applications such as, bone [10, 11], cartilage [12, 13], muscle [14, 15], neuronal [16, 17] and vascular [18, 19], among others. However, most of these polymeric fibers have been used just as cell support, not taking in to consideration the physical signal existing in the body. So, the use of polymer-based smart electrospun materials are attracting increasing interest once they can respond to different stimuli, making it an advantage for the development of suitable microenvironments for cell culture.

2 Smart Materials

Smart materials represent a key technology supporting novel products with unique capabilities [20]. The adjective "smart" implies that these materials are able to react to changes in their environment in a predetermined way [20, 21].

Thus, smart materials are able to respond to specific single or multiple external stimuli in a controllable, predictable and reproducible way [21]. Only some materials show certain intrinsic characteristics which can be exploited in products, systems or structures that in turn exhibit "smart" behavior [22]. Examples of this behavior include shape changes in response to an external stimulus; self-sensing, detection and/or quantification in response to an external stimulus; self-actuating, actuation in response to an external stimulus; self-healing, self-repair in response to damage; and self-diagnostic.

Shape-changing materials are based on shape-memory alloys, shape memory polymers and piezoelectric materials. They are capable of reversibly change their shape and/or dimensions when subjected to physical, chemical or biological external stimuli [20]. They play a key role in the biomedical field, including tissue engineering, diagnostic and therapeutic applications, as well as in robotics and artificial muscle actuators, among others [23]. Some of them are able to change their dimensions maintaining their shape, while others change their shape without changing their dimensions [24]. Shape changing materials are closely allied by self-actuation materials that develop a significant strain or displacement in response to an external stimulus. Composites containing piezoelectric elements, magnetostrictive elements,

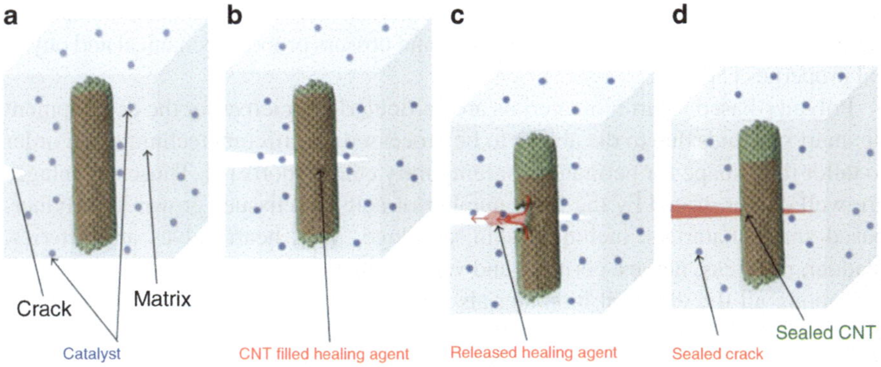

Fig. 1 Concept of the self-healing process using carbon nanotubes (From Aïssa [26]). Images a) to d) illustrate the differen steps of the self-healing process.

shape memory alloys and electrorheological fluids are examples of materials which exhibit this property [20].

Another possible function of smart materials is the introduction of self-healing characteristics that can be defined as the ability of a material to heal (recover/repair) damages automatically and autonomously. The different strategies of designing self-healing materials are release of healing agent, reversible cross-links and miscellaneous technologies [25]. An example of this approach involves the use of microcapsules containing a healing agent [26] (Fig. 1): when a crack reaches the microcapsule, the capsule breaks and the monomer bleeds into the crack, where it can polymerize and initiate the repair.

Self-sensing materials refer to materials sensing themselves and therefore the structure they are part of. An example of such behavior is the evaluation of structural damage through the electrical resistivity of a carbon fiber composite [27].

All these categories of smart materials are not necessarily mutually exclusive and more than one function can be present in a specific material. For example, self-sensing may be combined with, or initiate, self-healing [27].

Smart materials and system have a wide range of applications due to their varied response to external stimuli. There are a number of types of smart materials and they can be classified depending on the stimulus to which they are sensitive (Table 1) [28].

These materials have an increasingly important role in different areas of applications [58], including automotive [59], aerospace [60], biomedical [61] and tissue engineering [2] applications, among others.

In particular, the literature describes a variety of smart materials playing an increasing role in a wide range of tissue engineering strategies. This strong increase of the development of smart materials for tissue engineering applications is due in order to improve the limitations of conventional scaffolds, which are just used in a passive way, as support for the cells and tissues. It is observed that for specific cells and tissues, the active behavior of the material used for the scaffold can be taken to advantage, providing the necessary stimuli for proper tissue regeneration [2].

Table 1 Smart material types depending of the external stimuli for different applications and their function

External stimuli	Group of materials	Principle of response	References
Heat	Thermochromic	Change in color	[29, 30]
	Thermoelectric	Convert temperature differences into an electrical signal and vice versa	[31, 32]
	Shape memory materials	Return to its pre-deformed shape when heated	[33, 34]
Stress/pressure (mechanical)	Smart gels	Respond to changes in environment by shape changes	[35]
	Piezoelectric	Produce a voltage when stressed and change shape/produce stress when a voltage is applied	[5, 36]
Electrical current/voltage	Electrochromic	Change color	[37, 38]
	Dielectric elastomers	Produce large (up to 300%) strains under the influence of an external electric field	[39, 40]
	Electrorheological fluids	Shape variation when an electrical signal is applied	[41]
	Electrostrictive	Change their dimensions or shape under the application of an electric field	[42]
	Electroactive polymers	Exhibit a shape variation when an electrical stimuli is applied	[43–45]
	Piezoelectric	Produce a voltage when stressed and change shape/produce stress when a voltage is applied	[5, 36]
	Thermoelectric	Convert temperature differences into an electrical signal and vice versa	[31, 32]
Magnetic field	Magnetorheological fluids	Produces a shape variation when a magnetic field is applied	[46, 47]
	Magnetostrictive	Materials that change their magnetization under stress or change their shape if subjected to a magnetic field	[48]
	Magnetocaloric	Reversible temperature variations are obtained under the application of a magnetic field	[49]
	Magnetorheological elastomers	Variation of the mechanical properties under the application of a magnetic field	[46, 50]
pH-change/ solvent moisture	Smart gels	Mechanical or morphological variations under the applications of the specific stimulus	[51]
	Halochromic	Environmental pH variation by means of change in color	[52]

(continued)

Table 1 (continued)

External stimuli	Group of materials	Principle of response	References
Light	Photochromic	Change color as a response to a light stimulus	[53, 54]
	Photomechanical	Change shape under exposure to light	[55]
Molecular or biomolecular	Glucose-responsive	Physical changes are triggered in response to a changing glucose concentration	[56]
	Enzyme responsive	Macroscopic property changes when triggered by the biocatalytic action of an enzyme	[57]

Adapted from [22]

Many stimuli can be found to play an important role in the development and function of the human body. Some of these stimuli can be artificially controlled, such as light, magnetic or electric field, or naturally promoted by a physiological environment, such as electrical, temperature or pH changes, within the body. Thus, variations of pH can be found in muscle cells after muscle activity or in the brain, and pH variations also promote many cellular processes [23]. Mechanical stimuli are also constantly happening in the life of human body cells, playing a particularly relevant role in bone formation and ECM production [5].

Among the smart materials most often used for tissue engineering applications, are piezoelectric polymers. Many tissues of human body such as bone [62], nervous [63] and muscle [14] are controlled by electrical signals, piezoelectric materials allowing to apply electrical signals to the cells by mechano-electrical transduction. They have shown suitability for tissue engineering due to their ability to vary their surface charge when a mechanical load is applied, without need of direct connection to an external voltage source (Fig. 2).

As mechanical stimuli, can be found in many human body activities, piezoelectric materials and mechano-electrical transduction demonstrate large potential for tissue engineering. Table 2 summarizes the main studies with different scaffold morphologies that can be used to produce piezoelectric materials in the form of films, fibers, porous membranes and 3D scaffolds [5] for tissue engineering applications.

Further, in cases when the patient is immobilized due to serious health conditions, or when the natural mechanical stimulus does not occur, decreasing the effectiveness of piezoelectric materials, another type of smart materials can be taken to advantage. Thus, magnetoelectric composite materials have attracting increasing attentions as they allow to use of an external magnetic field to remotely trigger mechanical and electrical stimuli to growing tissues [93].

Another category of smart material that have often being used in tissue engineering, are thermoresponsive polymers. These materials are mainly used as substrate for cell culture, regulating the attachment and detachment of the cells from the substrate [94]. With this type of smart materials, it is possible to generate a technol-

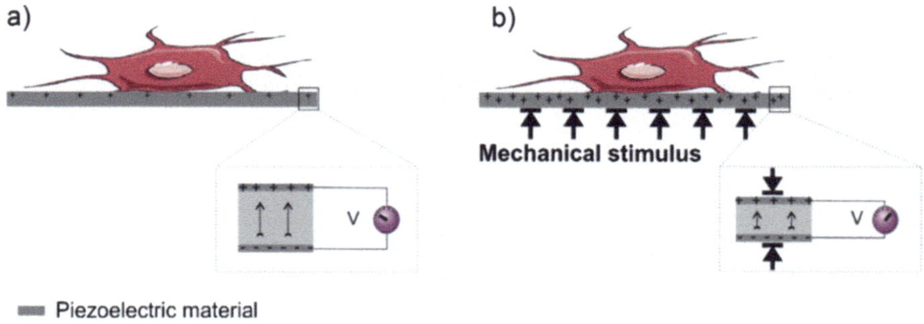

Fig. 2 Schematic representation of the piezoelectric effect (piezoelectric material representation at the bottom of the image) and corresponding cell culture on piezoelectric supports (**a**) without and (**b**) with mechanical stimulus, the later leading to an electrical potential variation of the materials which is, in turn, influences cell response (From Ribeiro et al. [5] with permission from the publisher (Elsevier))

ogy to fabricate different types of functional tissues for several tissue engineering applications such as bone, skin, cardiac and nerves, among others [23]. Further, thermoresponsive polymers are also commonly used as injectable gels, where the cells are encapsulated in a 3D structure within the body. In situ formation of cell/scaffold allows the delivery of encapsulated cells, nutrients and growth factors, minimizing the immune rejections in vivo. The main topography of thermoresponsive biomaterials for tissue engineering applications are hydrogels [95]. However, it is possible to obtain other structures such as specific 3D structures, interpenetrating networks, micelles, films and particles [94].

Nevertheless, it was verified that not only the stimulus is important to obtain promising results in regenerating, maintaining, or improving tissue functions of the human body. The architecture of the scaffolds used for tissue engineering also has critical importance. Intuitively, the best scaffold for an engineered tissue should be the extracellular matrix of the target tissue in its native state [96]. The fabrication technique for tissue engineering scaffolds depends almost entirely on the bulk and surface properties of the material and the proposed function of the scaffold. Each method presents distinct advantages and disadvantages, the appropriate technique being selected to meet the requirements for the specific type of tissue [8]. In the last years, significant advances have been achieved in the development of specific scaffolds geometries including hydrogels, three-dimensional polymeric scaffold, microspheres, acellular scaffolds and fibers scaffolds, among others [8].

Scaffolds should have an interconnected pore structure and high porosity to ensure cellular penetration and adequate diffusion of nutrients to cells within the construct and to the extracellular matrix formed by these cells [6]. The development of nanofibers has enhanced the scope for fabricating scaffolds that can potentially mimic the architecture of natural human tissue at the nanometer scale. There are three main techniques available for the synthesis of nanofibers: electrospinning, self-assembly and phase separation [8]. Electrospinning (ES) has attracted much

Table 2 Scaffold designs from piezoelectric materials for different tissue engineering applications

Applications	Scaffold design	Piezoelectric materials	References
Bone regeneration or bone tissue engineering	Films	Poly(vinylidene fluoride) (PVDF)/ copolymers	[62, 64–68]
		Poly(L-lactic acid) (PLLA)	[69]
		Poly(hydroxybutyrate) (PHB)/ copolymers	[70]
	Fibers	PVDF/copolymers	[71]
		PLLA	[72]
		PHB/copolymer	[70, 73]
		Composite poly(lactic acid) (PLA)/ demineralized bone powders	[74]
	3D blends/blends membranes (porous)	PVDF/copolymers	[75]
		PHB/copolymers	[76]
		Composite PVDF/starch/natural rubber (NR)	[75]
		Composite poly(vinylidene tri-fluoroethylene) (PVDF-TrFe)/ starch/NR	[75]
	Membranes	PVDF-TrFe/barium titanate (BT)	[77]
	3D porous scaffold	PLLA covered with bonelike apatite	[78]
		Apatite/collagen	[78]
Muscle regeneration	Films	PVDF	[14]
	Fibers	PVDF	[14, 79]
		Composite Au-PLLA	[80]
	Meshes	PVDF	[81]
Nerve regeneration	Films	PVDF	[82–84]
		PVDF-TrFE	[82, 85]
	Fibers	PVDF-TrFE	[16, 85]
		Collagen	[86]
	Blends membranes (porous)	PVDF	[87]
	Channels/tubes	PVDF	[88, 89]
		PVDF-TrFe	[90]
	3D porous scaffold	PLLA	[91]
	3D gel matrices	Collagen	[92]

Adapted from Ref. [5]

attention not only because of its versatility in spinning a wide variety of polymeric fibers but also because of its consistency in producing fibers with submicron thickness [97]. Electrospun fiber scaffolds have a high surface to volume ratio and combined with their microporous structure enhance cell adhesion, migration, proliferation and differentiated function, all of which are highly desired properties for tissue engineering applications [98].

Table 3 Some relevant electrospun smart materials for different tissue engineering

Effect of smart material	Tissue engineering applications	Polymers	References
Piezoelectric	Bone	Collagen	[99]
		Cellulose	[100]
		Chitin	[101]
		Keratin	[102]
		PVDF	[71]
		PHB	[103]
		PLLA	[104]
	Muscle	Collagen	[105]
		PVDF	[14]
	Nerve	PHB	[106]
		PLLA	[107]
	Other applications (skin substitutes or wound healing, cartilage)	PHB	[108]
		PLLA	[109]
		Collagen	[110, 111]
Conductive scaffolds	Cardiac	Poly(aniline) (PANI)	[112, 113]

Various engineering materials, including ceramics, metals (alloys), polymers and composites, have been developed to replace/support the function of biological materials. Metal, alloys and ceramics materials are being replaced by polymers because the developing scaffolds with the optimal characteristics, such as the strength, rate of degradation, porosity, and microstructure, as well as their shapes and sizes, are more readily and reproducibly controlled in polymeric scaffolds. Polymers are light weight, inexpensive, mechanically and electrically tough and show excellent compatibility with other organic and inorganic materials for the development of multifunctional hybrid systems [96]. Polymers can be easily processed by electrospinning and consequently, this potential leaves smart electrospun polymers materials as viable candidate for bone, neuronal, cardiac, and muscle, among others, tissue engineering applications. Table 3 summarizes some electrospun polymers with specific smart function, together with the respectively application.

In some cases, the scaffolds structures should be achieved with the few natural and synthetic materials available for a specific smart response, reinforced by other (nano)materials in order to achieve multifunctional smart materials. Some of these materials are for example the inclusion of magnetic sensitivity by using magnetic fillers and electrical sensitivity/conductivity by using electrically conducting fillers such as carbon or silver nanotubes or nanowires, among others [22].

A wide variety of electrospun smart materials have revealed interesting properties for applications in tissue engineering, as will be presented in detail in the following section.

3 Smart Piezoelectric and Conductive Polymer Electrospun Mats for Tissue Engineering

Different electrospun scaffolds based on smart materials have been processed in recent years mainly due to their morphological similarity to the extracellular matrix of the most tissues, large surface area to volume/mass ratio, possibility for surface functionalization and enhanced specific mechanical performance.

Natural and synthetic smart polymers have been electrospun in order to obtain fibers scaffolds with the ability to play a key role in signal transduction. As the most used smart materials have been the electrically active ones, this section is devoted to the main works on piezoelectric, conductive and magnetoelectric electrospun polymers for tissue engineering applications. The present report does not pretend to be exhaustive but to highlight the main materials, effects, cells and achievements obtained with these types of electrospun smart materials.

3.1 Piezoelectric Electrospun Nanofibers

Electroactive electrospun scaffolds, namely based on piezoelectric polymers, are among the most used smart materials for tissue engineering applications. Collagen, cellulose, chitin, amylose, keratin and fibrin are recognized as natural piezoelectric polymers [5]. Also synthetic polymers such as PLA, PHB, PVDF, PVDF-TrFe, and poly(amide) (PA) have been widely used to produce electrospun scaffolds [5].

Collagen is the most abundant protein in mammalian cells and the principal structural element of the mammalian ECM [114] and it has been used as biomaterial for tissue engineering applications. Collagen was primarily electrospun into fibers [115] using a collagen-poly(ethylene oxide) (PEO) solution and obtaining fibers with a diameter range from 100 to 150 nm. Pure collagen fibers of type I and III were successfully prepared from 1,1,1,3,3,3-hexafluoro-2-propanol (HFIP) solvent with average diameters ranging from 100 to 730 nm [115]. The electrospinning of collagen into fibers using different solvents (Fig. 3) [116] has been also reported, as well as their applicability in tissue engineering.

The major drawback associated to collagen electrospun scaffolds is the lack of mechanical properties upon hydration [115]. The fibers of collagen can be crosslinked to stabilize fiber structure and to keep the fibers from dissolving in an aqueous buffer. Further, the processing of collagen composite scaffolds with natural or synthetic polymers improve the properties of collagen-based scaffolds [115].

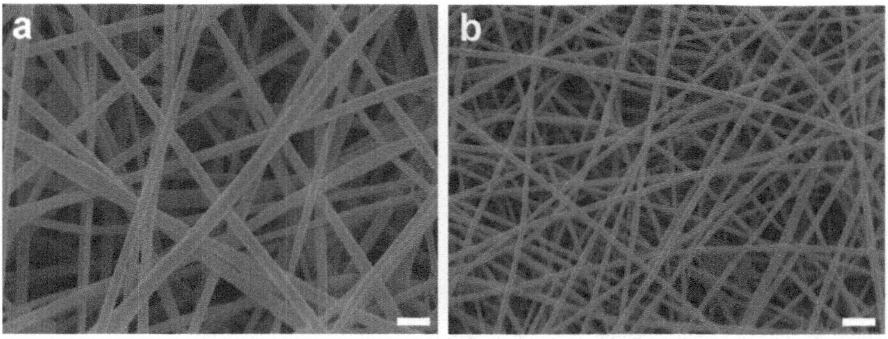

Fig. 3 SEM micrographs of as-spun and crosslinked scaffolds: (**a**) electrospun-trifluoroethanol (TFE); (**b**) electrospun-acetic acid (AcOH) (From Fiorani et al. [116] with permission from the publisher (Springer))

Electrospun collagen scaffolds promote stem cell adhesion, proliferation and differentiation of several cell types. The biological properties of electrospun collagen fibers were explored in wound dressing applications [115] and to repair and regenerate bone defects and injuries [117]. B. Shekhar Jha et al. [114], observed that electrospun collagen fibers have unique biological activity in a wide variety of tissue engineering applications. Cell adhesion experiments revealed that both native and denatured collagen chains show distinctive biological properties. Endothelial cells (ECs) were cultured in collagen scaffolds with an average cross-sectional fiber diameter lower than 1.0–1.50 μm and it was observed that the physical arrangement of fibers, namely the pore characteristics, allows ECs infiltration into the scaffolds [114]. Furthermore, it was also observed that electrospun collagen fibers induce osteoblast differentiation and subsequent formation of hydroxyapatite crystals [114]. Also collagen based composites found applications in tissue engineering: J. Venugopal et al. [105] found that human coronary artery smooth muscle cells (SMCs) seeded in poly(caprolactone) (PCL) nanofibers scaffolds coated with collagen migrated inside to the nanofibers scaffold forming a smooth muscle tissue. Electrospun collagen-blended poly(L-lactic acid)-co-poly(epsilon-caprolactone) (P(LLA-CL)), nanofibers enhanced the viability, spreading, and attachment of human coronary artery ECs preserving the phenotype [111]. Furthermore, when encapsulated with bovine serum albumin and recombinant human transforming growth factor-β3 (rhTGF-β3) it represents a promising tissue engineered scaffold for tracheal cartilage regeneration [111]. The deposition of collagen fibers on top of PLA fibers followed by a final layer of PLA fibers find applicability in vascular graft applications as it confers improved mechanical strength, the proper alignment of the fibers promoting the arterial SMCs orientation in the direction of the main stress directions [118]. It has been also reported that scaffolds based on collagen combined with proteins such as elastin allow developing more natural blood vessel [105].

Others electroactive scaffolds with a piezoelectric properties have been processed by electrospinning, including chitin, cellulose, amylose, keratin and fibrin.

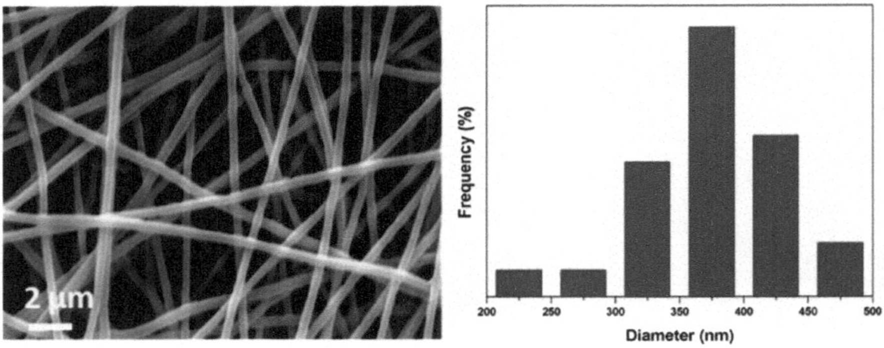

Fig. 4 SEM micrographs of HEC/PVA electrospun nanofibers (10:90) (From Chahal et al. [120] with permission from the publisher (Elsevier))

Cellulose is a linear polysaccharide and the most abundant natural resource on earth [119]. It is biodegradable and biocompatible [119, 120]. There have been few studies based on cellulose electrospun nanofibers mainly due to the difficulties during the electrospinning process, as it is difficulty to dissolve cellulose in the most common solvents [119]. *N*-methyl-morpholine N-oxide solution/water (NMMO/H_2O), lithium chloride/dimethyl acetamide (LiCl/DMAc) and ionic liquids are the usual solvents used to dissolve cellulose for electrospinning, due to the dielectric constant requirements for successful electrospinning [119]. An alternative approach emerged by developing scaffolds from cellulose derivatives including cellulose acetate, hydroxyl propyl, and others, obtained via reactions with hydroxyl side groups on the cellulose backbone [121, 122]. The electrospinning conditions, polymers, cellulose type and derivatives, solvents and average diameters were summarized by Kousaku Ohkawa [122]. Among the cellulose derivatives, cellulose acetate is the most promising for tissue engineering applications, particularly for bone tissue engineering, due to its electrospinnability and good mechanical properties [123]. Scaffolds based on cellulose acetate support osteoblast growth in vitro and immature form of bone formation in vivo. Also, cellulose based blends can be electrospun into fibers. Uniform, bead-free, porous, and non-woven morphology of cellulose derived hydroxyethyl cellulose (HEC)/poly(vinyl alcohol) (PVA) blends (10:90) (Fig. 4) with an average diameter of 524 ± 31 nm and with a tensile strength of 3.5 ± 1.01 MPa have been reported [120], the fiber diameter of HEC/PVA decreasing with increasing HEC content [120].

Chitin is another type of piezoelectric material that can be applied as scaffolds for tissue engineering [124, 125]. Chitin is a polysaccharide composed by *N*-acetylglucosamine and glucosamine repeating units in its structure and derives from a range of natural resources [101]. When the number of *N*-acetyl-glucosamine units is higher than 50%, the biopolymer is termed chitin. Its nontoxicity, biocompatibility and biodegradability make this polymer a strong candidate for applications in tissue engineering such as bone reparation and wound healing [126]. Few reports are known on the electrospinning of chitin, chitin derivatives or chitin

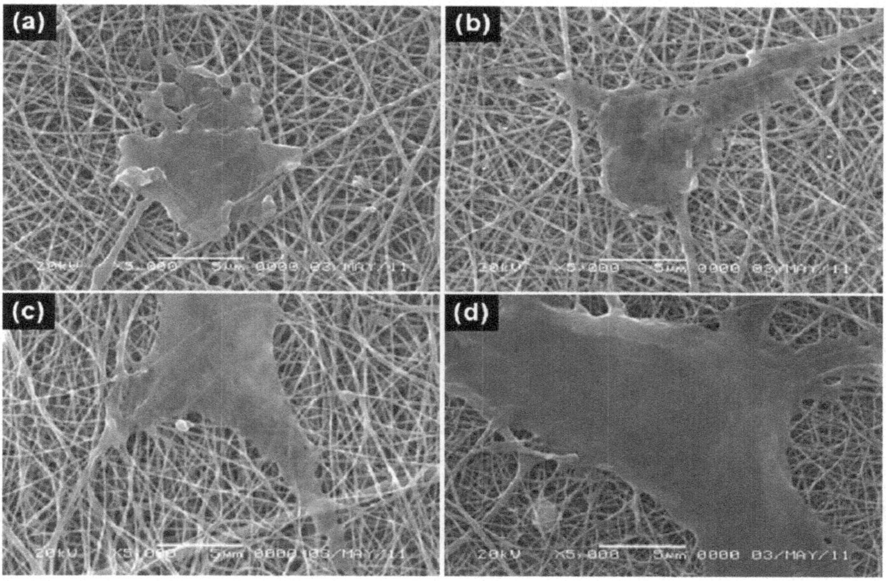

Fig. 5 Morphology of osteoblast cells on pristine nylon-6 mat after (**a**) 1 day and (**b**) 3 days, and on composite nylon-6/CB (N2) mat after (**c**) 1 day and (**d**) 3 days cell culture (From Pant et al. [124] with permission from the publisher (Elsevier))

composites with other polymers. This fact has is explained by its inherent poor electrospinnability, intractability, and insolubility [124]. H.R. Pant et al. [124] reported that chitin butyrate (CB) coated nylon-6 nanofibers prepared using single-spinneret electrospinning of blend solution increases its compatibility with osteoblast cells and bone formation ability (Fig. 5).

A. Pangon et al. [101] reported the development of chitosan/chitin whisker (CTWK) bionanocomposites in the form of hydroxyapatite (HA)-hybridized nanofibers. The nanofiber scaffold was nontoxic to osteoblast cells and cell viability and proliferation could be enhanced by increasing CTWK [101]. Y. Ji et al. evaluated the effect of the incorporation of chitin nanofibril into PCL matrix on cell attachment and proliferation by seeding human dermal fibroblasts on the fiber mat scaffolds and observed that the addition of chitin nanofibrils into PCL matrix facilitated fibroblasts attachment and infiltration [125].

Keratin electrospun fibers, also present a strong potential for tissue engineering [127–129]. Keratin can be found in hair and can be used in a variety of biomedical applications due to their biocompatibility and biodegradability [127]. Its amino acid sequences are the main responsible for its outstanding cytocompatibility [127]. However, due to its poor mechanical properties, several researchers have combined keratin with synthetic biodegradable polymers such as poly(3-hydroxybutyrate-co-3-hydroxyvalerate) (PHBV) and PLA [130, 131], among others. Y. Li et al. [127] reported the potential of co-electrospun of keratin blended with PCL to produce biodegradable cytocompatible mats to improve the cytocompatibility and hydrophilicity

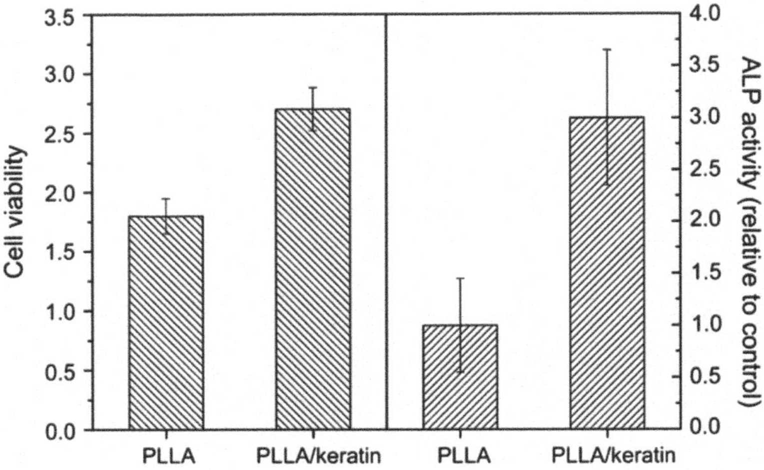

Fig. 6 MTT assay and ALP activity of osteoblast cells seeded on PLLA membranes with and without keratin. Error bars represent means ± standard deviation for n = 3 ($p < 0.05$) (From Li et al. [134] with permission from the publisher (Elsevier))

of PCL. PCL/keratins fiber scaffolds with averages diameters ranging of 477.3 ± 68.2 nm, 433.2 ± 65.1 nm and 330.6 ± 72.1 nm for different PCL/Keratins ratios were obtained. These composite nanofiber mats are characterized by good blood compatibility being a good candidate as scaffold for vascular tissue engineering [127]. Also Z.S. Thompson et al. [132] explored the potential of PCL/keratin fibers for tissue engineering and observed that the PCL/keratin composite scaffold presented excellent mechanical properties such as Young's modulus and failure point. The cell culture experiments showed that the blended scaffold promotes fibroblast cells adhesion and proliferation [132]. Keratin/PEO blends were developed by J. Fan et al. [128]. The cell culture performed with mouse fibroblasts revealed that the keratin/PEO blend nanofiber mats promote cell adhesion, spreading and growth [128]. Keratin and fibroin composites with different ratios were electrospun into fibers by Yen et al. [133] to tissue-engineered blood vessel in order to mimic the structure and morphology of the natural vascular structure. Moreover, Keratin/fibroin scaffolds offer temporary supports and guides for vascular tissue re-organization and regeneration. Human umbilical vein ECs were attached to the scaffold and showed no cytotoxicity and the phenotype of the keratin-fibroin membrane was not altered. The study reveal that a fibroin/keratin scaffold (8:2) can be considered as a tissue-engineered vascular grafts due to its similarity to native vessels [133]. A novel wound dressing based on nanofibrous PHBV–keratin mats was developed in [129]. The introduction of keratin enhanced cell proliferation, accelerating wound regeneration relatively to the control PHBV scaffold as shown by the histological examination results (Fig. 6). The study of viability and proliferation of osteoblast cells on the surface of PLLA and PLLA/keratin electrospun membranes also revealed more MC3T3-E1 pre-osteoblasts cells on PLLA/keratin scaffolds [134]. Similar results were obtained in [102] indicating that keratin can promote cell attachment and proliferation.

Fibrin is a biopolymer resulting from enzymatic reactions by the thrombin mediated cleavage of fibrin peptide A and B from the polypeptide Aα and Bβ chains of fibrinogen [135]. Fibrin plays an important role during wound healing and skin regeneration in skin injuries. Due to its properties, such as, blood clotting, cellular and matrix interactions, inflammatory response, wound healing and angiogenesis, fibrin has been used as a versatile scaffold for a variety of tissue engineering applications such as cardiovascular, liver, skin, bone and cartilage [135]. Furthermore, fibrin can be also used as a coating material for vascular constructs. M. Bacakova et al. [136] developed fibrin coated electrospun PLA nanofibers and results suggests that fibrin surrounded the individual fibers in the PLLA membrane, forming a thin fiber mesh in different places of the membrane surface, influencing the behavior of human dermal fibroblasts being particularly advantageous for skin tissue engineering. When used as a coating of on a synthetic surface, fibrin can enhance the endothelialization [135].

Also as keratin, fibrin possess poor mechanical properties limiting their use as an ideal scaffold [135]. P.R. Sreerekha et al. [135], developed a fibrin based multiscale hybrid scaffold by sequential electrospinning fibrin and PCL. The combination of the advantages of fibrin (natural wound healing matrix) with the biodegradable PCL, with excellent mechanical properties, resulted in nanofibers with an average diameter of 50–500 nm. The viability of human mesenchymal stem cells (MSCs) revealed that the scaffold is non-toxic and supports cell attachment, spreading and proliferation [135]. The multiscale scaffold enhances the growth and differentiation of human MSCs into cardiomyocytes [137].

Novel biocompatible materials based on electroactive electrospun synthetic polymers with favorable biocompatibility, good mechanical strength, known composition and simple design to minimize immune response were also electrospun in nanofibers scaffolds for tissue engineering. Among all polymers, semi- crystalline PVDF and its co-polymer PVDF-TrFe are recognized as the polymers with the larger piezoelectric constants (−24 to −34 and −38 pC/N, respectively) [5]. Due to their biocompatibility, non-toxicity and chemical stability, they have been explored as scaffold materials for exploring the concept of mechano-electrical transductions in bone tissue and nerve or neural and muscle regeneration [5]. Several studies reported the processing of these polymers by electrospinning into non-woven fiber mats randomly/oriented nanofibers. P. Martins et al. [14] demonstrated that PVDF electrospun random and aligned nanofibers are suitable for skeletal muscle tissue engineering. It was observed that fiber orientation influences the cell behavior into the scaffold, C2C12 myoblast cells aligning along the direction of the fibers within the matrix comparatively (Fig. 7) [14].

P. Hitscherich et al. [138] produced promising aligned PVDF-TrFe fibers for cardiovascular tissue constructs. The culture of cardiomyocytes and vascular endothelial cells derived from mouse embryonic stem cells on PVDF-TrFe scaffold reveal that the scaffold efficiently promotes the attachment and survival of both cardiovascular cell types. PVDF-TrFe scaffold can also be applied in neural repair applications [16],

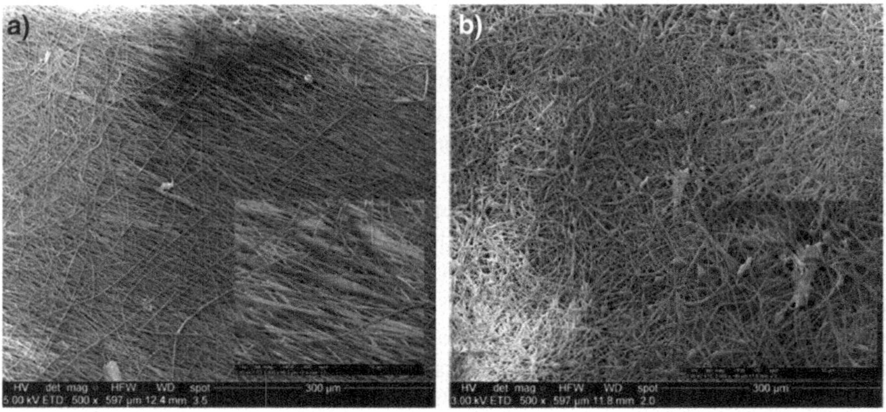

Fig. 7 Cell morphology of C2C12 myoblasts seeded on aligned PVDF fibers (**a**) and randomly oriented PVDF fibers (**b**). Adapted from (From Martins et al. [14])

promoting neuronal cell growth and neurite extension for neural applications [16]. Recently, novel PVDF nanofibers with ionic liquids bis(trifluoromethanesulfonyl) amide ([C_2mim][NTf_2]) emerged for tissue engineering applications, allowing the introduction of large strain mechanical actuation on the cells [139].

Due to its biocompatibility, non-toxicity, biodegradability, good solubility in some organic solvents and non-toxic degradation products, PLLA gained special attention for tissue engineering applications, being widely used for bone tissue engineering [104, 140]. PLLA nanofibers also find applications in musculoskeletal, nervous, cardiovascular, and cutaneous tissue engineering [5, 141]. M. Ali Derakhshan et al. [142] developed PLLA randomly oriented nanofiber scaffolds to reconstruct or replace the problematic bladder smooth muscle. Human bladder SMCs efficiently attach and interact with the random nanofibers scaffold and spread into the nanofibers. As Fig. 8 shows, the cells achieved highly elongated morphology and aligned direction in randomly arranged nanofibers [142].

H. Qi et al. [140] prepared smooth PLLA/PCL/HA composite scaffolds with average diameters ranging from 106 ± 18 nm to 195 ± 24 nm. Cell proliferation and osteogenic differentiation were evaluated with osteoblast cells and results suggested that the presence of HA in the composite is advantageous for osteoblast proliferation and osteogenic differentiation of MC3T3-E1 cells [140]. Biocompatible and non-toxic alginate/PLA nanofibers with an average diameter about 250 ± 90 nm produced by emulsion electrospinning emerged as an innovative type of tissue engineering material [143]. These fiber composite promotes the cell adhesion and growth of cells on nanofibers and the subsequently proliferation and differentiation [143]. PLLA fibers reinforced with strontium-bioactive borosilicate glass microparticles also can induce the osteogenic differentiation of bone marrow MSCs [104]. M. Salehi et al. [144] fabricated PLLA/collagen nanofiber scaffolds with sustained release of aloe vera gel using a chitosan layer for wound healing applications. These hybrid scaffold enhanced the low tensile strength of PLLA after coating with a chitosan layer. The in vitro tests performed by the interaction of the scaffold with

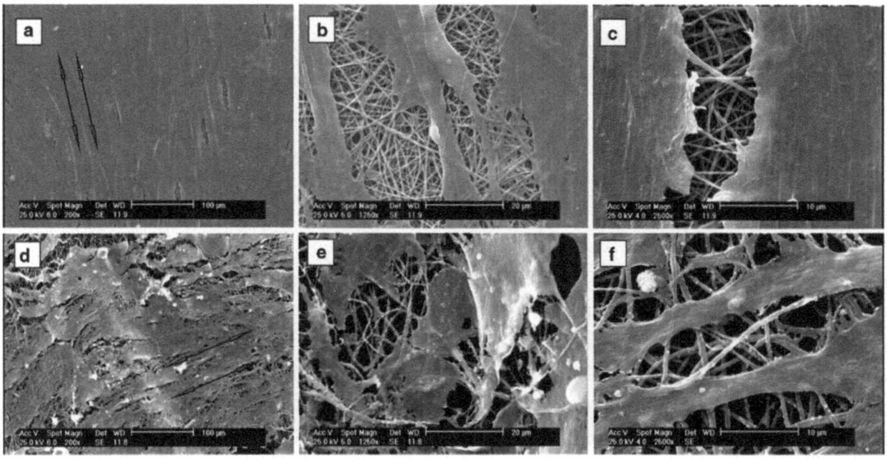

Fig. 8 The hBSMCs 30 days after seeding on smaller-PLLA (**a–c**) and also on the thicker-PLLA nanofibers (**d–f**) in different magnifications. The arrows show the direction of cell elongations (**a–d**) (From Derakhshan et al. [142] with permission from the publisher (Springer))

mouse fibroblasts demonstrated that the scaffold supports attachment, viability, and proliferation of the cells. Silk fibroin (SF)/PLLA scaffolds were produced by W. Liu et al. [145] with excellent biocompatibility for chondrocyte attachment with a good spreading and ECM secretion of chondrocytes on the surface of the fibers SF/PLLA scaffolds representing a promising material for potential application in cartilage tissue engineering.

PHB is a biocompatible polymer, non-toxic and shows high mechanical strength (comparatively to natural polymers). The main disadvantage of this polymer is associated with its low hydrophilicity and slow mass loss rate. As in the case of some natural polymers, to overcome this drawback, PHB can be blended with other polymers such as poly(hydroxybutyrate-hydroxyhexanoate) (PHBHHx), PHBV, PANI, PCL and chitosan [108]. S. Kosé et al. [106] explored whether random-PHB and aligned-PHB nanofiber PHB membranes changed the phenotype and features of human MSCs cells. The study confirmed that fiber orientation influenced the phenotype and biological behavior of human MSCs. While cells expanded in an aligned-PHB scaffold showed fibroid-like morphology, cells interacting with the random-PHB membrane being placed homogeneously and demonstrating polygonal morphology. Moreover, the adhesion and proliferation of human MSCs was higher on the aligned-PHB membrane than on the randomly oriented one [106]. These nanofibers can be also used for bladder tissue repair [146].

Coaxially electrospun fibers of PHB blended with gelatin and also with PVA find applicability in skin regeneration, supporting the growth of human dermal fibroblasts and keratinocytes with normal morphology [147, 148]. Other blends with natural polymers such as collagen demonstrated their applicability as a good scaffold for tissue engineering [149]. D. Sadeghi et al. [108] processed blended PHB electrospun scaffolds with chitosan and evaluated their potential for tissue engineering, namely the production of hydrophilic fibers scaffolds with higher mass loss rates for cartilage

Table 4 Relevant works on electrical stimulation in tissue engineering strategies based on conductive electrospun scaffolds

Conductive scaffolds	DC current (A), potential (V)	Duration of the stimulus	Ref.
PPy/PLA	DC, 100 mV	Lasting 30 min/day for 5 days	[160]
PANI/PLCL	DC, 0–200 mA	48 h	[161]
PANI/PCL/gelatin	DC, 100 mV	1 h	[162]

Adapted from Ref. [5]

tissue engineering applications. The blended scaffolds induced chondrocytes attachment to the surface [108]. Furthermore, PHB scaffolds nanofibers can also be used for nerve tissue engineering due to their ability to induce neural progenitor cells and embryonal cell lines to neuronal differentiation [150, 151].

In vitro assay with pre osteoblasts and PHB blended with HA nanofibers revealed that PHB/HA scaffolds induce the pre-osteoblasts and MSCs cell viability and cell spreading on the fibers scaffold [103, 152, 153]. The combination of PHB/HA fibers with modified gelatin/HA hydrogel in a single tri-layered scaffold also provides mechanical strength and a suitable environment for mouse calvarial pre-osteoblasts (MC3T3-E1) cell encapsulation and proliferation [154].

Also PHB/PCL/58S Sol-gel nanofiber scaffolds can be applied in bone engineering [155]. Furthermore, D. Daranarong et al. [156] developed PHB/PCL nanofibers as scaffolds for nerve repair due to its ability to support the attachment and proliferation of murine olfactory ensheathing cells. Other in vitro studies with PHB/PCL and PHB/PHBV scaffolds have carried out and are suggested for application in tissue engineering [156–158].

3.2 *Conductive Electrospun Smart Scaffolds*

Conductive polymers are a class of electroactive materials with applicability in tissue engineering mainly due to their electrical conducting properties, but also to the ability to tailor their chemical, electrical and physical properties by incorporating antibodies, enzymes and other biological moieties [43]. Several conductive polymers have been electrospun into fibers for tissue engineering. Table 4 summarizes the main works based on the electrical stimulation for tissue engineering strategies based on conductive electrospun scaffolds.

Conductive polymers can be electrospun alone but the resulting fibers are unsuitable for biological applications. In this sense, conductive polymers are typically combined with spinnable polymers such as PEO, polystyrene, PLA, and PCL, coating the electrospun fibers with the respective conductive polymer [43, 159].

The application of conductive nanofibers in tissue engineering is very recent and in this sense, few studies have been reported yet. The most commonly conductive polymers processed into fibers by the electrospinning are poly(pyrrole) (PPy) and PANI.

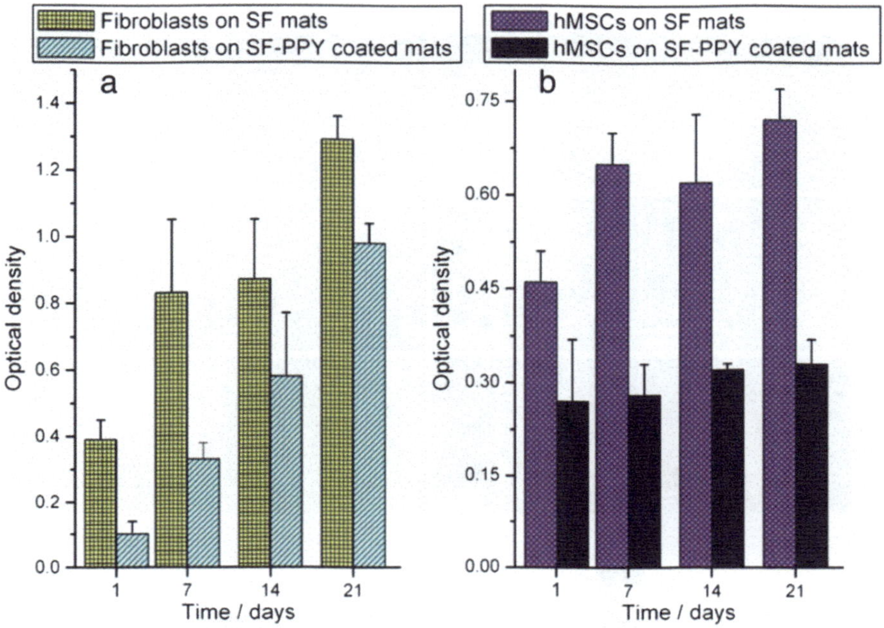

Fig. 9 (**a**) MTT results for fibroblasts growing on SF and SF-PPy coated mats at different times (seeding density 2×10^5 cells cm^{-2}). (**b**) MTT results for human MSCs growing on SF and SF-PPy coated mats at different times (seeding density 2×10^5 cells cm^{-2}) (From Aznar-Cervantes et al. [164] with permission from the publisher (Elsevier))

PPy [5] promotes cell growth and proliferation, in particular for axon growth in vitro and in vivo experiments [5, 163]. S. Aznar-Cervantes et al. [164] developed conductive electrospun SF scaffolds by chemical polymerization coating with PPy with higher ability (comparatively to the SF scaffold) to support the adhesion and proliferation of MSCs and fibroblasts (Fig. 9).

PPy coated electrospun poly(lactic-co-glycolic acid) (PLGA) nanofibers may be suitable applicability for neuronal tissue scaffolds. These conductive nanofibers induced longer neurites formation when stimulated with a potential of 10 mV/cm comparatively to unstimulated PC12 cells [165]. Other approach to fabricate conductive scaffolds is to is to blend them with a carrier material before the electrospinning process.

To the best of our knowledge, few studies have been reported based on PPy electrospun scaffolds for tissue engineering. J. Zhou et al. [160, 166] developed a conductive and biocompatible PLA/PPy aligned nanofibers scaffold with a conductivity of 1×10^{-5} s/cm (for a PPy content above 10%) with potential application in peripheral nerve regeneration. The elongation of human umbilical cord MSCs and Schwann cells onto the PLA/PPy scaffold was parallel to the direction of the fibers, being the nanofibers capable to tune the cell orientation (Fig. 10) [166].

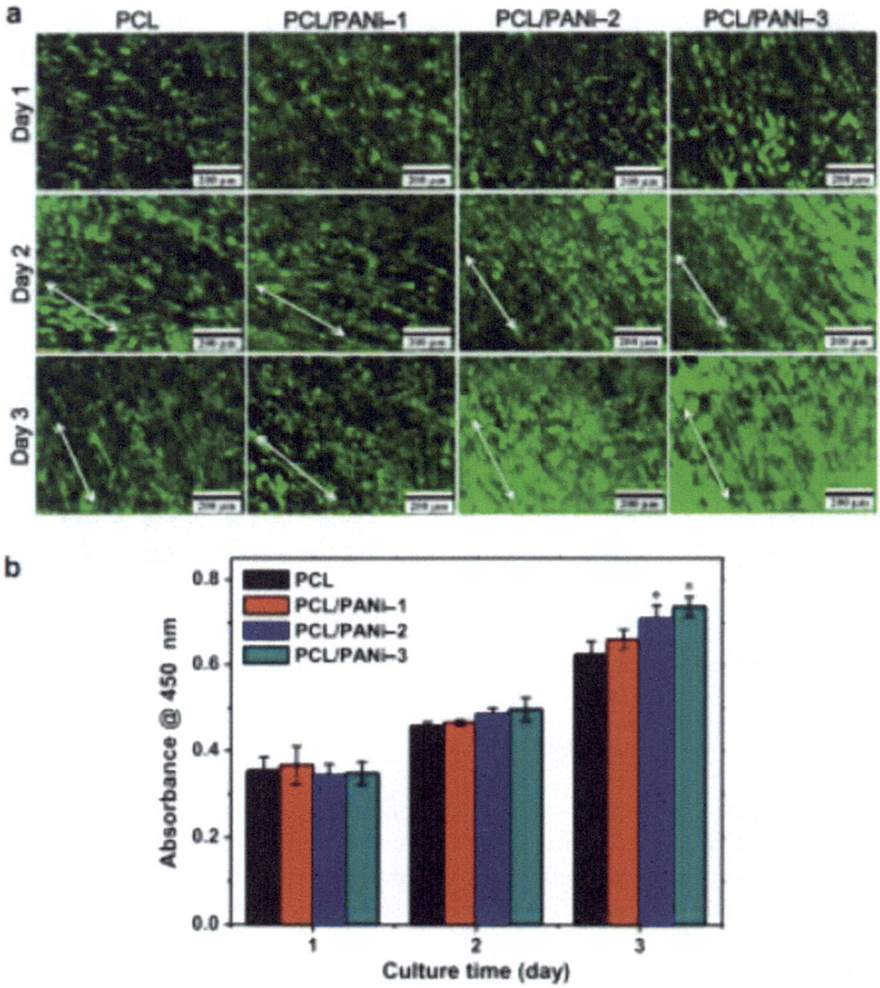

Fig. 10 (**a**) Representative fluorescent images of C2C12 myoblasts cultured on the aligned PCL and aligned PCL/PANI nanofibers for 1, 2 and 3 days and stained using a live/dead cell viability kit (green, viable cells; red, dead cells). Arrows indicate the major directions of the aligned fibers in the images. (**b**) Quantification of the cell viability of C2C12 myoblasts in (**a**) using a CCK-8 kit. *Significantly different compared with pure PCL nanofibers ($p < 0.05$, n = 5) (From Chen et al. [168] with permission from the publisher (Elsevier))

Human umbilical cord MSCs seeded on aligned PLA/PPy fibers can be electrically stimulated with a DC power supply of 100 mV/mm inducing the differentiation of human umbilical cord MSCs into nerve cells [160].

PANI blended with PCL was electrospun into fibers with potential application as substrate for skeletal muscle [167, 168] and cardiac tissue engineering [112] due to the synergistically stimulation of the cells, increasing the electrical conductivity of the PANI/PCL scaffolds with increasing PANI content [168]. Nanofibers scaffolds based on PANI/poly(L-lactide-co-e-caprolactone) (PLCL) also presents potential as substrate to support the adhesion of NIH-3T3 fibroblasts and C2C12 myoblasts cells [161], cell growth being controlled by the application of an electrical stimulation [161]. PANI and blended poly(methyl methacrylate) (PMMA) as a supporting polymer has been also implemented for the regulation of the differentiation and proliferation of nerve cells [169]. Poly(acrylonitrile) (PAN)-PANI blends have been also demonstrated as a suitable scaffolds for muscle tissue engineering once promotes cell proliferation and also differentiation of the satellite cells [170]. PANI blended with gelatin demonstrated its potentiality to support the attachment, migration, and proliferation of H9C2 rat cardiac myoblasts [113]. With increasing PANI the average fiber size is reduced by approximately 1 order of magnitude to <100 nm and the conductivity increases [113]. The electrical stimulation of PANI/PCL/gelatin electrospun nanofibers also enhanced significantly cell proliferation and neurite growth when compared to non-stimulated scaffolds [162].

3.3 Magnetically Active Electrospun Smart Scaffolds

A novel approach in smart materials for tissue engineering applications has been achieved by electrospun fibers based on materials with a magnetoelectric (ME) response [93]. ME emerges as a class of materials allowing the use of an external magnetic field to remotely control tissue stimulation without the need of patient movement, based on the combination of magnetostrictive and piezoelectric materials [93]. The potential application of ME materials in tissue engineering was evaluated in Terfenol-D/PVDF-TrFe composite films able to provide mechanical and electrical stimuli to MC3T3-E1 pre-osteoblast cells being the stimuli remotely triggered by an applied magnetic field films [93]. ME electrospun fibers based on cobalt ferrite ($CoFe_2O_4$)/PVDF [171] have been also reported, with suitable ME response for tissue engineering applications.

4 Conclusions

Electrospun materials based on smart polymers has been gaining increasing interest for tissue engineering applications, emerging as promising substrates with the ability to mimic the properties and structural characteristics of the ECM and allowing to contribute to the development of suitable cell microenvironments. Electrospun mats present interesting characteristics for tissue engineering applications, including large surface area to volume/mass ratio, large variety of possibilities for surface functionalization and improved mechanical performance. Further, when electrospinning is performed from smart and responsive materials, they can be used as biomimetic templates with the ability to respond to an appropriate environment and to assist specific cell responses, such as biological, physical and mechanical functions of native tissues.

This chapter has reported shown that smart materials based on electrospinning of piezoelectric natural polymers such as collagen, cellulose, chitin, amylose, keratin and fibrin and synthetic polymers such as PLA, PHB, PVDF, PVDF-TrFe, and PA have gained an increase interest for tissue applications. In vitro and in vivo studies reveal that fibroblasts, myoblasts and chondrocytes cultured into electrospun scaffolds (random and aligned fibers) based on these smart materials are a promising approach to wound dressings, vascular dressings, and to repair and regenerate bone defects and injuries. Also, conductive polymers, such as PPy and PANI, are recognized as interesting materials for biomedical applications. The synergistically stimulation of the cells, increasing the electrical conductivity promotes a significant enhancement of cell proliferation and growth when compared to non-stimulated scaffolds. Further, ME materials allow the use of an external magnetic field to remotely control tissue stimulation combining magnetostrictive and piezoelectric materials have been explored. Other smart materials effects, such as thermoresponsive or pH responsive, among others, have been scarcely investigated yet. Even for the most applied smart materials, most of the conducted tests have been performed under static conditions, proving the suitability of the developed material for tissue engineering but not proving the effect of the smart material active response (e.g. mechano-electrical transduction in piezoelectric, magneto-mechanical or magneto-electrical in magneto-electrics or the effect or the electrical signal in conductive polymer scaffolds) on cell behavior.

Thus, in order to properly evaluate the great potential of electrospun polymers for tissue engineering applications, not only a larger variety of materials with tailored specific active properties have to be processes, but also suitable bioreactors must be developed to fully take advantage of the materials during the cell culture procedure. This will allow the proper evaluation of the effect of the active response, to develop suitable microenvironments suitable for cell and tissue oriented tissue engineering strategies and, finally, to advance in the true challenge of understanding the cell differentiation and proliferation process and to develop proper functional tissue regeneration strategies.

Acknowledgments This work was supported by the Portuguese Foundation for Science and Technology (FCT) in the framework of the Strategic Funding UID/FIS/04650/2013 and grants SFRH/BD/111478/2015 (S.R.), SFRH/BPD/121526/2016 (D.C.) and SFRH/BPD/90870/2012 (C.R.). The authors acknowledge funding by the Spanish Ministry of Economy and Competitiveness (MINECO) through the project MAT2016-76039-C4-3-R (AEI/FEDER, UE) and from the Basque Government Industry Department under the ELKARTEK program.

References

1. Kazanci M (2003) A review of polymeric smart materials for biomedical applications. Mater Technol 18(2):87–93. https://doi.org/10.1080/10667857.2003.11753019
2. Ravichandran R, Sundarrajan S, Venugopal JR, Mukherjee S, Ramakrishna S (2012) Advances in polymeric systems for tissue engineering and biomedical applications. Macromol Biosci 12(3):286–311. https://doi.org/10.1002/mabi.201100325

3. Ikada Y (2006) Challenges in tissue engineering. J R Soc Interface 3(10):589–601. https://doi.org/10.1098/rsif.2006.0124
4. Bierbaum S, Hintze V, Scharnweber D (2012) Functionalization of biomaterial surfaces using artificial extracellular matrices. Biomatter 2(3):132–141. https://doi.org/10.4161/biom.20921
5. Ribeiro C, Sencadas V, Correia DM, Lanceros-Méndez S (2015) Piezoelectric polymers as biomaterials for tissue engineering applications. Colloids Surf B Biointerfaces 136:46–55. https://doi.org/10.1016/j.colsurfb.2015.08.043
6. O'Brien FJ (2011) Biomaterials & scaffolds for tissue engineering. Mater Today 14(3):88–95. https://doi.org/10.1016/S1369-7021(11)70058-X
7. Seidi A, Ramalingam M, Elloumi-Hannachi I, Ostrovidov S, Khademhosseini A (2011) Gradient biomaterials for soft-to-hard interface tissue engineering. Acta Biomater 7(4):1441–1451. https://doi.org/10.1016/j.actbio.2011.01.011
8. Dhandayuthapani B, Yoshida Y, Maekawa T, Kumar DS (2011) Polymeric scaffolds in tissue engineering application: a review. Int J Polym Sci. https://doi.org/10.1155/2011/290602
9. Ingavle GC, Leach JK (2014) Advancements in electrospinning of polymeric nanofibrous scaffolds for tissue engineering. Tissue Eng Part B Rev 20(4):277–293. https://doi.org/10.1089/ten.teb.2013.0276
10. Khajavi R, Abbasipour M, Bahador A (2016) Electrospun biodegradable nanofibers scaffolds for bone tissue engineering. J Appl Polym Sci. https://doi.org/10.1002/app.42883
11. Gönen SÖ, Erol Taygun M, Küçükbayrak S (2016) Fabrication of bioactive glass containing nanocomposite fiber mats for bone tissue engineering applications. Compos Struct 138:96–106. https://doi.org/10.1016/j.compstruct.2015.11.033
12. Neves SC, Moreira Teixeira LS, Moroni L, Reis RL, Van Blitterswijk CA, Alves NM, Karperien M, Mano JF (2011) Chitosan/Poly(ε-caprolactone) blend scaffolds for cartilage repair. Biomaterials 32(4):1068–1079. https://doi.org/10.1016/j.biomaterials.2010.09.073
13. Pham QP, Sharma U, Mikos AG (2006) Electrospinning of polymeric nanofibers for tissue engineering applications: a review. Tissue Eng 12(5):1197–1211. https://doi.org/10.1089/ten.2006.12.1197
14. Martins PM, Ribeiro S, Ribeiro C, Sencadas V, Gomes AC, Gama FM, Lanceros-Mendez S (2013) Effect of poling state and morphology of piezoelectric poly(vinylidene fluoride) membranes for skeletal muscle tissue engineering. RSC Adv 3(39):17938–17944. https://doi.org/10.1039/c3ra43499k
15. Karam JP, Muscari C, Montero-Menei CN (2012) Combining adult stem cells and polymeric devices for tissue engineering in infarcted myocardium. Biomaterials 33(23):5683–5695. https://doi.org/10.1016/j.biomaterials.2012.04.028
16. Lee Y-S, Collins G, Arinzeh TL (2011) Neurite extension of primary neurons on electrospun piezoelectric scaffolds. Acta Biomater 7(11):3877–3886. https://doi.org/10.1016/j.actbio.2011.07.013
17. Seidlits SK, Lee JY, Schmidt CE (2008) Nanostructured scaffolds for neural applications. Nanomedicine 3(2):183–199. https://doi.org/10.2217/17435889.3.2.183
18. Zhu Y, Cao Y, Pan J, Liu Y (2010) Macro-alignment of electrospun fibers for vascular tissue engineering. J Biomed Mater Res B Appl Biomater 92(2):508–516. https://doi.org/10.1002/jbm.b.31544
19. Ercolani E, Del Gaudio C, Bianco A (2015) Vascular tissue engineering of small-diameter blood vessels: reviewing the electrospinning approach. J Tissue Eng Regen Med 9(8):861–888. https://doi.org/10.1002/term.1697
20. Bogue R (2014) Smart materials: a review of capabilities and applications. Assem Autom 34(1):16–22. https://doi.org/10.1108/aa-10-2013-094
21. Xia F, Jiang L (2008) Bio-inspired, smart, multiscale interfacial materials. Adv Mater 20(15):2842–2858. https://doi.org/10.1002/adma.200800836
22. Ferreira ADBL, Nóvoa PRO, Marques AT (2016) Multifunctional material systems: a state-of-the art review. Compos Struct 151:3–35. https://doi.org/10.1016/j.compstruct.2016.01.028

23. Cardoso VF, Ribeiro C, Lanceros-Mendez S (2017) Metamorphic biomaterials. In: Rodrigues L, Mota M (eds) Bioinspired materials for medical applications. Woodhead Publishing, Philadelphia, pp 69–99
24. Rasmussen MK, Pedersen EW, Petersen MG, Hornbæk K (2012) Shape-changing interfaces: a review of the design space and open research questions. Conference on human factors in computing systems – proceedings. https://doi.org/10.1145/2207676.2207781
25. Ghosh SK (2009) Self-healing materials: fundamentals, design strategies, and applications. Springer, Weinheim. https://doi.org/10.1002/9783527625376.ch1
26. Aïssa B, Therriault D, Haddad E, Jamroz W (2012) Self-healing materials systems: overview of major approaches and recent developed technologies. Adv Mater Sci Eng. https://doi.org/10.1155/2012/854203
27. Chung DDL (2007) Damage detection using self-sensing concepts. Proc Inst Mech Eng G J Aerosp Eng 221(4):509–520. https://doi.org/10.1243/09544100JAERO203
28. Kamila S (2013) Introduction, classification and applications of smart materials: an overview. Am J Appl Sci 10(8):876–880. https://doi.org/10.3844/ajassp.2013.876.880
29. Seyfouri MM, Binions R (2017) Sol-gel approaches to thermochromic vanadium dioxide coating for smart glazing application. Sol Energy Mater Sol Cells 159:52–65. https://doi.org/10.1016/j.solmat.2016.08.035
30. Yang YS, Zhou Y, Chiang FBY, Long Y (2016) Temperature-responsive hydroxypropylcellulose based thermochromic material and its smart window application. RSC Adv 6(66):61449–61453. https://doi.org/10.1039/c6ra12454b
31. Musho T (2015) Predicting the figure of merit of nanostructured thermoelectric materials. J Mater Res 30(17):2628–2637. https://doi.org/10.1557/jmr.2015.256
32. Hudak NS, Amatucci GG (2008) Small-scale energy harvesting through thermoelectric, vibration, and radio frequency power conversion. J Appl Phys. https://doi.org/10.1063/1.2918987
33. Chan BQY, Low ZWK, Heng SJW, Chan SY, Owh C, Loh XJ (2016) Recent advances in shape memory soft materials for biomedical applications. ACS Appl Mater Interfaces 8(16):10070–10087. https://doi.org/10.1021/acsami.6b01295
34. Choudhary N, Kaur D (2016) Shape memory alloy thin films and heterostructures for MEMS applications: a review. Sensors Actuators A Phys 242:162–181. https://doi.org/10.1016/j.sna.2016.02.026
35. Yu XD, Chen LM, Zhang MM, Yi T (2014) Low-molecular-mass gels responding to ultrasound and mechanical stress: towards self-healing materials. Chem Soc Rev 43(15):5346–5371. https://doi.org/10.1039/c4cs00066h
36. Khan A, Abas Z, Kim HS, Oh IK (2016) Piezoelectric thin films: an integrated review of transducers and energy harvesting. Smart Mater Struct. https://doi.org/10.1088/0964-1726/25/5/053002
37. Wang Y, Runnerstrom EL, Milliron DJ (2016) Switchable materials for smart windows. In: Prausnitz JM (eds) Annual review of chemical and biomolecular engineering. Annual review of chemical and biomolecular engineering, pp 283–304
38. Cai GF, Wang JX, Lee PS (2016) Next-generation multifunctional electrochromic devices. Acc Chem Res 49(8):1469–1476. https://doi.org/10.1021/acs.accounts.6b00183
39. Zhu FB, Zhang CL, Qian J, Chen WQ (2016) Mechanics of dielectric elastomers: materials, structures, and devices. J Zhejiang Univ Sci A 17(1):1–21. https://doi.org/10.1631/jzus.A1500125
40. Shankar R, Ghosh TK, Spontak RJ (2007) Dielectric elastomers as next-generation polymeric actuators. Soft Matter 3(9):1116–1129. https://doi.org/10.1039/b705737g
41. Chen YG, Yan H (2011) The performance analysis of electro-rheological damper. In: Zhu G (eds) Materials science and engineering. Advanced materials research, pp 443–448
42. Ganet F, Le MQ, Capsal JF, Lermusiaux P, Petit L, Millon A, Cottinet PJ (2015) Development of a smart guide wire using an electrostrictive polymer: option for steerable orientation and force feedback. Sci Rep. https://doi.org/10.1038/srep18593
43. Balint R, Cassidy NJ, Cartmell SH (2014) Conductive polymers: towards a smart biomaterial for tissue engineering. Acta Biomater 10(6):2341–2353. https://doi.org/10.1016/j.actbio.2014.02.015

44. De Luca V, Digiamberardino P, Di Pasquale G, Graziani S, Pollicino A, Umana E, Xibilia MG (2013) Ionic electroactive polymer metal composites: fabricating, modeling, and applications of postsilicon smart devices. J Polym Sci Part B Polym Phys 51(9):699–734. https://doi.org/10.1002/polb.23255
45. Shi ZJ, Gao X, Ullah MW, Li SX, Wang Q, Yang G (2016) Electroconductive natural polymer-based hydrogels. Biomaterials 111:40–54. https://doi.org/10.1016/j.biomaterials.2016.09.020
46. Choi SB, Li WH, Yu M, Du HP, Fu J, Do PX (2016) State of the art of control schemes for smart systems featuring magneto-rheological materials. Smart Mater Struct 25(4):043001. https://doi.org/10.1088/0964-1726/25/4/043001
47. Vicente J, Klingenberg DJ, Hidalgo-Alvarez R (2011) Magnetorheological fluids: a review. Soft Matter 7(8):3701–3710. https://doi.org/10.1039/c0sm01221a
48. Levi DS, Kusnezov N, Carman GP (2008) Smart materials applications for pediatric cardiovascular devices. Pediatr Res 63(5):552–558. https://doi.org/10.1203/PDR.0b013e31816a9d18
49. Yu C, Wang CF, Chen S (2015) Facile access to versatile hydrogels via interface-directed frontal polymerization derived from the magnetocaloric effect. J Mater Chem A 3(33):17351–17358. https://doi.org/10.1039/c5ta03811a
50. Sutrisno J, Purwanto A, Mazlan SA (2015) Recent progress on magnetorheological solids: materials, fabrication, testing, and applications. Adv Eng Mater 17(5):563–597. https://doi.org/10.1002/adem.201400258
51. Koetting MC, Peters JT, Steichen SD, Peppas NA (2015) Stimulus-responsive hydrogels: theory, modern advances, and applications. Mater Sci Eng R Rep 93:1–49. https://doi.org/10.1016/j.mser.2015.04.001
52. Jesus MR, Sarika R, Shankaran DR (2016) Curcumin labeled PMMA halochromic smart flexible films. Mater Focus 5(2):177–181. https://doi.org/10.1166/mat.2016.1299
53. Zhang JJ, Zou Q, Tian H (2013) Photochromic materials: more than meets the eye. Adv Mater 25(3):378–399. https://doi.org/10.1002/adma.201201521
54. Zhang JJ, Wang JX, Tian H (2014) Taking orders from light: progress in photochromic biomaterials. Mater Horiz 1(2):169–184. https://doi.org/10.1039/c3mh00031a
55. Sanchez-Arevalo FM, Garnica-Palafox IM, Jagdale P, Hernandez-Cordero J, Rodil SE, Okonkwo AO, Hernandez FCR, Tagliaferro A (2015) Photomechanical response of composites based on PDMS and carbon soot nanoparticles under IR laser irradiation. Opt Mater Exp 5(8):1792–1805. https://doi.org/10.1364/ome.5.001792
56. Wu Q, Wang L, Yu HJ, Wang JJ, Chen ZF (2011) Organization of glucose-responsive systems and their properties. Chem Rev 111(12):7855–7875. https://doi.org/10.1021/cr200027j
57. Rica R, Aili D, Stevens MM (2012) Enzyme-responsive nanoparticles for drug release and diagnostics. Adv Drug Deliv Rev 64(11):967–978. https://doi.org/10.1016/j.addr.2012.01.002
58. Hu J, Meng H, Li G, Ibekwe SI (2012) A review of stimuli-responsive polymers for smart textile applications. Smart Mater Struct 21(5):053001. https://doi.org/10.1088/0964-1726/21/5/053001
59. Manz H, Breitbach EJ (2001) Application of smart materials in automotive structures. Proc SPIE Int Soc Opt Eng 4332:197–204. https://doi.org/10.1117/12.429657
60. Gopalakrishnan S (2014) Smart materials technology for aerospace applications. In: Vinoy KJ, Ananthasuresh GK, Pratap R, Krupanidhi SB (eds) Micro and smart devices and systems. Springer, New Delhi, pp 423–437
61. Thévenot J, Oliveira H, Sandre O, Lecommandoux S (2013) Magnetic responsive polymer composite materials. Chem Soc Rev 42(17):7099–7116. https://doi.org/10.1039/c3cs60058k
62. Ribeiro C, Moreira S, Correia V, Sencadas V, Rocha JG, Gama FM, Ribelles JLG, Lanceros-Mendez S (2012) Enhanced proliferation of pre-osteoblastic cells by dynamic piezoelectric stimulation. RSC Adv 2(30):11504–11509. https://doi.org/10.1039/c2ra21841k
63. Seil JT, Webster TJ (2010) Electrically active nanomaterials as improved neural tissue regeneration scaffolds. Wiley Interdiscip Rev Nanomed Nanobiotechnol 2(6):635–647. https://doi.org/10.1002/wnan.109

64. Frias C, Reis J, Silva FCE, Potes J, Simoes J, Marques AT (2010) Polymeric piezoelectric actuator substrate for osteoblast mechanical stimulation. J Biomech 43(6):1061–1066. https://doi.org/10.1016/j.jbiomech.2009.12.010
65. Reis J, Frias C, Castro CCE, Botelho ML, Marques AT, Simoes JAO, Silva FCE, Potes J (2012) A new piezoelectric actuator induces bone formation in vivo: a preliminary study. J Biomed Biotechnol. https://doi.org/10.1155/2012/613403
66. Rodrigues MT, Gomes ME, Mano JF, Reis RL (2008) β-PVDF membranes induce cellular proliferation and differentiation in static and dynamic conditions. In: Marques AT, Silva AF, Baptista APM et al (eds) Advanced materials forum. Materials science forum, pp 72–76
67. Parssinen J, Hammaren H, Rahikainen R, Sencadas V, Ribeiro C, Vanhatupa S, Miettinen S, Lanceros-Mendez S, Hytonen VP (2015) Enhancement of adhesion and promotion of osteogenic differentiation of human adipose stem cells by poled electroactive poly(vinylidene fluoride). J Biomed Mater Res A 103(3):919–928. https://doi.org/10.1002/jbm.a.35234
68. Ribeiro C, Parssinen J, Sencadas V, Correia V, Miettinen S, Hytonen VP, Lanceros-Mendez S (2015) Dynamic piezoelectric stimulation enhances osteogenic differentiation of human adipose stem cells. J Biomed Mater Res A 103(6):2172–2175. https://doi.org/10.1002/jbm.a.35368
69. Ikada Y, Shikinami Y, Hara Y, Tagawa M, Fukada E (1996) Enhancement of bone formation by drawn poly(L-lactide). J Biomed Mater Res 30(4):553–558. https://doi.org/10.1002/(SICI)1097-4636(199604)30:4<553::AID-JBM14>3.0.CO;2-I
70. Wang Y-W, Wu Q, Chen J, Chen G-Q (2005) Evaluation of three-dimensional scaffolds made of blends of hydroxyapatite and poly(3-hydroxybutyrate-co-3-hydroxyhexanoate) for bone reconstruction. Biomaterials 26(8):899–904. https://doi.org/10.1016/j.biomaterials.2004.03.035
71. Damaraju SM, Wu S, Jaffe M, Arinzeh TL (2013) Structural changes in PVDF fibers due to electrospinning and its effect on biological function. Biomed Mater 8(4):045007. https://doi.org/10.1088/1748-6041/8/4/045007
72. Prabhakaran MP, Venugopal J, Ramakrishna S (2009) Electrospun nanostructured scaffolds for bone tissue engineering. Acta Biomater 5(8):2884–2893. https://doi.org/10.1016/j.actbio.2009.05.007
73. Paşcu EI, Stokes J, McGuinness GB (2013) Electrospun composites of PHBV, silk fibroin and nano-hydroxyapatite for bone tissue engineering. Mater Sci Eng C 33(8):4905–4916. https://doi.org/10.1016/j.msec.2013.08.012
74. Ko EK, Jeong SI, Rim NG, Lee YM, Shin H, Lee B-K (2008) In vitro osteogenic differentiation of human mesenchymal stem cells and in vivo bone formation in composite nanofiber meshes. Tissue Eng A 14(12):2105–2119. https://doi.org/10.1089/ten.tea.2008.0057
75. Marques L, Holgado LA, Simoes RD, Pereira J, Floriano JF, Mota L, Graeff CFO, Constantino CJL, Rodriguez-Perez MA, Matsumoto M, Kinoshita A (2013) Subcutaneous tissue reaction and cytotoxicity of polyvinylidene fluoride and polyvinylidene fluoride-trifluoroethylene blends associated with natural polymers. J Biomed Mater Res Part B 101(7):1284–1293. https://doi.org/10.1002/jbm.b.32941
76. Mota C, Wang S-Y, Puppi D, Gazzarri M, Migone C, Chiellini F, Chen G-Q, Chiellini E (2014) Additive manufacturing of poly[(R)-3-hydroxybutyrate-co-(R)-3-hydroxyhexanoate] scaffolds for engineered bone development. J Tissue Eng Regen Med 11(1):175–186. https://doi.org/10.1002/term.1897
77. Gimenes R, Zaghete MA, Bertolini M, Varela JA, Coelho LO, Silva NF (2004) Composites PVDF-TrFE/BT used as bioactive membranes for enhancing bone regeneration. In: BarCohen Y (eds) Smart structures and materials: electroactive polymer actuators and devices proceedings of the society of photo-optical instrumentation engineers, pp 539–547
78. Chen Y, Mak AFT, Wang M, Li J, Wong MS (2006) PLLA scaffolds with biomimetic apatite coating and biomimetic apatite/collagen composite coating to enhance osteoblast-like cells attachment and activity. Surf Coat Technol 201(3-4):575–580. https://doi.org/10.1016/j.surfcoat.2005.12.005

79. Inui A, Kokubu T, Makino T, Nagura I, Toyokawa N, Sakata R, Kotera M, Nishino T, Fujioka H, Kurosaka M (2010) Potency of double-layered poly L-lactic acid scaffold in tissue engineering of tendon tissue. Int Orthop 34(8):1327–1332. https://doi.org/10.1007/s00264-009-0917-8
80. McKeon-Fischer KD, Freeman JW (2011) Characterization of electrospun poly(L-lactide) and gold nanoparticle composite scaffolds for skeletal muscle tissue engineering. J Tissue Eng Regen Med 5(7):560–568. https://doi.org/10.1002/term.348
81. Jansen PL, Klinge U, Anurov M, Titkova S, Mertens PR, Jansen M (2004) Surgical mesh as a scaffold for tissue regeneration in the esophagus. Eur Surg Res 36(2):104–111. https://doi.org/10.1159/000076650
82. Valentini RF, Vargo TG, Gardella JA Jr, Aebischer P (1992) Electrically charged polymeric substrates enhance nerve fibre outgrowth in vitro. Biomaterials 13(3):183–190. https://doi.org/10.1016/0142-9612(92)90069-Z
83. Valentini RF, Vargo TG, Gardella JA, Aebischer P (1994) Patterned neuronal attachment and outgrowth on surface modified, electrically charged fluoropolymer substrates. J Biomater Sci Polym Ed 5(1–2):13–36. https://doi.org/10.1163/156856294x00626
84. Royo-Gascon N, Wininger M, Scheinbeim JI, Firestein BL, Craelius W (2013) Piezoelectric substrates promote neurite growth in rat spinal cord neurons. Ann Biomed Eng 41(1):112–122. https://doi.org/10.1007/s10439-012-0628-y
85. Lee YS, Arinzeh TL (2012) The influence of piezoelectric scaffolds on neural differentiation of human neural stem/progenitor cells. Tissue Eng A 18(19–20):2063–2072. https://doi.org/10.1089/ten.TEA.2011.0540
86. de Guzman RC, Loeb JA, VandeVord PJ (2010) Electrospinning of matrigel to deposit a basal lamina-like nanofiber surface. J Biomater Sci Polym Ed 21(8/9):1081–1101. https://doi.org/10.1163/092050609x12457428936116
87. Young T-H, Chang H-H, Lin D-J, Cheng L-P (2010) Surface modification of microporous PVDF membranes for neuron culture. J Membr Sci 350(1–2):32–41. https://doi.org/10.1016/j.memsci.2009.12.009
88. Aebischer P, Valentini RF, Dario P, Domenici C, Galletti PM (1987) Piezoelectric guidance channels enhance regeneration in the mouse sciatic nerve after axotomy. Brain Res 436(1):165–168. https://doi.org/10.1016/0006-8993(87)91570-8
89. Delaviz H, Faghihi A, Delshad AA, Bahadori MH, Mohamadi J, Roozbehi A (2011) Repair of peripheral nerve defects using a polyvinylidene fluoride channel containing nerve growth factor and collagen gel in adult rats. Cell J 13(3):137–142
90. Fine EG, Valentini RF, Bellamkonda R, Aebischer P (1991) Improved nerve regeneration through piezoelectric vinylidenefluoride-trifluoroethylene copolymer guidance channels. Biomaterials 12(8):775–780. https://doi.org/10.1016/0142-9612(91)90029-A
91. Evans GRD, Brandt K, Niederbichler AD, Chauvin P, Hermann S, Bogle M, Otta L, Wang B, Patrick CW (2000) Clinical long-term in vivo evaluation of poly(L-lactic acid) porous conduits for peripheral nerve regeneration. J Biomater Sci Polym Ed 11(8):869–878. https://doi.org/10.1163/156856200744066
92. O'Shaughnessy TJ, Lin HJ, Ma W (2003) Functional synapse formation among rat cortical neurons grown on three-dimensional collagen gels. Neurosci Lett 340(3):169–172. https://doi.org/10.1016/S0304-3940(03)00083-1
93. Ribeiro C, Correia V, Martins P, Gama FM, Lanceros-Mendez S (2016) Proving the suitability of magnetoelectric stimuli for tissue engineering applications. Colloids Surf B Biointerfaces 140:430–436. https://doi.org/10.1016/j.colsurfb.2015.12.055
94. Gandhi A, Paul A, Sen SO, Sen KK (2015) Studies on thermoresponsive polymers: phase behaviour, drug delivery and biomedical applications. Asian J Pharm Sci 10(2):99–107. https://doi.org/10.1016/j.ajps.2014.08.010
95. Klouda L (2015) Thermoresponsive hydrogels in biomedical applications: a seven-year update. Eur J Pharm Biopharm 97:338–349. https://doi.org/10.1016/j.ejpb.2015.05.017.

96. Chan BP, Leong KW (2008) Scaffolding in tissue engineering: general approaches and tissue-specific considerations. Eur Spine J 17(4):467–479. https://doi.org/10.1007/s00586-008-0745-3
97. Bhardwaj N, Kundu SC (2010) Electrospinning: a fascinating fiber fabrication technique. Biotechnol Adv 28(3):325–347. https://doi.org/10.1016/j.biotechadv.2010.01.004
98. Cui WG, Zhou Y, Chang J (2010) Electrospun nanofibrous materials for tissue engineering and drug delivery. Sci Technol Adv Mater 11(1):014108. https://doi.org/10.1088/1468-6996/11/1/014108
99. Venugopal J, Low S, Choon AT, Sampath Kumar TS, Ramakrishna S (2008) Mineralization of osteoblasts with electrospun collagen/hydroxyapatite nanofibers. J Mater Sci Mater Med 19(5):2039–2046. https://doi.org/10.1007/s10856-007-3289-x
100. Singh BN, Panda NN, Mund R, Pramanik K (2016) Carboxymethyl cellulose enables silk fibroin nanofibrous scaffold with enhanced biomimetic potential for bone tissue engineering application. Carbohydr Polym 151:335–347. https://doi.org/10.1016/j.carbpol.2016.05.088
101. Pangon A, Saesoo S, Saengkrit N, Ruktanonchai U, Intasanta V (2016) Hydroxyapatite-hybridized chitosan/chitin whisker bionanocomposite fibers for bone tissue engineering applications. Carbohydr Polym 144:419–427. https://doi.org/10.1016/j.carbpol.2016.02.053
102. Zhao X, Lui YS, Choo CKC, Sow WT, Huang CL, Ng KW, Tan LP, Loo JSC (2015) Calcium phosphate coated keratin-PCL scaffolds for potential bone tissue regeneration. Mater Sci Eng C 49:746–753. https://doi.org/10.1016/j.msec.2015.01.084
103. Sadat-Shojai M (2016) Electrospun polyhydroxybutyrate/hydroxyapatite nanohybrids: microstructure and bone cell response. J Mater Sci Technol 32(10):1013–1020. https://doi.org/10.1016/j.jmst.2016.07.007
104. Fernandes JS, Gentile P, Martins M, Neves NM, Miller C, Crawford A, Pires RA, Hatton P, Reis RL (2016) Reinforcement of poly-l-lactic acid electrospun membranes with strontium borosilicate bioactive glasses for bone tissue engineering. Acta Biomater 44:168–177. https://doi.org/10.1016/j.actbio.2016.08.042
105. Venugopal J, Ma LL, Yong T, Ramakrishna S (2005) In vitro study of smooth muscle cells on polycaprolactone and collagen nanofibrous matrices. Cell Biol Int 29(10):861–867. https://doi.org/10.1016/j.cellbi.2005.03.026
106. Kosé S, Aerts Kaya F, Denkbas EB, Korkusuz P, Cetinkaya FD (2016) Evaluation of biocompatibility of random or aligned electrospun polyhydroxybutyrate scaffolds combined with human mesenchymal stem cells. Turk J Biol 40:410–419. https://doi.org/10.3906/biy-1508-18
107. Yu Y, Meng D, Man L, Wang X (2016) The interactions between aligned poly(L-lactic acid) nanofibers and SH-SY5Y cells in vitro. J Nanosci Nanotechnol 16(6):6407–6413. https://doi.org/10.1166/jnn.2016.1088.
108. Sadeghi D, Karbasi S, Razavi S, Mohammadi S, Shokrgozar MA, Bonakdar S (2016) Electrospun poly(hydroxybutyrate)/chitosan blend fibrous scaffolds for cartilage tissue engineering. J Appl Polym Sci 133(47). https://doi.org/10.1002/app.44171
109. Mohiti-Asli M, Saha S, Murphy SV, Gracz H, Pourdeyhimi B, Atala A, Loboa EG (2015) Ibuprofen loaded PLA nanofibrous scaffolds increase proliferation of human skin cells in vitro and promote healing of full thickness incision wounds in vivo. J Biomed Mater Res B Appl Biomater 105(2):327–339. https://doi.org/10.1002/jbm.b.33520
110. Jha BS, Ayres CE, Bowman JR, Telemeco TA, Sell SA, Bowlin GL, Simpson DG (2011) Electrospun collagen: a tissue engineering scaffold with unique functional properties in a wide variety of applications. J Nanomater. https://doi.org/10.1155/2011/348268
111. Wang J, Sun B, Tian L, He X, Gao Q, Wu T, Ramakrishna S, Zheng J, Mo X (2017) Evaluation of the potential of rhTGF- β3 encapsulated P(LLA-CL)/collagen nanofibers for tracheal cartilage regeneration using mesenchymal stems cells derived from Wharton's jelly of human umbilical cord. Mater Sci Eng C 70(Part 1):637–645. https://doi.org/10.1016/jmsec201609.044

112. Borriello A, Guarino V, Schiavo L, Alvarez-Perez MA, Ambrosio L (2011) Optimizing PANi doped electroactive substrates as patches for the regeneration of cardiac muscle. J Mater Sci Mater Med 22(4):1053–1062. https://doi.org/10.1007/s10856-011-4259-x
113. Li M, Guo Y, Wei Y, MacDiarmid AG, Lelkes PI (2006) Electrospinning polyaniline-contained gelatin nanofibers for tissue engineering applications. Biomaterials 27(13):2705–2715. https://doi.org/10.1016/j.biomaterials.2005.11.037
114. Jha BS, Ayres CE, Bowman JR, Telemeco TA, Sell SA, Bowlin GL, Simpson DG (2011) Electrospun collagen: a tissue engineering scaffold with unique functional properties in a wide variety of applications. J Nanomater 2011:1–15. https://doi.org/10.1155/2011/348268
115. Fullana MJ, Wnek GE (2012) Electrospun collagen and its applications in regenerative medicine. Drug Deliv Transl Res 2(5):313–322. https://doi.org/10.1007/s13346-012-0087-x
116. Fiorani A, Gualandi C, Panseri S, Montesi M, Marcacci M, Focarete ML, Bigi A (2014) Comparative performance of collagen nanofibers electrospun from different solvents and stabilized by different crosslinkers. J Mater Sci Mater Med 25(10):2313–2321. https://doi.org/10.1007/s10856-014-5196-2
117. Dhand C, Ong ST, Dwivedi N, Diaz SM, Venugopal JR, Navaneethan B, Fazil M, Liu S, Seitz V, Wintermantel E, Beuerman RW, Ramakrishna S, Verma NK, Lakshminarayanan R (2016) Bio-inspired in situ crosslinking and mineralization of electrospun collagen scaffolds for bone tissue engineering. Biomaterials 104:323–338. https://doi.org/10.1016/j.biomaterials.2016.07.007
118. Stitzel JD, Pawlowski KJ, Wnek GE, Simpson DG, Bowlin GL (2001) Arterial smooth muscle cell proliferation on a novel biomimicking, biodegradable vascular graft scaffold. J Biomater Appl 16(1):22–33. https://doi.org/10.1106/u2uu-m9qh-y0bb-5gyl
119. He X, Cheng L, Zhang X, Xiao Q, Zhang W, Lu C (2015) Tissue engineering scaffolds electrospun from cotton cellulose. Carbohydr Polym 115:485–493. https://doi.org/10.1016/j.carbpol.2014.08.114
120. Chahal S, Hussain FSJ, Kumar A, Rasad MSBA, Yusoff MM (2016) Fabrication, characterization and in vitro biocompatibility of electrospun hydroxyethyl cellulose/poly (vinyl) alcohol nanofibrous composite biomaterial for bone tissue engineering. Chem Eng Sci 144:17–29. https://doi.org/10.1016/j.ces.2015.12.030
121. Frey MW (2008) Electrospinning cellulose and cellulose derivatives. Polym Rev 48(2):378–391. https://doi.org/10.1080/15583720802022281.
122. Ohkawa K (2015) Nanofibers of cellulose and its derivatives fabricated using direct electrospinning. Molecules 20(5):9139–9154. https://doi.org/10.3390/molecules20059139
123. Atila D, Keskin D, Tezcaner A (2016) Crosslinked pullulan/cellulose acetate fibrous scaffolds for bone tissue engineering. Mater Sci Eng C 69:1103–1115. https://doi.org/10.1016/j.msec.2016.08.015
124. Pant HR, Kim HJ, Bhatt LR, Joshi MK, Kim EK, Kim JI, Abdal-hay A, Hui KS, Kim CS (2013) Chitin butyrate coated electrospun nylon-6 fibers for biomedical applications. Appl Surf Sci B 285:538–544. https://doi.org/10.1016/j.apsusc.2013.08.089
125. Ji Y, Liang K, Shen X, Bowlin GL (2014) Electrospinning and characterization of chitin nanofibril/polycaprolactone nanocomposite fiber mats. Carbohydr Polym 101:68–74. https://doi.org/10.1016/j.carbpol.2013.09.012
126. Jayakumar R, Chennazhi KP, Srinivasan S, Nair SV, Furuike T, Tamura H (2011) Chitin scaffolds in tissue engineering. Int J Mol Sci 12(3):1876–1887. https://doi.org/10.3390/ijms12031876
127. Li Y, Wang Y, Ye J, Yuan J, Xiao Y (2016) Fabrication of poly(ε-caprolactone)/keratin nanofibrous mats as a potential scaffold for vascular tissue engineering. Mater Sci Eng C 68:177–183. https://doi.org/10.1016/j.msec.2016.05.117
128. Fan J, Lei T-D, Li J, Zhai P-Y, Wang Y-H, Cao F-Y, Liu Y (2016) High protein content keratin/poly (ethylene oxide) nanofibers crosslinked in oxygen atmosphere and its cell culture. Mater Des 104:60–67. https://doi.org/10.1016/j.matdes.2016.05.022

129. Yuan J, Geng J, Xing Z, Shim K-J, Han I, Kim J-C, Kang I-K, Shen J (2015) Novel wound dressing based on nanofibrous PHBV-keratin mats. J Tissue Eng Regen Med 9(9):1027–1035. https://doi.org/10.1002/term.1653
130. Yuan J, Xing Z-C, Park S-W, Geng J, Kang I-K, Yuan J, Shen J, Meng W, Shim K-J, Han I-S, Kim J-C (2009) Fabrication of PHBV/keratin composite nanofibrous mats for biomedical applications. Macromol Res 17(11):850–855. https://doi.org/10.1007/bf03218625
131. Yuan J, Shen J, Kang I-K (2008) Fabrication of protein-doped PLA composite nanofibrous scaffolds for tissue engineering. Polym Int 57(10):1188–1193. https://doi.org/10.1002/pi.2463
132. Thompson ZS, Rijal NP, Jarvis D, Edwards A, Bhattarai N (2016) Synthesis of keratin-based nanofiber for biomedical engineering. JoVE J Vis Exp. https://doi.org/10.3791/53381.
133. Yen K-C, Chen C-Y, Huang J-Y, Kuo W-T, Lin F-H (2016) Fabrication of keratin/fibroin membranes by electrospinning for vascular tissue engineering. J Mater Chem B 4(2):237–244. https://doi.org/10.1039/c5tb01921d
134. Li J, Li Y, Li L, Mak AFT, Ko F, Qin L (2009) Preparation and biodegradation of electrospun PLLA/keratin nonwoven fibrous membrane. Polym Degrad Stab 94(10):1800–1807. https://doi.org/10.1016/j.polymdegradstab.2009.06.004
135. Sreerekha PR, Menon D, Nair SV, Chennazhi KP (2013) Fabrication of fibrin based electrospun multiscale composite scaffold for tissue engineering applications. J Biomed Nanotechnol 9(5):790–800. https://doi.org/10.1166/jbn.2013.1585
136. Bacakova M, Musilkova J, Riedel T, Stranska D, Brynda E, Zaloudkova M, Bacakova L (2016) The potential applications of fibrin-coated electrospun polylactide nanofibers in skin tissue engineering. Int J Nanomedicine 11:771–789. https://doi.org/10.2147/ijn.s99317
137. Sreerekha PR, Menon D, Nair SV, Chennazhi KP (2013) Fabrication of electrospun poly (Lactide-co-Glycolide)-fibrin multiscale scaffold for myocardial regeneration in vitro. Tissue Eng A 19(7-8):849–859. https://doi.org/10.1089/ten.tea.2012.0374
138. Hitscherich P, Wu S, Gordan R, Xie L-H, Arinzeh T, Lee EJ (2016) The effect of PVDF-TrFE scaffolds on stem cell derived cardiovascular cells. Biotechnol Bioeng 113(7):1577–1585. https://doi.org/10.1002/bit.25918
139. Dias JC, Correia DC, Lopes AC, Ribeiro S, Ribeiro C, Sencadas V, Botelho G, Esperanca JMSS, Laza JM, Vilas JL, Leon LM, Lanceros-Mendez S (2016) Development of poly(vinylidene fluoride)/ionic liquid electrospun fibers for tissue engineering applications. J Mater Sci 51(9):4442–4450. https://doi.org/10.1007/s10853-016-9756-3
140. Qi H, Ye Z, Ren H, Chen N, Zeng Q, Wu X, Lu T (2016) Bioactivity assessment of PLLA/PCL/HAP electrospun nanofibrous scaffolds for bone tissue engineering. Life Sci 148:139–144. https://doi.org/10.1016/j.lfs.2016.02.040
141. Santoro M, Shah SR, Walker JL, Mikos AG (2016) Poly(lactic acid) nanofibrous scaffolds for tissue engineering. Adv Drug Deliv Rev 107:206–212. https://doi.org/10.1016/j.addr.2016.04.019
142. Derakhshan MA, Pourmand G, Ai J, Ghanbari H, Dinarvand R, Naji M, Faridi-Majidi R (2016) Electrospun PLLA nanofiber scaffolds for bladder smooth muscle reconstruction. Int Urol Nephrol 48(7):1097–1104. https://doi.org/10.1007/s11255-016-1259-2
143. Xu W, Shen R, Yan Y, Gao J (2017) Preparation and characterization of electrospun alginate/PLA nanofibers as tissue engineering material by emulsion eletrospinning. J Mech Behav Biomed Mater 65:428–438. https://doi.org/10.1016/j.jmbbm.2016.09.012
144. Salehi M, Farzamfar S, Bastami F, Tajerian R (2016) Fabrication and characterization of electrospun PLLA/collagen nanofibrous scaffold coated with chitosan to sustain release of aloe vera gel for skin tissue engineering. Biomed Eng Appl Basis Commun. https://doi.org/10.4015/s1016237216500356
145. Liu WW, Li ZQ, Zheng L, Zhang XY, Liu P, Yang T, Han B (2016) Electrospun fibrous silk fibroin/poly(L-lactic acid) scaffold for cartilage tissue engineering. Tissue Eng Regen Med 13(5):516–526. https://doi.org/10.1007/s13770-016-9099-9

146. Karahaliloğlu Z, Demirbilek M, Şam M, Sağlam N, Mızrak AK, Denkbaş EB (2016) Surface-modified bacterial nanofibrillar PHB scaffolds for bladder tissue repair. Artif Cells Nanomed Biotechnol 44(1):74–82. https://doi.org/10.3109/21691401.2014.913053
147. Nagiah N, Madhavi L, Anitha R, Anandan C, Srinivasan NT, Sivagnanam UT (2013) Development and characterization of coaxially electrospun gelatin coated poly (3-hydroxybutyric acid) thin films as potential scaffolds for skin regeneration. Mater Sci Eng C 33(7):4444–4452. https://doi.org/10.1016/j.msec.2013.06.042
148. Asran AS, Razghandi K, Aggarwal N, Michler GH, Groth T (2010) Nanofibers from blends of polyvinyl alcohol and polyhydroxy butyrate as potential scaffold material for tissue engineering of skin. Biomacromolecules 11(12):3413–3421. https://doi.org/10.1021/bm100912v
149. Meng W, Kim S-Y, Yuan J, Kim JC, Kwon OH, Kawazoe N, Chen G, Ito Y, Kang I-K (2007) Electrospun PHBV/collagen composite nanofibrous scaffolds for tissue engineering. J Biomater Sci Polym Ed 18(1):81–94. https://doi.org/10.1163/156856207779146114
150. Khorasani MT, Mirmohammadi SA, Irani S (2011) Polyhydroxybutyrate (PHB) scaffolds as a model for nerve tissue engineering application: fabrication and in vitro assay. Int J Polym Mater Polym Biomater 60(8):562–575. https://doi.org/10.1080/00914037.2010.531809.
151. Genchi GG, Ciofani G, Polini A, Liakos I, Iandolo D, Athanassiou A, Pisignano D, Mattoli V, Menciassi A (2015) PC12 neuron-like cell response to electrospun poly(3-hydroxybutyrate) substrates. J Tissue Eng Regen Med 9(2):151–161. https://doi.org/10.1002/term.1623
152. Ramier J, Bouderlique T, Stoilova O, Manolova N, Rashkov I, Langlois V, Renard E, Albanese P, Grande D (2014) Biocomposite scaffolds based on electrospun poly(3-hydroxybutyrate) nanofibers and electrosprayed hydroxyapatite nanoparticles for bone tissue engineering applications. Mater Sci Eng C 38:161–169. https://doi.org/10.1016/j.msec.2014.01.046
153. Guan D, Chen Z, Huang C, Lin Y (2008) Attachment, proliferation and differentiation of BMSCs on gas-jet/electrospun nHAP/PHB fibrous scaffolds. Appl Surf Sci 255(2):324–327. https://doi.org/10.1016/j.apsusc.2008.06.093
154. Sadat-Shojai M, Khorasani M-T, Jamshidi A (2016) A new strategy for fabrication of bone scaffolds using electrospun nano-HAp/PHB fibers and protein hydrogels. Chem Eng J 289:38–47. https://doi.org/10.1016/j.cej.2015.12.079
155. Ding Y, Li W, Müller T, Schubert DW, Boccaccini AR, Yao Q, Roether JA (2016) Electrospun polyhydroxybutyrate/poly(ε-caprolactone)/58S sol–gel bioactive glass hybrid scaffolds with highly improved osteogenic potential for bone tissue engineering. ACS Appl Mater Interfaces 8(27):17098–17108. https://doi.org/10.1021/acsami.6b03997
156. Daranarong D, Chan RTH, Wanandy NS, Molloy R, Punyodom W, Foster LJR (2014) Electrospun polyhydroxybutyrate and poly(L-lactide-co-ε-caprolactone) composites as nanofibrous scaffolds. Biomed Res Int. https://doi.org/10.1155/2014/741408
157. Ding Y, Roether JA, Boccaccini AR, Schubert DW (2014) Fabrication of electrospun poly (3-hydroxybutyrate)/poly (ε-caprolactone)/silica hybrid fibermats with and without calcium addition. Eur Polym J 55:222–234. https://doi.org/10.1016/j.eurpolymj.2014.03.020
158. Zonari A, Novikoff S, Electo NRP, Breyner NM, Gomes DA, Martins A, Neves NM, Reis RL, Goes AM (2012) Endothelial differentiation of human stem cells seeded onto electrospun polyhydroxybutyrate/polyhydroxybutyrate-co-hydroxyvalerate fiber mesh. PLoS One 7(4):e35422. https://doi.org/10.1371/journal.pone.0035422
159. Xie J, MacEwan MR, Willerth SM, Li X, Moran DW, Sakiyama-Elbert SE, Xia Y (2009) Conductive core–sheath nanofibers and their potential application in neural tissue engineering. Adv Funct Mater 19(14):2312–2318. https://doi.org/10.1002/adfm.200801904
160. Zhou J, Cheng L, Sun X, Wang X, Jin S, Li J, Wu Q (2016) Neurogenic differentiation of human umbilical cord mesenchymal stem cells on aligned electrospun polypyrrole/polylactide composite nanofibers with electrical stimulation. Front Mater Sci 10(3):260–269. https://doi.org/10.1007/s11706-016-0348-6
161. Jeong SI, Jun ID, Choi MJ, Nho YC, Lee YM, Shin H (2008) Development of electroactive and elastic nanofibers that contain polyaniline and poly(L-lactide-co-ε-caprolactone)

for the control of cell adhesion. Macromol Biosci 8(7):627–637. https://doi.org/10.1002/mabi.200800005

162. Ghasemi-Mobarakeh L, Prabhakaran MP, Morshed M, Nasr-Esfahani MH, Ramakrishna S (2009) Electrical stimulation of nerve cells using conductive nanofibrous scaffolds for nerve tissue engineering. Tissue Eng A 15(11):3605–3619. https://doi.org/10.1089/ten.tea.2008.0689
163. Ghasemi-Mobarakeh L, Prabhakaran MP, Morshed M, Nasr-Esfahani MH, Baharvand H, Kiani S, Al-Deyab SS, Ramakrishna S (2011) Application of conductive polymers, scaffolds and electrical stimulation for nerve tissue engineering. J Tissue Eng Regen Med 5(4):17–35. https://doi.org/10.1002/term.383.
164. Aznar-Cervantes S, Roca MI, Martinez JG, Meseguer-Olmo L, Cenis JL, Moraleda JM, Otero TF (2012) Fabrication of conductive electrospun silk fibroin scaffolds by coating with polypyrrole for biomedical applications. Bioelectrochemistry 85:36–43. https://doi.org/10.1016/j.bioelechem.2011.11.008
165. Lee JY, Bashur CA, Goldstein AS, Schmidt CE (2009) Polypyrrole-coated electrospun PLGA nanofibers for neural tissue applications. Biomaterials 30(26):4325–4335. https://doi.org/10.1016/j.biomaterials.2009.04.042
166. Zhou J-F, Wang Y-G, Cheng L, Wu Z, Sun X-D, Peng J (2016) Preparation of polypyrrole-embedded electrospun poly(lactic acid) nanofibrous scaffolds for nerve tissue engineering. Neural Regen Res 11(10):1644–1652. https://doi.org/10.4103/1673-5374.193245
167. SH K, Lee SH, Park CB (2012) Synergic effects of nanofiber alignment and electroactivity on myoblast differentiation. Biomaterials 33(26):6098–6104. https://doi.org/10.1016/j.biomaterials.2012.05.018
168. Chen M-C, Sun Y-C, Chen Y-H (2013) Electrically conductive nanofibers with highly oriented structures and their potential application in skeletal muscle tissue engineering. Acta Biomater 9(3):5562–5572. https://doi.org/10.1016/j.actbio.2012.10.024
169. Castagna R, Tunesi M, Saglio B, Della Pina C, Sironi A, Albani D, Bertarelli C, Falletta E (2016) Ultrathin electrospun PANI nanofibers for neuronal tissue engineering. J Appl Polym Sci 133(35). https://doi.org/10.1002/app.43885
170. Hosseinzadeh S, Mahmoudifard M, Mohamadyar-Toupkanlou F, Dodel M, Hajarizadeh A, Adabi M, Soleimani M (2016) The nanofibrous PAN-PANi scaffold as an efficient substrate for skeletal muscle differentiation using satellite cells. Bioprocess Biosyst Eng 39(7):1163–1172. https://doi.org/10.1007/s00449-016-1592-y
171. Gonçalves R, Martins P, Moya X, Ghidini M, Sencadas V, Botelho G, Mathur ND, Lanceros-Mendez S (2015) Magnetoelectric $CoFe_2O_4$/polyvinylidene fluoride electrospun nanofibres. Nanoscale 7(17):8058–8061. https://doi.org/10.1039/c5nr00453e

The manufacturer's authorised representative in the EU is Springer Nature Customer Service Centre GmbH, Europaplatz 3, 69115 Heidelberg, Germany. If you have any concerns regarding our products, please contact ProductSafety@springernature.com

Printed and bound by CPI Group (UK) Ltd, Croydon, CR0 4YY
30/09/2026
02225657-0002